Saas-Fee Advanced Course 35

D. Jewitt A. Morbidelli H. Rauer

Trans-Neptunian Objects and Comets

Saas-Fee Advanced Course 35

Swiss Society for Astrophysics and Astronomy
Edited by K. Altwegg, W. Benz and N. Thomas

With 132 Figures, 18 in Color

Springer

David Jewitt
University of Hawaii
Institute for Astronomy
2680 Woodlawn Drive
Honolulu, HI 96822, USA
jewitt@hawaii.edu

Alessandro Morbidelli
Observatoire de la Côte
dŠAzur/CNRS
B.P. 4229
06304 Nice Cedex 4, France
morby@obs-nice.fr

Heike Rauer
DLR/Institut für Planetenforschung
Rutherfordstr. 2
12489 Berlin, Germany
and
TU Berlin/Zentrum
für Astronomie und Astrophysik
Hardenbergstr. 36
10623 Berlin, Germany
heike.rauer@dlr.de

Volume Editors:
Kathrin Altwegg
Willy Benz
Nicolas Thomas
Universität Bern
Physikalisches Institut
Sidlerstrasse 5
3012 Bern, Switzerland

This series is edited on behalf of the Swiss Society for Astrophysics and Astronomy:
Société Suisse d'Astrophysique et d'Astronomie
Observatoire de Genève, ch. des Maillettes 51, 1290 Sauverny, Switzerland

---

Cover picture: See chapter by D. Jewitt, Fig. 12.

Library of Congress Control Number: 2007934029

ISBN 978-3-540-71957-1 Springer Berlin Heidelberg New York

Springer is a part of Springer Science+Business Media
springer.com

Typesetting: by the authors and Integra using a Springer LaTeX macro package
Cover design: WMXDesign GmbH, Heidelberg

Printed on acid-free paper SPIN: 11808169 55/Integra 5 4 3 2 1 0

Preface

The 35th Saas Fee Winter School was held on 13–18 March 2005 in the skiing village of Mürren in the Berner Oberland. In view of the excitement generated over the past 15 years by the discovery of the Kuiper Belt and Trans-Neptunian Objects and also by the ongoing Rosetta mission to comet Churyumov-Gerasimenko, it was decided to combine discussion of these primitive objects into one winter school under the title, "Trans-Neptunian Objects and Comets." The aim was to provide an overview of these objects, to discuss their relationships, and to identify directions for future research. The school attracted over 60 students from all over the world. We were fortunate that not merely were the students able to hear a set of outstanding lectures but were also able to enjoy marvellous weather in one of the most beautiful parts of Switzerland.

The organizers thank the lecturers, Dave Jewitt, Alessandro Morbidelli, and Heike Rauer, for the tremendous effort they made in preparing the lectures and the text for this volume. Stephan Graf, Annette Jäckel, and Jonathan Horner provided reviews, checked the text and references, and assisted in the production. We also thank Frau Staehli and the staff of the Hotel Eiger in Mürren for the warm welcome and their generosity. We also thank Ms. Kathrin Weyeneth and Ms. Edith Hertig from the Physikalisches Institut for their secretarial support for the school.

Financial assistance provided by the Swiss Society for Astrophysics and Astronomy and the European Space Agency is gratefully acknowledged.

Kathrin Altwegg
Willy Benz
Nicolas Thomas

Contents

List of Previous Saas-Fee Advanced Courses

!! 2005 Trans-Neptunian Objects and Comets
D. Jewitt, A. Morbidelli, H. Rauer

!! 2004 The Sun, Solar Analogs and the Climate
J.D. Haigh, M. Lockwood, M.S. Giampapa

!! 2003 Gravitation Lensing: Strong, Weak and Micro
P. Schneider, C. Kochanek, J. Wambsganss

!! 2002 The Cold Universe
A.W. Blain, F. Combes, B.T. Draine

!! 2001 Extrasolar Planets
T. Guillot, P. Cassen, A. Quirrenbach

!! 2000 High-Energy Spectroscopic Astrophysics
S.M. Kahn, P. von Ballmoos, R.A. Sunyaev

!! 1999 Physics of Star Formation in Galaxies
F. Palla, H. Zinnecker

!! 1998 Star Clusters
B.W. Carney, W.E. Harris

!! 1997 Computational Methods for Astrophysical Fluid Flow
R.J. LeVeque, D. Mihalas, E.A. Dorfi, E. Müller

!! 1996 Galaxies Interactions and Induced Star Formation
R.C. Kennicutt, F. Schweizer, J.E. Barnes

!! 1995 Stellar Remnants
S.D. Kawaler, I. Novikov, G. Srinivasan

!! 1994 Plasma Astrophysics
J.G. Kirk, D.B. Melrose, E.R. Priest

!! 1993 The Deep Universe
A.R. Sandage, R.G. Kron, M.S. Longair

!! 1992 Interacting Binaries
S.N. Shore, M. Livio, E.J.P. van den Heuvel

!! 1991 The Galactic Interstellar Medium
W.B. Burton, B.G. Elmegreen, R. Genzel

!! 1990 Active Galactic Nuclei
R. Blandford, H. Netzer, L. Woltjer

* 1989 The Milky Way as a Galaxy
G. Gilmore, I. King, P. van der Kruit

! 1988 Radiation in Moving Gaseous Media
H. Frisch, R.P. Kudritzki, H.W. Yorke

! 1987 Large Scale Structures in the Universe
A.C. Fabian, M. Geller, A. Szalay

! 1986 Nucleosynthesis and Chemical Evolution
J. Audouze, C. Chiosi, S.E. Woosley

! 1985 High Resolution in Astronomy
R.S. Booth, J.W. Brault, A. Labeyrie

! 1984 Planets, Their Origin, Interior and Atmosphere
D. Gautier, W.B. Hubbard, H. Reeves

! 1983 Astrophysical Processes in Upper Main Sequence Stars
A.N. Cox, S. Vauclair, J.P. Zahn

* 1982 Morphology and Dynamics of Galaxies
J. Binney, J. Kormendy, S.D.M. White

! 1981 Activity and Outer Atmospheres of the Sun and Stars
F. Praderie, D.S. Spicer, G.L. Withbroe

* 1980 Star Formation
J. Appenzeller, J. Lequeux, J. Silk

* 1979 Extragalactic High Energy Physics
F. Pacini, C. Ryter, P.A. Strittmatter

* 1978 Observational Cosmology
J.E. Gunn, M.S. Longair, M.J. Rees

* 1977 Advanced Stages in Stellar Evolution
I. Iben Jr., A. Renzini, D.N. Schramm

* 1976 Galaxies
K. Freeman, R.C. Larson, B. Tinsley

* 1975 Atomic and Molecular Processes in Astrophysics
A. Dalgarno, F. Masnou-Seeuws, R.V.P. McWhirter

* 1974 Magnetohydrodynamics
L. Mestel, N.O. Weiss

* 1973 Dynamical Structure and Evolution of Stellar Systems
G. Contopoulos, M. Hénon, D. Lynden-Bell

* 1972 Interstellar Matter
N.C. Wickramasinghe, F.D. Kahn, P.G. Metzger

* 1971 Theory of the Stellar Atmospheres
D. Mihalas, B. Pagel, P. Souffrin

---

* Out of print

! May be ordered from Geneva Observatory
Saas-Fee Courses
Geneva Observatory
CH-1290 Sauverny
Switzerland

!! May be ordered from Springer and/or are available online at springerlink.com.

Kuiper Belt and Comets: An Observational Perspective

D. Jewitt

Note to the Reader
These notes outline a series of lectures given at the Saas Fee Winter School held in Mürren, Switzerland, in March 2005. As I see it, the main aim of the Winter School is to communicate (especially) with young people in order to inflame their interests in science and to encourage them to see ways in which they can contribute and maybe do a better job than we have done so far. With this in mind, I have written up my lectures in a less than formal but hopefully informative and entertaining style, and I have taken a few detours to discuss subjects that I think are important but which are usually glossed-over in the scientific literature.

1 Preamble

Almost exactly 400 years ago, planetary astronomy kick-started the era of modern science, with a series of remarkable discoveries by Galileo concerning the surfaces of the Moon and Sun, the phases of Venus, and the existence and motions of Jupiter's large satellites. By the early 20th century, the focus of astronomical attention had turned to objects at larger distances, and to questions of galactic structure and cosmological interest. At the start of the 21st century, the tide has turned again. The study of the Solar system, particularly of its newly discovered outer parts, is one of the hottest topics in modern astrophysics with great potential for revealing fundamental clues about the origin of planets and even the emergence of life. New technology has been crucial to each of these steps. Galileo's refractor gave a totally new view of the sky. A hundred years ago, photographic plates and large telescopes allowed the first spectroscopic observations of distant galaxies revealing, through Hubble's law, the third dimension of distance into the plane of the sky. In our own time, highly sensitive, wide-field electronic detectors have enabled the discovery and the exploration of the Kuiper Belt, while fast computers allow us to make numerical simulations with a degree of sophistication that was previously unimaginable.

D. Jewitt, *Kuiper Belt and Comets*. In: K. Altwegg et al., Trans-Neptunian Objects and Comets, Saas-Fee Advanced Courses, pp. 1–78 (2008)
DOI 10.1007/978-3-540-71958-8_1

As a result of all this, our view of the Solar system is in the middle of a great change. Our appreciation of the different types of objects (planets, asteroids, comets, etc) orbiting the Sun is changing in response to new observations. Our understanding of their evolutionary connections with each other and with the formation epoch is changing as we develop more and more elaborate schemes to synthesize the new data. Additionally, our perception of the Solar system in the bigger context of the galactic disk is changing, particularly as we detect planets encircling other stars (in systems that are, for the most part, dynamically not very like our own). All of this makes it a great time to review what we know about the Solar system in the context of the Saas Fee winter school series, one of very few Saas Fee lectures to be dedicated to the universe at $z \sim 0$.

This article parallels five lectures given in Mürren, Switzerland, in March 2005, as part of the Saas Fee Lecture Series entitled "Trans-Neptunian Objects and Comets." Some of these lectures were given "off the cuff," and I have tried to reconstruct them from memory and a few notes. The degree to which this succeeds is unknown and it does not matter: the participants, like this lecturer, have no doubt forgotten most of what was said while readers who were not in Saas Fee for the Lecture Series never knew. The style of the write-up is deliberately informal.

1.1 The Conduct of Research into the Subject

In this section, I want to take advantage of the open format of the Saas Fee lecture series to briefly discuss the conduct of modern science, particularly as it relates to the new study of the Solar system. Partly, this is for fun and for my own entertainment, but I also have a serious purpose: there are real misconceptions about what is happening (as opposed to what should happen), sometimes even in the minds of the best scientists. Most of us probably possess vaguely Popperian [124] notions about the conduct of science. Essentially, Popper argued that we advance in science by the falsification of hypotheses. Observations suggest hypotheses that make predictions, which can be confirmed or refuted by new observations, and so on. But not all of us work within this framework, and there are few clues as to the real methods or motivations of scientists in the stylized and frequently dry presentations that are demanded for publication in the refereed journals. It is the absence of discussion about the realities of the practice of science that has allowed false ideas to spread unchecked. The Saas Fee participants, especially those likely to become major figures in the future exploration of the Solar system, are the main targets of my remarks.

Observations

Observationally, the goal is to determine objective reality through careful studies that are unbiassed (or at least well calibrated), fully understood, independently reproducible and motivated by the desire to test a hypothesis. Several things must be said about this idealized goal.

- Real science is much more affected by chance discoveries than one would guess from the simple description of Popper's scheme, above. Sometimes, the biggest advance comes from simply looking, not from testing a hypothesis.
- The flip-side of this is that the human brain is rarely able to perceive or assimilate things that it does not expect to see, and so, fundamental discoveries made by chance are very rare (but disproportionately important). We are like ants in the city: comfortable with the dirt in front of us but unable to perceive the buildings above.
- Although it seems that it should be otherwise, taking good observations is incredibly hard. Too many things can go wrong; there are many sources of error both random and systematic, and it is often difficult or impossible to accurately quantify these uncertainties. As a result, observations that seem secure (or "statistically significant" as we say with a misleading air of detachment) are often wrong, leading us up blind alleys that can take years to escape.
- An equally serious problem is that it is easy to take the "wrong" measurement, by which I mean a measurement that has no great impact on our perception of the big picture. In fact, most observers, including this one, spend most of their time taking measurements that are unimportant. The simple reason is that we usually cannot see clearly enough to predetermine which measurements will be of the greatest value. Theories and models are supposed to help us here: usually they do not.

As observers, we are swimming in mud (Fig. 1): it is hard work, we cannot see where we are going but sometimes we bump into interesting things as we crawl our way along.

Theories and Models

The purpose of theories and models is to use available data together with established physical laws to make observationally testable predictions. Predictions provide an objective and indispensable way to test the theories and models. Unfortunately, theory rarely works this way, because the systems under consideration are very complicated and a large number of processes interact in a way that is difficult to treat. Making observationally testable predictions is difficult because a given model, with changes to one or two of its many free parameters, can usually accommodate a wide range of outcomes, regardless of whether the model is correct. Making predictions that are falsifiable is the hard part of making models, which is why many modelers do not do it.

Here are some problems with theory and theorists.

Fig. 1. Observers, doing what they do best. Photo courtesy Talisman Creations

- The main problem for theorists and modelers is that the world is very complex, and most problems are observationally under-constrained. Analytical approaches offer real insight and understanding but are mostly confined to the study of highly simplified approximations. Numerical approaches provide a way to deal with the complexity, but at the expense of adding typically large numbers of under-determined model parameters and initial conditions.
- It has become common to present models that fit the available data but which offer no observationally testable predictions, leaving the reader to speculate about what predictions the model might make if only the authors had written them down. The reason for this is clear enough: making observationally testable predictions is difficult (and scary too: you could be wrong!). But without predictions the models have no scientific value. Some have argued that the mere fact that a model can fit many and varied observations in a self-consistent way is evidence in itself for the correctness of the model. Nonsense!
- The meaning of the word "predict" is also under attack. Sometimes, the authors say that their model "predicts" some quantity or property, but closer inspection shows that the thing has already been measured. One cannot predict something which is already known! What the modelers mean is that they can fit the data, not predict new data. There is a big difference.
- Models are frequently over-sold Fig. 2. It is almost *de rigueur* for modelers to add comforting phrases like "our conclusions are insensitive to the parameters assumed in the model..." and "our model has only one free parameter..." whether or not these statements are true!

Fig. 2. The theorist, spotlessly clean, whose theory explains everything and has no free parameters. The halo and the facial expression signify his wisdom and purity. Courtesy Virginia A. Tikan

Of course, it is the interaction between the observers and the theorists that gives our subject its extraordinary vitality and power. Science without observations would collapse into dull paralysis within months. Science without models would soon degenerate into stamp collecting. But this does not mean that we have to accept either the observations or the models uncritically. In particular, we should not accept models that fail to make observationally testable predictions. They may offer beautiful descriptions of what we observe but, without predictions, we will never know if they have deeper meaning.

The Kuiper belt is still very much in the discovery phase, and we should not expect a scientifically compelling picture of its formation and evolution to emerge overnight. With this warning of a turbulent and uncertain background, we are ready to launch into an overview of the modern Solar system.

2 The Modern Solar System

2.1 Protoplanetary Disk

Scale Constraints

The most noticeable feature of the Solar system is that the planets follow nearly circular orbits about the Sun in roughly the same plane. This architecture strongly suggests that the planets formed by accretion in a circum-Solar disk. The properties of this disk, now long-gone, can be inferred only approximately from the modern-day system.

The extent of the solar nebula is not tightly observationally constrained, but again we can set limits. At the inner edge, it is reasonable to suppose that the disk extended practically to the surface of the protosun. Indeed, material flowed through the disk into the Sun as part of its formation. At the outer edge, we surmise that the disk extended to roughly the outer extremity of the well-established part of the Kuiper belt (roughly 50 AU). Observations of disks around other stars show that disks are commonly much larger, extending to hundreds of AU around stars of Solar mass. The timescales for the growth of solid bodies scale with heliocentric distance, R, as R^3, give or take one power of R. One possibility is that the protoplanetary disk may initially have been hundreds of AU in extent but that no large bodies grew in the outer parts. In this case, deeper survey observations should reveal smaller bodies beyond the ~50 AU edge, something that seems not to be true. Another possibility is that the small size of the Kuiper belt (specifically of the classical belt) results from tidal truncation by a passing star, as argued by Ida et al. [66] and others since.

Structure Constraints

The current mass of the objects in the Solar system (excluding the Sun) is about 10^{-3} $M_\odot$, most of which is in Jupiter. Obviously, this sets a strong lower limit to the initial mass of the disk. A more realistic limit is set by careful consideration of the compositions of the planets and the (probably good) assumption that the disk had a basically cosmic composition. For instance, consider the Earth. Its mass consists mostly of heavy elements (called "metals" by terminology-bending astrophysicists), whereas, in a mixture containing a cosmic proportion of H and He, the "metals" carry only ~0.01 of the mass. Therefore, the so-called augmented mass of the Earth (the mass of material of cosmic composition containing an Earth mass of metals) is about 100 $M_\oplus$. This same treatment of the other planets leads to a best estimate of the minimum disk mass of order 0.01 $M_\odot$. Models with this mass are known as MMSN models: Minimum Mass solar nebula models.

The distribution of mass and temperature within the protoplanetary disk are usually approximated by power laws

$$\Sigma(R) = \Sigma(R_0)\left[\frac{R_0}{R}\right]^p \tag{1}$$

$$T(R) = T(R_0)\left[\frac{R_0}{R}\right]^q \tag{2}$$

where $\Sigma(R)$ [kg m^{-2}] and $T(R)$ are the column density and temperature of the disk at radius R, R_0 is a reference radius, often taken as 1 AU (the orbit of Earth) or 10 AU (orbit of Saturn), and the indices p and q describe the radial fall-off of the density and temperature, respectively. Estimates of Σ_0 and p can be obtained by studying the distribution of mass within the Solar system. If we smear the augmented masses of the planets over annuli extending half way to the nearest planet (e.g., Saturn would be smeared from 7.5 to 15 AU) we obtain $p \sim 3/2$ (with an uncertainty of at least $\pm 1/2$) and $\Sigma(R_0) \sim 50\,\mathrm{kg\,m^{-2}}$ at $R_0 = 10\,\mathrm{AU}$. This is the total (gas plus dust) surface density. The dust surface density is about 100 times smaller. The temperature of a blackbody in radiative equilibrium with sunlight is described by (2) with $T(R_0 = 10) = 88\,\mathrm{K}$ and $q = 0.5$.

The values of disk parameters so derived are not particularly accurate, given the uncertainties in computing augmented masses from current masses and given the likelihood that the orbits of the planets were not always where we now find them. Still, the above give a reasonable starting guess for the structure of the disk. It is natural to think that observations of disks around young stars should provide independent constraints on likely disk parameters. Unfortunately, most existing data generally lack angular resolution high enough for the disk spatial parameters to be directly measured. Instead, the disk parameters are inferred from measurements of the spectral energy distribution using models in which the number of free parameters is larger than the number of observational constraints. Assuming $p = 3/2$, measurements give mean values $q = 0.6 \pm 0.1$ and $T(10\,AU) = 45 \pm 21\,\mathrm{K}$ [4], which fit well with the nominal values. The dust mass inferred from disk observations averages $M_\mathrm{d} = 4 \times 10^{-3}\,\mathrm{M_\odot}$ ([4]; from 67 classical T-Tauri stars, likely analogs of the young Sun). The dust mass is really a lower limit to the mass in solids: particles much larger than the millimeter wavelengths of observation contribute little to the measured radiation and go undetected. Augmented to cosmic composition, the implied average disk mass is $\sim 0.4\,\mathrm{M_\odot}$. This is substantially larger than MMSN but the scatter in disk masses is large, as are the uncertainties, and there are presumably observational biases against the measurement of lower disk masses.

Constraints on Disk Timescales and Environment

The most important observational constraints on timescales in the protoplanetary disk are provided by measurements of the products of radioactive decay of short-lived elements in meteorites. The latter are rocks derived by shattering collisions amongst the asteroids and delivered to Earth by gravitational scattering after their orbits become planet-crossing. Minerals in many meteorites incorporate the decay products of short-lived nuclei, showing that the minerals formed on timescales comparable to the half-lives of the decaying elements. The quintessential example is provided by ^{26}Al, which β-decays into ^{26}Mg with a half-life $t_{1/2} = 0.7\,\mathrm{Myr}$ [90]. When ^{26}Mg is found incorporated within the mineral structure of a meteorite, we may conclude that ^{26}Al was originally present. To be captured in abundance, ^{26}Al must have been incorporated into the meteorites within a few half-lives of its formation. Element formation occurs naturally in the explosion of massive stars as supernovae, but the significance of ^{26}Al has sometimes been questioned because it can be also formed by spallation reactions with particles accelerated to energies >MeV [91]. Such particles might have been emitted by the magnetically super-active young sun. Recent measurements of ^{60}Ni, which is produced by the decay of ^{60}Fe with a half-life of 1.5 Myr [116], do not suffer this ambiguity because there is no route to its production through spallation. We conclude with confidence that macroscopic solid bodies formed in the asteroid belt on timescales of a few Myr.

Other timescale constraints come from observations of circumstellar matter in disks around nearby Solar-mass stars. These observations show that circumstellar gas has a lifetime that is less than 10 Myr [10, 161] and potentially just a few Myr. Dust emission from stars also declines rapidly with age (Fig. 3). The initial decline is probably due to growth into particles that are much larger than the wavelength of observation (typically ∼1 mm). There is evidence for thermal excess above the emission from the stellar photospheres in stars as old as ∼0.5 Gyr, and this dust is probably produced in recent times by collisions among unseen bodies in the circumstellar disks, or released by unseen comets. The general decline in the dustiness of nearby stars is occasionally punctuated by objects with surprising dust emission excess. This could be showing that the stars are, for some reason, intrinsically more dusty than others of similar age. An alternative explanation is that the dust has been

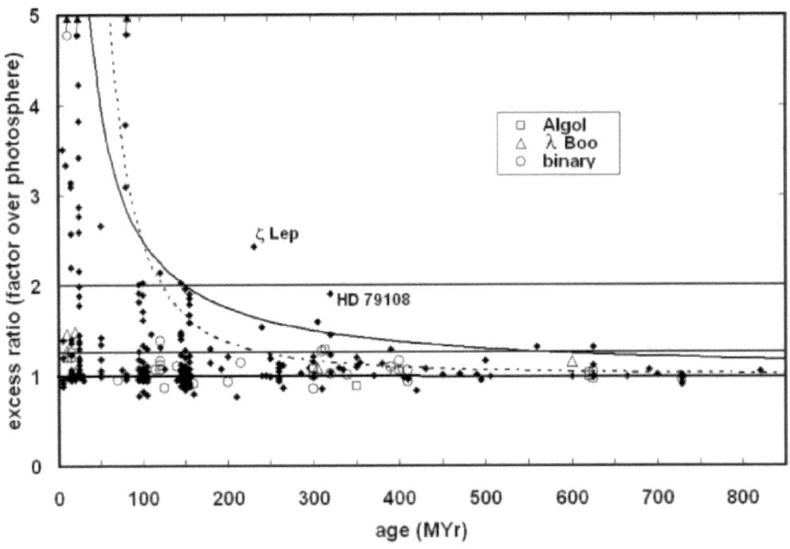

Fig. 3. Dust emission from nearby stars at 24 μm wavelength expressed as a ratio to the flux density expected from the photosphere alone. Values >1 indicate excess emission, most likely from circumstellar dust heated by starlight. The emission generally declines with stellar age, but, at any given age, there is a range of thermal excesses, with occasional dramatic spikes, as at ζ Lep and HD 79108. The solid curve shows a 1/(time) dependence. Ages of the stars are estimated from cluster membership and from models of their spectra, and are accurate to about a factor of two. One interpretation of the spikes is that dust is impulsively created by collisions between massive bodies. Figure reproduced from [131]

recently created, perhaps by impact and shattering of massive planetesimals in the unseen circumstellar disks [131].

Two pieces of evidence suggest that the Sun formed in a star cluster.

First, some of the short-lived radionuclides (notably ^{60}Fe) must have been produced, in an exploding star, only shortly before their incorporation into minerals and meteorite parent bodies (asteroids), otherwise, they would have already decayed to insignificance. Supernovae are very rare (the galactic rate is only ~one per 50 years) and typically distant so that the likelihood of having one occur nearly simultaneously with the formation of solid bodies in the disk is small. The simplest interpretation is that the Sun was part of a cluster of stars in which nearby high mass members exploded upon reaching the ends of their stable main-sequence lifetimes. An estimate of the cluster population can be made based on the dual requirements that the cluster must have been populated enough to contain a massive star capable of reaching supernova status but yet not so populated that gravitational perturbations would have noticably disturbed the orbits of the planets. A cluster containing $\sim 2000 \pm 1100$ stars seems capable of meeting both conditions [2].

Second, the truncated outer edge of the classical Kuiper belt and the excited dynamical structure of the belt in general suggests to some that the protoplanetary disk might have been tidally truncated by a passing star [66,114]. Numerical simulations show that to truncate or seriously disturb the disk down to radius r [AU] implies a stellar impact parameter $\sim 3\,r$. The classical belt ends near 50 AU, requiring a Solar mass star to pass ~150 AU from the Sun. In its current environment, the sun and stars are separated by ~1 pc (200,000 AU), and the probability of two stars passing within 150 AU in the 4.6 Gyr age is negligible. Again, a plausible inference is that the mean distance between the Sun and nearby stars was once much smaller: the Sun was in a cluster.

2.2 The Three Domains

It is useful to consider the Solar system as divided into three domains, based on the compositions, masses, and radial distances of its constituents. These are as follows:

The Domain of the Terrestrial Planets

The primary objects are Mercury, Venus, Earth, and Mars, but the asteroids in the main-belt between Mars and Jupiter are also included (the largest asteroid is (1) Ceres; see Table 1). These objects are all distinguished by refractory (non-volatile) compositions dominated by metals [principally iron (Fe) and nickel (Ni)] and compounds of silicon (Si), oxygen (O), magnesium (Mg), and aluminium (Al). The bulk densities are high ($\rho = 3930\,\mathrm{kg\,m^{-3}}$ for Mars up to $5515\,\mathrm{kg\,m^{-3}}$ for Earth, the latter slightly enhanced by self-compression due to gravity), reflecting the lack of volatiles. Densities of many

Table 1. Terrestrial Planets

Object	Mass/$M_\oplus$	Radius/$R_\oplus$	ρ [kg m^{-3}]	a [AU]	e	i [deg]
Mercury	0.06	0.38	5430	0.387	0.206	7.0
Venus	0.81	0.95	5424	0.723	0.007	3.4
Earth	1	1	5520	1.000	0.017	0.0
Mars	0.11	0.53	3930	1.523	0.093	1.8
Ceres	1.6×10^{-4}	0.08	2080	2.766	0.078	10.6

asteroids are smaller, apparently because of porous internal structures created by impact fragmentation and reassembly of these bodies since their formation. The densities of stony meteorites, small fragments from the asteroid belt, are $\rho \sim 3000\,\mathrm{kg\,m^{-3}}$.

All these bodies appear to have formed by "binary accretion," the step-by-step growth occurring when two bodies collide and stick, starting from tiny dust particles in the original nebula about the Sun and reaching up to the sizes of the Earth and Venus. Indeed, the N-body models that are used to study the dynamics and growth of bodies in the outer Solar system have been honed to their highest levels of perfection in the study of terrestrial planet growth. Still, new data continue to surprise and unnerve us. For example, N-body accretion models show that Earth grew to its final mass on a timescale ~100–200 My [18, 129], and this long timescale has remained more or less unchanged for the past several decades, since detailed estimates were first made by G. Wetherill. It stands in contrast to new isotopic data from the Hafnium-Tungsten (Hf-W) decay [67]. Hafnium decays to Tungsten, $^{182}\mathrm{Hf} \rightarrow {}^{182}\mathrm{W}$, with a 9- Myr half life. The quantity of ^{182}W in the Earth's mantle (relative to the core) provides a measure of the amount of the unstable Hf isotope at the epoch of core formation, and so sets the timescale for Earth's differentiation. The W-Hf data show that the Earth accreted the bulk of its mass within 30 Myr, whereas major asteroids such as Vesta formed in an even shorter 3 Myr [67]. This is a half to one order of magnitude discrepancy with the N-body models and remains unexplained.

The relevance to us is that models can give very plausible but wholly incorrect solutions. Without the benefit of independent constraints from the isotopes, we would remain completely unaware that the N-body terrestrial planet growth models are too slow. In the outer Solar system (where independent constraints on the models from isotopes or other sources are unavailable), it is easy to see that we are skating on very, very thin ice.

The Domain of the Giant Planets

Gas Giants

Jupiter and Saturn (Figs. 4 and 5), in addition to being two orders of magnitude more massive than the Terrestrial planets (see Table 2), have very differ-

Fig. 4. Gas giant Jupiter from the Galileo spacecraft, showing its banded cloud structure and the Great Red Spot. Image from NASA

ent, much more volatile-rich compositions. Jupiter and Saturn are mass-wise dominated by hydrogen (H_2) and helium (He) and are known as "gas giants."

The formation of the giant planets is imperfectly understood. Prevailing ideas suggest that, in the Solar system, the gas giant planets formed by a process of nucleated instability, a bit like a rain drop forming by condensation of water molecules on a refractory aerosol. The model was developed by Mizuno and others [111,123]. Briefly, solid bodies collide and grow by binary accretion in the protoplanetary disk, much as they did in the domain of the Terrestrial planets. Upon reaching a critical mass, generally estimated to be $\sim$10 $M_\oplus$, the core precipitates the infall of surrounding nebular gas, producing a hydrody-

Fig. 5. Gas giant Saturn from the Cassini spacecraft. Courtesy NASA

Table 2. Giant Planets

Object	Mass/$M_\oplus$	Radius/$R_\oplus$	ρ [kg m^{-3}]	a [AU]	e	i [deg]
Jupiter	316	11.21	1330	5.203	0.048	1.3
Saturn	95	9.45	700	9.537	0.054	2.5
Uranus	14.5	4.01	1300	19.191	0.047	0.8
Neptune	16.6	3.88	2300	30.068	0.009	1.8

namic flow that results in very rapid mass growth of the planet. As the planet mass undergoes a runaway growth, tidal torques exerted by the planet on the protoplanetary disk open a "gap" around the orbit of the planet. Subsequent mass in-flow to the planet continues at a reduced rate.

Growth by nucleated instability clearly involves two distinct timescales. First, the core must grow to critical mass. Second, the nebular gas must be accreted by the core. Core growth, which occurs by binary accretion as for the terrestrial planets, is the slower process. It is the principal cause of concern with the nucleated instability model and so has been the subject of much attention. The key issue is that the core must grow on a timescale that is short compared with the timescale for the dissipation of the gas nebula. Observations of young stars with dusts disks generally fail to reveal attendant gas, leading to the inference that the gas is quickly removed, probably on timescales of a few Myr for sun-like stars and almost certainly on timescales <10 Myr [10]. This sets an upper limit to the core growth times and is a primary challenge to the core accretion model. One way in which core growth might have been accelerated is through an increase in the disk column density just beyond the snow-line, owing to the extra mass in solids added by the freeze-out of nebular water vapor [20]. Million year growth times at the orbit of Jupiter are not hard to obtain from current models, but more work is needed to induce Uranus and, especially, Neptune to grow on cosmically reasonable timescales.

A different giant planet growth scenario has been proposed in which the "slow step" of core accretion is side-stepped. In this model, the protoplanetary disk is supposed to have been intrinsically unstable to collapse under its own gravity. Disk instabilities clearly favor higher than MMSN disks (models typically assume disk masses ~10 times the MMSN in order to produce spontaneous collapse), but even MMSN models have been reported to be susceptible to collapse under some circumstances [8]. Formation of giant planets by spontaneous collapse does not suffer the timescale problem of the nucleated instability model (because there is no need to wait for a nucleus to form), but there are other problems related to the long-term stability of the collapsing planet. Investigators differ on this issue. The differences are not fully understood, but might relate to the accuracy with which cooling processes are represented [14].

Neither core accretion nor nebula collapse predicted the over-abundance of heavy elements measured in Jupiter by the Galileo entry probe ([120], see

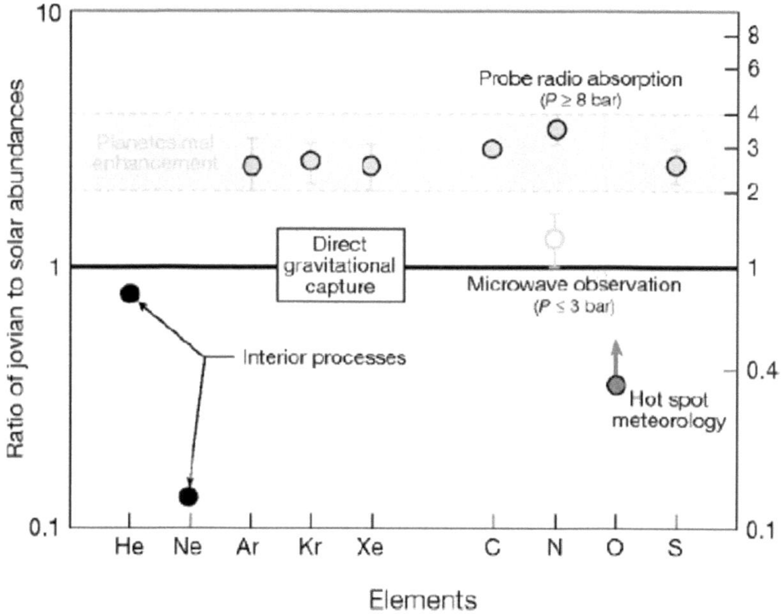

Fig. 6. Metal abundances in Jupiter relative to those in the Sun, as measured by the Galileo entry probe. Helium and Neon are low in abundance because they are partly dissolved in the metallic hydrogen core. Oxygen is low, probably because the probe entered Jupiter's atmosphere at an (unrepresentative) hot-spot location, where conditions were atypically dry. The other measured elements are over-abundant relative to their Solar proportions. From [120]

Fig. 6). In fact, pure collapse models implicitly contradict it because gravitational instabilities provide no way to selectively accrete elements according to their molecular weight. Pressure gradient forces might help to concentrate solids near growing planets [56], and one might conjecture that Jupiter's heavy elements were accreted by the capture of ice-rich planetesimals in the extended atmosphere of the newly formed planet. There are problems with providing enough planetesimals to deliver the mass of Jupiter's metal excess above Solar composition. This process further fails to explain N and Ar, which are over-abundant in Jupiter by factors of 3 or 4 (Fig. 6) but which are too volatile to be carried by asteroids or the known comets in any appreciable abundance. The suggestion advanced by Owen et al. [120] is that Jupiter's core grew by the accretion of ultra-cold ($\sim$30 K) planetesimals, in which N, Ar, and other volatiles were efficiently trapped (probably by adsorption within amorphous water ice). But 30 K is too cold to fit the protoplanetary disk at 5 AU (c.f. Equation 2, which gives $T = 125$ K at this distance). A convincing resolution of this puzzle has yet to be identified.

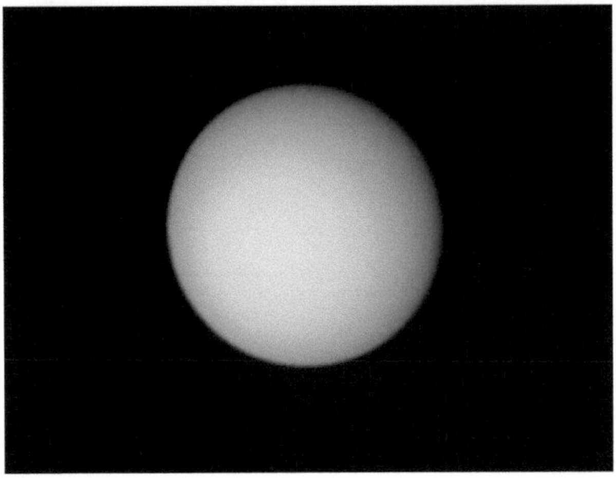

Fig. 7. Ice giant Uranus from the Voyager 2 spacecraft. Courtesy NASA

Ice Giants

Compared to Jupiter and Saturn, Uranus (Fig. 7) and Neptune (Fig. 8) are an order of magnitude less massive and also compositionally distinct, being depleted in H_2 and He. The bulk of their mass is contained in heavier elements that form ices at low temperatures, such as C, N, and O. Uranus and Neptune are known as "ice giants" for this reason. The difficulty in forming Uranus and Neptune on any reasonable timescale has motivated a number of novel, alternative suggestions. For example, in one well-publicized model, Uranus and Neptune are envisioned to have formed between Jupiter and Saturn, were then scattered outwards by mutual perturbations, and, finally, their orbits

Fig. 8. Ice giant Neptune from the Voyager 2 spacecraft. Courtesy NASA

were circularized by friction with an assumed massive disk [149]. To make all this happen, the authors placed the giant planets initially at 6.0, 7.4, 9.0, and 11.1 AU and assumed that they were initially each of $10\,M_\oplus$, with an additional $95\,M_\oplus$ of planetesimals between 12 AU and the assumed edge of the protoplanetary disk at 60 AU. In common with almost all other N-body Solar system simulations, they neglected collective interactions in the $95\,M_\oplus$ disk (these might be expected to generate waves that could be important in the redistribution of angular momentum in the disk [155]). Dynamical effects of the few $\times 10^4\,M_\oplus$ of nebular gas (which must also have been present in order to keep the overall disk composition in approximately cosmic proportions) were also neglected, except that some of this gas was used to feed the runaway growth of the gas giants. The authors assert that their scenario for Uranus and Neptune formation is insensitive to the above assumptions, and, indeed, it is easy to imagine that the first core to experience runaway mass growth should exert a strong gravitational influence on other cores nearby, perhaps scattering them outwards. On the other hand, the initial conditions may have been very different from the ones envisioned in [149]. Worst of all, it is not clear to me what new observations can be taken to test it.

An equally fascinating but rather different scenario for rapid ice giant formation assumes that these planets started out as gas giants and were then eroded down to their observed masses by intense fluxes of ionizing radiation from a nearby, massive star [9]. According to this model, the future ice giants are selectively depleted in mass relative to the surviving gas giants because they are more distant from the sun. Photoionized hydrogen (whose temperature is $\sim 10^4$ K and thermal velocity $\sim 10\,\mathrm{km\,s^{-1}}$) escapes more rapidly from heliocentric orbit at the distances of Uranus and Neptune than at Jupiter and Saturn, leaving the former two planets unprotected from the radiation while the latter two are heavily shielded. Again, the authors do not suggest observational tests of this model, although non-thermal loss of gases from planetary atmospheres often leads to isotopic fractionation effects that might be expected in this extreme case.

The Domain of the Comets

There are several useful definitions of what it is to be a comet, not all of them mutually consistent. The different definitions are used concurrently, sometimes without a clear understanding of the differences between them. The three different classification schemes are idealized in Fig. 9.

Observationally, a comet is any object showing a gravitationally unbound atmosphere, known as a “coma” (from the Greek for “hair”). The coma is a low-surface brightness region surrounding the central, mass-dominant nucleus. It owes its brightness to a combination of sunlight resonantly scattered from molecules and molecular fragments (radicals) and light scattered from tiny dust particles entrained in the outflowing gas. The visibility of the coma depends on the instrumental sensitivity and angular resolution.

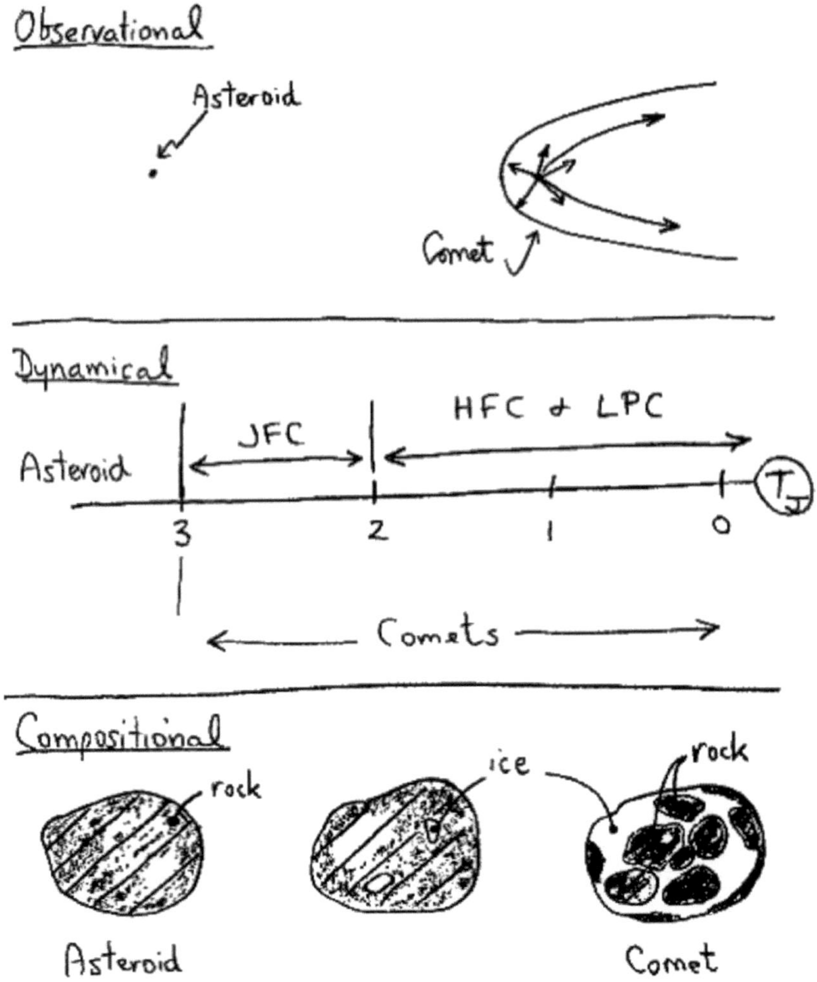

Fig. 9. Schematic diagram showing three different criteria for distinguishing comets from asteroids. Observationally, a comet is any body showing a coma (unbound atmosphere) at any point in its orbit. Dynamically, the distinction is made based on some model parameter, typically the Tisserand parameter, T_J. JFC, HFC, and LPC denote Jupiter-Family Comets, Halley-Family Comets, and Long-Period Comets. The Main-Belt Comets (MBCs) are located with the asteroids, in the middle panel of the figure. Compositionally, the distinction is based on the presence or absence of bulk ice in the body. The different definitions lead to the same classification in most cases, but there are growing numbers of bodies that are "cometary" by one definition but not the others

For this reason, objects that are discovered by survey telescopes as "asteroids" (i.e., bodies having no atmospheres) are commonly reclassified as comets based on the subsequent detection of comae by observers using more sensitive telescopes. Moreover, the strength of the coma diminishes rapidly with heliocentric distance, falling to invisibility beyond the orbit of Jupiter except in a few unusual cases. On longer timescales, cometary activity can evolve in response to evolutionary process on the surface, in a crust or "mantle" that throttles the release of escaping gas. What appears as a comet now might look completely asteroidal to observers of the twenty second century. Obviously, this observational definition of comet-hood is not at all a perfect one.

Compositionally, a comet may be defined as a small body in which a substantial part of the mass is contained in ice. Practically, we may expect all objects that condensed beyond the "snow-line" to contain bulk water ice. The snow-line is now near the orbit of Jupiter; all small bodies from the Jovian Trojans outward are likely to be compositional comets by this reasoning, whether or not they show comae. In the past, the snow-line may have been closer to the sun, meaning that ice could be present in many of the main-belt asteroids. These bodies are compositionally comets. Unfortunately, we have no meaningful way to estimate the bulk composition of a body without drilling into it, and this definition of comet-hood is consequently hard to apply.

Dynamically, a comet is any body with a Tisserand parameter measured with respect to Jupiter, $T_{\mathrm{J}} \leq 3$ (the main-belt asteroids have $T_{\mathrm{J}} > 3$). The Tisserand parameter is a constant of the motion in the restricted, circular three-body approximation, defined by

$$T_{\mathrm{J}} = \frac{a_{\mathrm{J}}}{a} + 2\left[(1-e^2)\frac{a}{a_{\mathrm{J}}}\right]^{1/2} cos(i) \tag{3}$$

where a_{J} is the semimajor axis of Jupiter's orbit (assumed circular); a, e, and i are the semimajor axis, eccentricity and inclination of the small body orbit. Bodies with $T_{\mathrm{J}} \leq 3$ strongly interact with the planet, indicating a short dynamical lifetime and a source elsewhere. Those with $T_{\mathrm{J}} > 3$ are effectively decoupled from the planet. This definition, although seemingly clean-cut, also suffers from ambiguity. Some main-belt asteroids can be scattered onto orbits with $T_{\mathrm{J}} \leq 3$. A few comets (the most famous is 2P/Encke) have $T_{\mathrm{J}} > 3$ (although only slightly so), making them dynamically asteroidal.

The timescale for the loss of volatiles from a body is just $\tau_{\mathrm{dv}} \sim M/(\mathrm{d}M/\mathrm{d}t))$, where M is the mass and $\mathrm{d}M/\mathrm{d}t$ the rate of loss of mass. Whipple and authors since have assumed that mass loss is predominantly by sublimation [?], at a rate that can be calculated from the assumption of radiative equilibrium on the nucleus. There is growing evidence that the mass loss in at least some comets may be dominated by disintegration of the nucleus, in which mass is shed in macroscopic blocks or chunks rather than molecule-by-molecule as in the process of sublimation. Neglecting this possibility for the moment, we write the energy balance equation for a sublimating ice patch as

$$\frac{L_\odot}{4\pi R^2}(1-A)cos(\theta) = \epsilon\sigma T^4 + L(T)\frac{\mathrm{d}m}{\mathrm{d}t} + f_\mathrm{c} + f_\mathrm{g}. \quad (4)$$

Here, $L_\odot$ is the luminosity of the Sun, R is the heliocentric distance, A and ϵ are the albedo and the emissivity of the surface, θ is the angle between the direction to the Sun and the surface normal, $L(T)$ is the latent heat of sublimation of the ice at temperature T, dm/dt is the mass loss rate per unit area and f_c represents the conducted energy flux from the surface while f_g is the flux of energy carried by gas flow into the nucleus. A few things should be noted. The quantity $L_\odot/(4\pi R^2)$ is the flux of sunlight falling on the projected surface. When evaluated at $R = 1\,\mathrm{AU}$, this quantity is called the Solar Constant, $F_\odot$, and has the value $F_\odot = 1360\,\mathrm{W\,m^{-2}}$. The first term on the right-hand side represents the power per unit area lost by radiation into space. The second term is the power per unit area consumed by sublimation. Physically this power is used to break the bonds connecting molecules together in the solid phase. The last term in the equation accounts for thermal conduction and can be either positive or negative, depending on the temperature gradient in the upper layers of the nucleus.

For a non-volatile ($L \rightarrow \infty$) black-body ($A = 0$, $\epsilon = 1$) material oriented perpendicular to the Sun ($\theta = 0$) and neglecting thermal conduction, the temperature is just

$$T = \left[\frac{F_\odot}{\sigma R_\mathrm{AU}^2}\right]^{1/4} \sim \frac{393}{R_\mathrm{AU}^{1/2}}. \quad (5)$$

This corresponds to the temperature at the sub-Solar point on a perfectly absorbing body. The average temperature on a spherical isothermal object will be reduced by a factor $4^{1/4}$, because the average value of $\cos(\theta)$ over the sunlit hemisphere is 1/4, giving $T \sim 278/R_\mathrm{AU}^{1/2}$.

For a sublimating surface, (3) cannot be solved without prior knowledge of the temperature dependence of the latent heat. The Clausius–Clapeyron equation (for the slope of the solid-gas phase boundary in pressure vs. temperature space) can be used or, more directly, measurements of the thermal pressure exerted by sublimating water ice as a function of temperature can be employed. For illustrative purposes, we here consider an extreme approximation.

When close to the Sun (say for $R_\mathrm{AU} < 1\,\mathrm{AU}$) water ice, the dominant cometary volatile, uses so much energy to sublimate that we may write

$$\frac{L_\odot}{4\pi R^2}(1-A)\cos(\theta) \sim L(T)\frac{\mathrm{d}m}{\mathrm{d}t}. \quad (6)$$

as a rough approximation to (4). Then, we see that the characteristic mass loss rate per unit area (again with $\theta = 0$) is just

$$\frac{\mathrm{d}m}{\mathrm{d}t} \sim \frac{F_\odot}{L(T)R_\mathrm{AU}^2} \quad (7)$$

and we have assumed for simplicity that the surface is perfectly absorbing, $A = 0$. Substituting $F_\odot = 1360\,\mathrm{W\,m^{-2}}$ and $L(T) = 2\times10^6\,\mathrm{J\,kg^{-1}}$ (for water ice), we have $\mathrm{d}m/\mathrm{d}t \sim 7\times10^{-4}/R_\mathrm{AU}^2$ [$\mathrm{kg\,s^{-1}\,m^{-2}}$].

The rate at which the sublimation surface recedes into the body of the nucleus is just

$$\frac{\mathrm{d}\ell}{\mathrm{d}t} = \rho^{-1}\frac{\mathrm{d}m}{\mathrm{d}t} \tag{8}$$

where ρ is the bulk density. With $\rho \sim 1000\,\mathrm{kg\,m^{-3}}$, we estimate $\mathrm{d}\ell/\mathrm{d}t \sim 0.7\,\mu\mathrm{m\,s^{-1}}$ at $R_{\mathrm{AU}} = 1\,\mathrm{AU}$. The sublimation lifetime of a nucleus of radius r_{n} is then

$$\tau_{\mathrm{dv}} \sim \frac{r_{\mathrm{n}}}{\mathrm{d}\ell/\mathrm{d}t} \sim \frac{\rho r_{\mathrm{n}}}{\mathrm{d}m/\mathrm{d}t} \tag{9}$$

and, with the standard values as above, we obtain

$$\tau_{\mathrm{dv}} \sim 50\left(\frac{r_{\mathrm{n}}}{1\,\mathrm{km}}\right)\ [\mathrm{yr}_1] \tag{10}$$

In this equation, the unit of time is denoted yr_1 to emphasize that it is the number of years of equivalent exposure to sunlight at 1 AU.

Of course, no real comets circle the Sun in the orbit of the Earth. Instead, they follow eccentric orbits with larger semimajor axes and are hot enough to sublimate only when they dip in to perihelion. Still, the approximation described above nicely illustrates the fact that sublimation can potentially limit the active lifetimes of the comets to very small values, certainly values that are tiny compared with the 4.6 Gyr age of the Solar system.

Less approximate solutions of the energy balance equation are plotted in Fig. 10. There I show the average value of $\mathrm{d}m/\mathrm{d}t$ computed around the orbits of comets having eccentricities $e = 0$, 0.5 and 0.9, as a function of the semimajor axis. At a given semimajor axis, the net effect of non-zero eccentricity is to increase the orbitally averaged mass loss rate relative to the circular orbit approximation, because sublimation grows fast enough near perihelion to overwhelm the long period of inactivity as the comet sails out to and back from aphelion. Figure 10 shows that, for a typical short-period comet having $a = 4\,\mathrm{AU}$ and $e = 0.3$, the orbitally averaged mass loss rate is $\mathrm{d}m/\mathrm{d}t \sim 10^{-7}$ $[\mathrm{kg\,s^{-1}\,m^{-2}}]$, giving $\mathrm{d}\ell/\mathrm{d}t \sim 10^{-10}\,\mathrm{m\,s^{-1}}$ and $\tau_{\mathrm{dv}} \sim 3{\times}10^5\,\mathrm{yr}$.

All of the above is simplistic and intended merely to make a point, namely that sublimation can destroy nuclei quickly. We will have more to say about this later. For now, we use it to assert that the active comets must be derived from inactive source regions, if they are (as we believe) as old as the Solar system.

Source Regions

Three distinct source regions of the comets are now recognized. One, the Oort Cloud, was identified half a century ago [119] and is well known as the source of the long-period comets. The second, the Kuiper belt, was discovered in 1992 [73] and has played a major role in revamping our understanding of the Solar system. It is the source of the Jupiter Family Comets. The third, the

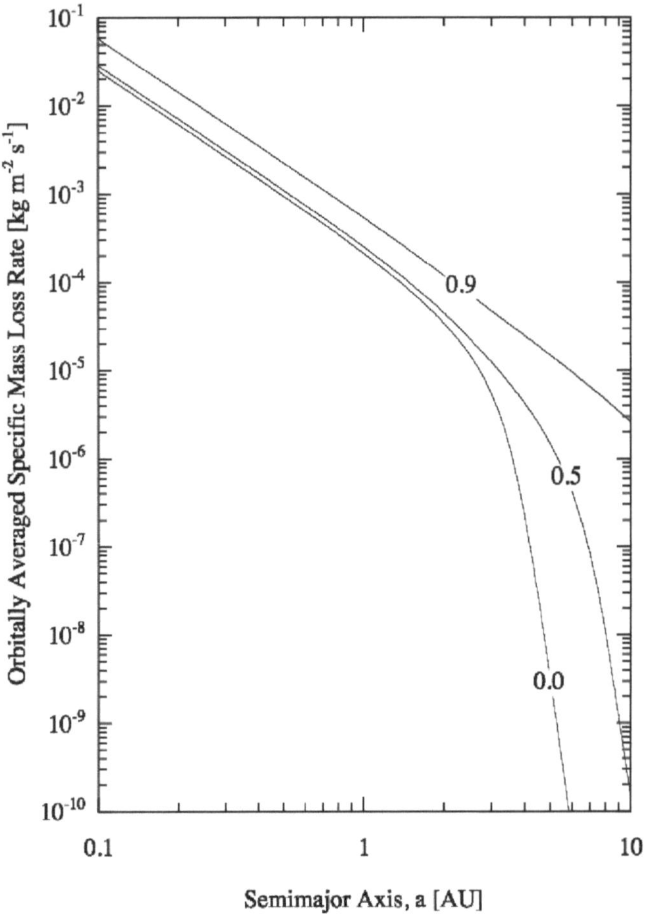

Fig. 10. sublimation rate as a function of semimajor axis for comets having orbital eccentricities as marked. At a given semimajor axis, the sublimation rate averaged around the orbit increases with orbital eccentricity. This is because the enhanced sublimation near perihelion in an eccentric orbit more than compensates for the long period of inactivity surrounding aphelion. From [79]

Main-belt source, was discovered after the Saas Fee workshop [63] and is being written about here for the first time. Comets in this region are unique in being activated not by increased insolation resulting from inward dynamical evolution but by the transient exposure of near-surface ices, probably by collisions with other main-belt objects. Relations between the source regions and various small-body populations in the Solar system are summarized in Fig. 11.

Oort Cloud Source

The Oort Cloud was identified from observations of long-period comets, whose orbits appear randomly (isotropically) distributed over the sky and whose

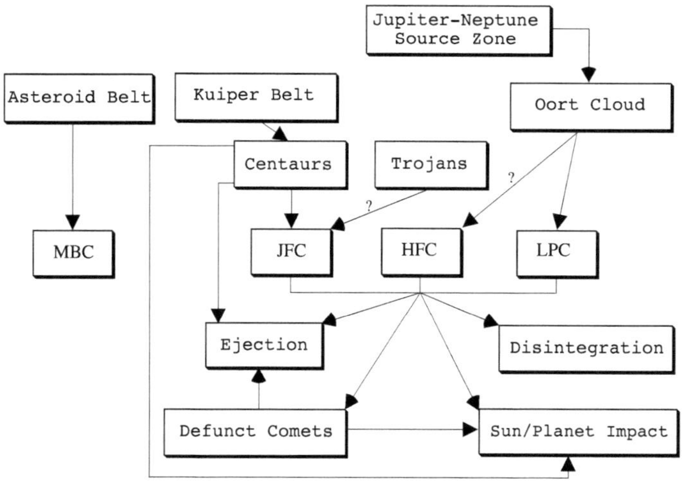

Fig. 11. Flow diagram for the Solar system. This chart shows, at the top, the Kuiper belt and Oort cloud reservoirs. Arrows indicate dynamical flow-down into other populations, including the Jupiter family comets (JFCs), Halley family comets (HFCs), and other long-period comets (LPCs). Escaped Trojans would resemble JFCs. Although no specific cases are known, I have indicated the Trojans as a possible source by an arrow marked "?". The reservoir from which the HFCs are derived is not well understood, but most researchers believe that a source in the inner Oort cloud is likely. This is indicated by another arrow with a "?". On the left is shown the newly identified MBC class, co-located with their source region in the asteroid belt. At the bottom are four processes that represent the demise of the comets

semimajor axes are clustered at large values. The key observation made by Oort was that the orbital energies of many long-period comets (which Oort expressed by the inverse semimajor axes of their orbits) are smaller than the characteristic value of the energy change resulting from gravitational perturbations exerted by Jupiter in a single passage [119]. He concluded that comets were falling into the planetary region from large (but finite) distances, and that many of the long-period comets had not been through the planetary region before, for otherwise they would already have been scattered out of the narrow (bound) energy peak in which they sit. This basic conclusion remains unchanged, to the undying credit of Mr. Oort. Likewise, available data, much improved in quantity and quality since Oort's time, continue to show that the cloud is closely spherical in shape, albeit with a characteristic diameter (~100,000 AU) that is about half the value he calculated (see [158] and [51] for refreshingly written overviews of the observational constraints on the Oort cloud).

Other features of Oort's model are more puzzling. He found very few examples of comets that have been scattered out of the Oort peak (to more tightly bound, smaller orbits), relative to the number of comets in the peak. Three possibilities exist to explain this mismatch between the dynamical model and the data: (1) the model could be wrong, or incomplete, (2) incoming comets could become intrinsically fainter (and therefore harder to detect) once they have passed through the inner Solar system, or (3) a large fraction of the incoming comets could vanish after their first few journeys through the Solar system. There seems to be no great enthusiasm amongst dynamicists for concluding that Oort's dynamical model is wrong or incomplete. Indeed, no dynamical explanation could be found (by Oort in 1950 nor by Wiegert and Tremaine [158] in a careful analysis some 50 years later). Like Oort [119], all researchers have assumed that the disagreement between the data and the model is best explained by fading or disintegration of the incoming comets. However, the nature and reality of the fading remain unidentified. The low rate of detection of weakly active or completely inactive long-period comets has been interpreted as evidence that objects from the Oort cloud do not merely run out of gas but physically disintegrate [94]. This conclusion rests on a poorly known relation between the brightness of active long-period comets and the sizes of their underlying nuclei. For example, if the nuclei are much smaller than assumed in [94], then they might escape detection without disintegrating.

The population and mass of the Oort cloud are also uncertain. The population is derived from measurements of the rate of arrival of new comets from the Oort cloud coupled with models of the rate of erosion of the cloud by external perturbers. Oort considered passing stars to be the main external perturbers. The asymmetric tide of the Milky Way is now thought to be a larger perturber [58]. In addition, the rate of arrival of new comets is subject to observational biases that are difficult to quantify. Until recently, published population estimates relied on the work of visual observers [43, 64], most of whose survey techniques and other details went unpublished. A recent attempt to use data from the LINEAR survey (whose parameters are better, but still not completely, known) gives $\sim 5 \times 10^{11}$ comets with absolute magnitude $H \leq 11$ [51], about 10 times smaller than estimated previously.

Lastly, the relation of the Halley family comets to the Oort Cloud is unclear. These objects have distinctly non-random distributions of inclinations (with some retrograde members but many more prograde ones) and orbital periods, by definition, <200 year. The most likely source is the inner Oort cloud, but the location and population of this region remain poorly constrained.

Kuiper Belt Source

The Kuiper belt became real with the discovery of 1992 QB1 [73]. Before that time, its only observed member was Pluto, misleadingly given planetary status for a host of mostly socio-scientific reasons. In fact, if Pluto had been accurately interpreted in 1930, our study of the structure of the Solar system could have advanced by many decades over the actual case (e.g., many bright KBOs have been "precovered" in photographic plates taken in the 1950s. They were not *discovered* using plates because astronomers did not think to look for them until after the discovery of 1992 QB1). Indeed, at least one astronomer correctly recognized in 1930 that Pluto must be just one of many trans-Neptunian objects [92] based on the dubiousness of the proposition that Tombaugh had been lucky enough to find the only one so soon after starting his survey. This reasoned position was drowned out by the assertion that Pluto must be the long-sought "Planet X," predicted by Percival Lowell on the basis of a model of (what turned out to be unreal) deviations in the motion of Uranus. Still, everything is obvious in hindsight, and it is too easy to see what should have been done knowing what we know, and too difficult to reconstruct the full state of confusion that reigned only a few decades ago. For example, Edgeworth in 1943 [40] speculated about "clusters" in the trans-Plutonian region (clusters were his idea for the structure of comets) while Kuiper (for whom the belt is somewhat ironically named) in 1951 [85] considered that this region should be empty, having been cleared of objects by strong perturbations from "massive" Pluto. Later, Fernandez in 1980 [45] reasoned that a flat disk source was needed to explain the inclination distribution of the short-period comets. Before this time, most researchers had been happy with the contention that short-period comets were somehow dynamically evolved versions of long-period comets. Later still, in 1988, Duncan and collaborators [38] showed, using numerical methods, the correctness of Fernandez' argument.

The dynamics of the Kuiper Belt are extensively and masterfully discussed in the Saas Fee lectures by Alessandro Morbidelli [112].

Main Belt Source

Main belt comets (MBCs) have orbits in the main asteroid belt between Mars and Jupiter. At the time of writing, three MBCs have been identified ([63]; see Fig. 12). The best known is asteroid 7968 also known as 133P/Elst-Pizarro, first observed to be accompanied by a dust trail in 1996. Initially interpreted as an impact-produced dust cloud [150], the reappearance of the trail near perihelion in 2003 showed that another explanation is required [62]. The newest examples are comet P/2005 U1 (Read) and (118401) 1999 RE70, both of which show persistent dust emission over timescales of months. These three objects have similar semimajor axes located beyond 3 AU, in the outer regions of the main belt. Their orbital inclinations are also all small, but the similar a and i are at least in part results of observational bias, because our surveys have targeted exactly these types of object. The MBC orbits are decoupled from both Mars and Jupiter and appear to be dynamically stable on billion year timescales, like those of the main-belt asteroids that occupy exactly the same region of orbital element space (Fig. 13).

Could the MBCs be comets captured from other regions, for example from the Jupiter family comet (JFC) or long-period comet (LPC) populations? As

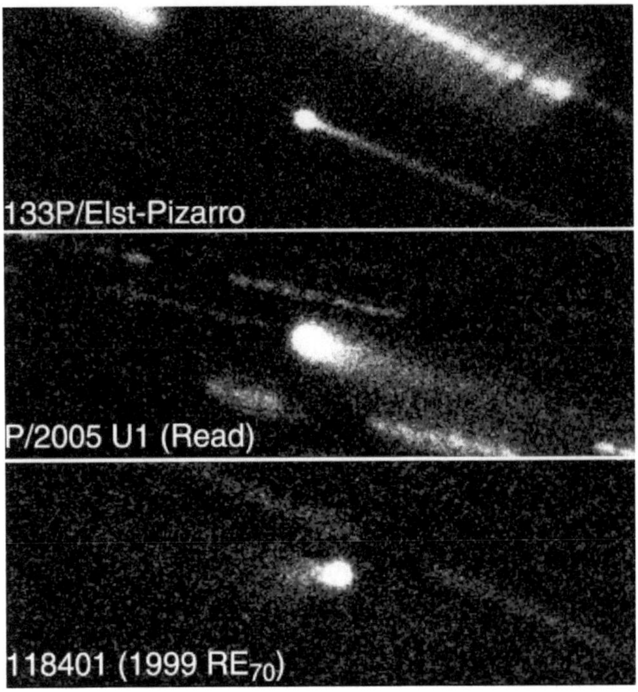

Fig. 12. Three main-belt comets (MBCs) in deep CCD images from Mauna Kea. These objects emit dust like comets but have orbits that are like those of outer main-belt asteroids. Background stars and galaxies appear trailed owing to the non-sidereal motions of the MBCs. From [63]

observers, we are open to this possibility, but dynamical simulations of the motions of comets suggest that this is very unlikely. In fact, pure dynamical calculations completely fail to inject comets into MBC-like orbits even when the perturbations of the Terrestrial planets are included [47, 95]. Some work has been done on the effects of non-gravitational accelerations (caused by asymmetric mass loss from cometary nuclei), but again, MBC-like objects are not produced. Failing some dramatic revision of the dynamics, we are forced to the conclusion that the MBCs are what they appear to be: asteroids that outgas like comets.

Several lines of argument indicate that the mass loss from MBCs is driven by sublimation, probably of near-surface water ice. First, mass loss from 133P has been observed at consecutive perihelia but not in between. This is exactly as expected for sublimation-driven activity. The sunward "nose" of the coma of P/2005 U1 (Read) is well resolved, with an apex scale of several arcseconds. This implies that the particles are ejected from the nucleus at considerable speed ($>100\,\mathrm{m\,s^{-1}}$), as expected for water ice sublimating at $\sim 3\,\mathrm{AU}$. Other explanations for mass loss seem less viable. The nucleus of 133P is rapidly rotating, and it is possible that centripetal effects assist the launching of dust

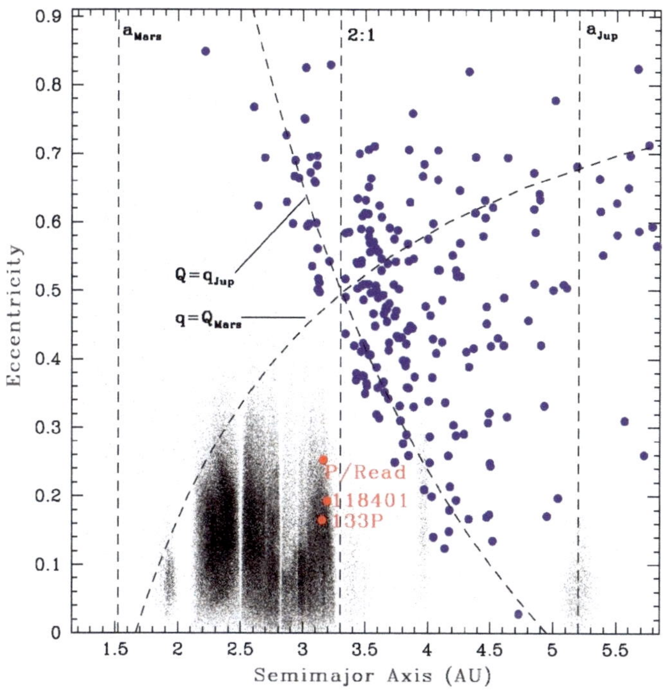

Fig. 13. Semimajor axis vs. orbital eccentricity for asteroids (small black dots), Jupiter family comets (blue circles) and the known MBCs (red circles). Vertical dashed lines mark the semimajor axes of the orbits of Mars and Jupiter and the 2:1 mean-motion resonance with Jupiter, which practically defines the outer edge of the main belt. Curved dashed lines show the locus of orbits which are just Mars and Jupiter crossing. Objects below these two curves cross neither Mars nor Jupiter, like essentially all of the main-belt asteroids. The MBCs fall within the domain occupied by stable main-belt asteroids and far from the periodic comets. From [63]

particles from its surface. However, centripetal effects alone cannot explain the observation that activity is confined to perihelion. Neither do we find evidence for rapid rotation in P/2005 U1 (Read) or 1999 RE70: these objects spin so slowly that rotation can play no role in the mass loss. Electrostatic levitation of dust grains has been observed in the terminator regions of the moon, where the derived velocities of the dust grains are $\sim 1\,\mathrm{m\,s^{-1}}$. Such low speeds are incompatible with the extended coma of P/2005 U1 (Read) and, furthermore, it is hard to see how electrostatic ejection of grains could be episodic (as on 133P), or why it would be confined to only three of several hundred asteroids examined in detail by our ongoing survey.

For these reasons, it appears that the MBCs are really comets in a special population where the source reservoir and the current locations are one and the same. Unlike the long- and short-period comets, the MBCs are not activated by being brought from cold storage locations into the hot inner Solar system. Instead, we suspect that they are activated collisionally. For example, the mass loss from P/2005 U1 (Read) corresponds to sublimation from an exposed patch of dirty water ice having a diameter of only ~20 m. Such a patch could be exposed by the impact of a meter-scale boulder into the nucleus surface. The mass loss rate at 3 AU is about 10^{-5} kg s^{-1} m^{-2} (Fig. 10) and, with density $\rho \sim 1000\,\mathrm{kg\,m^{-3}}$, the surface recession rate is $d\ell/dt \sim 10^{-8}\,\mathrm{m\,s^{-1}}$. An ice patch 20 m in diameter would sublimate to a depth equal to its diameter on timescale $\tau \sim 2 \times 10^9$ s (50 years), thereafter declining into inactivity from self-shadowing. Triggering collisions involving the impact of 1-m scale boulders should not be overly rare: we expect to find many MBCs in planned all-sky surveys such as Pan STARRS.

Is ice in the asteroid belt surprising? It should not be. Some meteorites show textural and geochemical evidence that they have been aqueously altered, probably by being bathed in liquid water at temperatures not far above the triple point [84]. This evidence includes the presence of clay minerals and serpentines that most naturally form with water, as well as carbonates and mineral deposits in veins that cross-cut other structures in the meteorites (showing that the vein materials were emplaced after formation). Spectrally, about half of the outer belt asteroids show absorption features attributed to water of hydration in minerals (not free ice, but water bound chemically within other materials such as clays [16]). At both smaller and larger distances, the prevalence of these hydration features decreases. One interpretation that fits the available data is that, at smaller distances, the asteroids were too hot for liquid water to have survived while at larger distances the ice was so cold as to never be melted, foreclosing the possibility of hydration reactions that could produce water of hydration bands [82].

The greatest excitement behind the MBCs lies in the potential relation between these objects and the oceans (and, through water, life). Earth probably formed too hot to trap much water, and so, it is widely believed that a separate source is required. Possible sources include the comets (but the measured deuterium/hydrogen (D/H) ratios in the three that have been measured seems higher than in the oceans [109]) or watery asteroids like the MBCs [113]. MBCs are so close to Earth that we should soon be able to visit them with a mass spectrometer, to measure their D/H (and other isotope ratios, including $^{16}O/^{17}O/^{18}O$) abundances, and so to make a direct comparison with the oceans.

3 Cometary Nuclei

The nucleus is the fundamental component of any comet because it contains most of the mass. Unfortunately, it is also the hardest to study, because most of the cross-section is carried by dust and gas ejected from the nucleus and not by the nucleus itself. As a result, physical studies of comets have, until recent times, been biassed toward the study of gas and dust released from the nucleus by its sublimation. These are the subjects of Heike Rauer's [128] lectures in this Saas Fee workshop. A great deal of important information about comets has been gleaned, for example, from the study of molecular fragments from dissociated parent molecules. In this section, though, I want to focus on what we have learned about the nucleus itself.

The first well-established detections of nuclei were achieved from the ground in 1984, quickly followed by close-up images of the nucleus of 1P/Halley in 1986 (Fig. 14). Before that time, direct observation of the nucleus was held by many to be impossible because of contamination of the nuclear signal by scattering from nearby dust and gas. A common misperception is that the nucleus is invisible from the ground because it is shielded from view by near-nucleus dust. This is almost never the case for a very simple reason: dust is ejected from the nucleus by the drag forces exerted on it by sublimated ice. If the coma were to become optically thick, the source of heat driving the sublimation would be shut down, reducing the dust opacity. Feedback, then, stabilizes the optical depth along a line of sight from the nucleus to the Sun, to be smaller than unity. Transient exceptions to this feedback control can

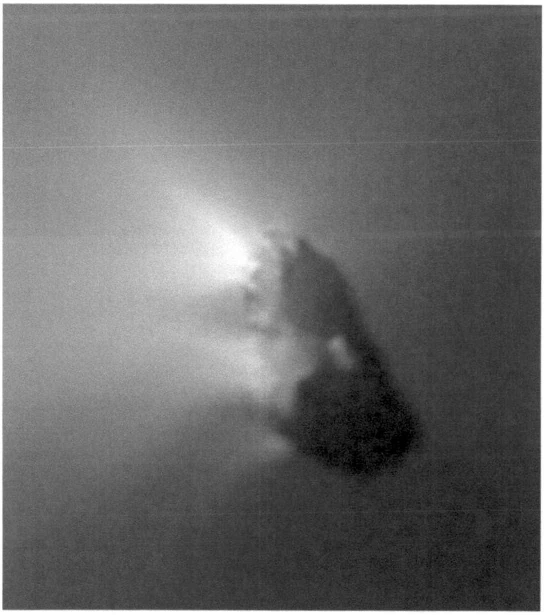

Fig. 14. Nucleus of 1P/Halley imaged from the ESA Giotto spacecraft. This classic image was the first to show a nucleus at high spatial resolution. While various surface features can be discerned, it is obvious that important structure lurks beneath the resolution of the data. Dust jets are seen to emanate primarily from the sun-facing side of the nucleus. Courtesy Giotto camera PI H. U. Keller and ESA

be imagined and might occur, but few or none of the observed properties of comets require large broadband optical depths to be understood. Cometary comae are, to a good level of approximation, optically thin. Since 1986, the nuclei of comets Borrelly (Fig. 15), Wild 2 (Fig. 16), and Tempel 1 (Fig. 17) have been imaged by spacecraft.

Nucleus size

Cometary nuclei subtend minuscule angular diameters (milliarcseconds) and are unresolved in optical ground-based data. No occultation of a field star by a nucleus has ever been observed: most nucleus sizes must be inferred by indirect means. The size of the cometary nucleus can be inferred from the "classical" technique first used by Dave Allen [3] in which simultaneous optical (scattered) and infrared (thermally emitted) flux densities are compared. This method is so important to the study of small bodies that it is worth describing in more detail: in essence it is very simple. Photons from the Sun strike a body and are either reflected or absorbed. The fraction reflected is called the "Bond albedo," A. Photons not reflected are absorbed, raising the temperature of the body and producing thermally emitted photons at longer wavelength. The fraction of the incident photons that is absorbed is $(1 - A)$.

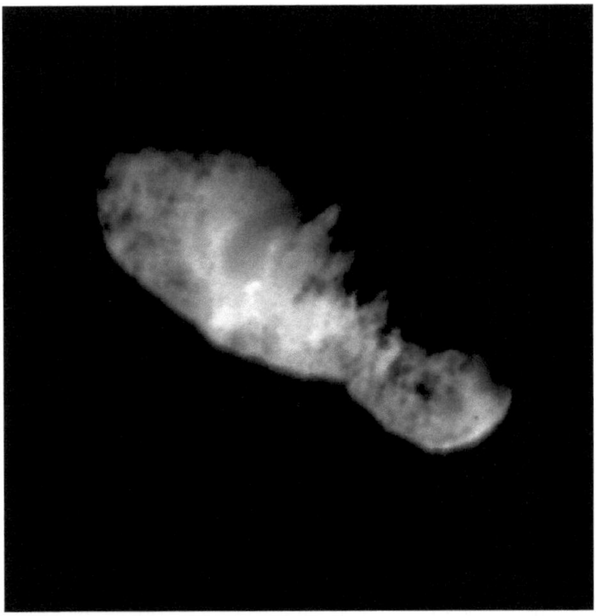

Fig. 15. Nucleus of P/Borrelly imaged from NASA's Deep Space 1 spacecraft. The effective radius is ~2.2 km and surface albedo ~0.03. Note the lobed structure of the nucleus (perhaps caused by a composite structure consisting of two major bodies in contact) and the smooth "pond" material above the waist. Courtesy NASA

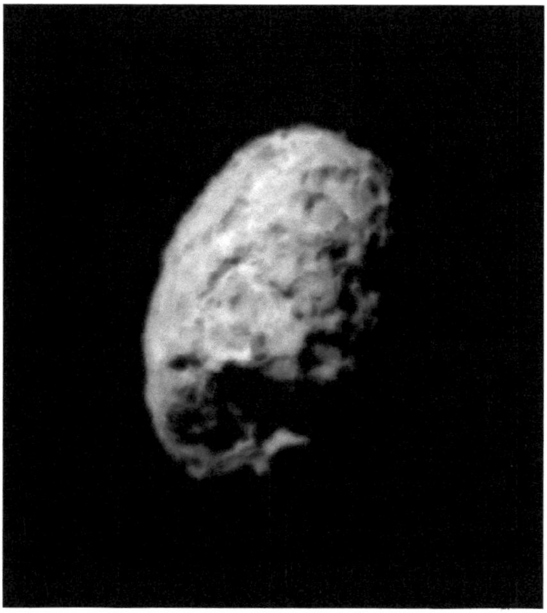

Fig. 16. Nucleus of P/Wild 2 imaged from the NASA Stardust spacecraft. The effective radius is ~2.1 km and surface albedo ~0.03. Note the remarkably smooth shape of the nucleus, which resembles that of a rotational figure of equilibrium. Courtesy Don Brownlee and NASA

The optical flux density scattered from a body is proportional to the product $C_e p$, where C_e is the cross-section, while the thermally emitted flux density is proportional to $C_e(1-p)$. Here, p is the "geometric albedo," which is related to the Bond albedo by $A = pq$, where q is a measure of the angular dependence of the scattering function called the "phase function." Provided q is known, measurements at optical and thermal wavelengths permit us to solve for the two unknowns C_e and p (cf. Fig. 18). The measurements should be simultaneous because small bodies are usually not spherical, causing C_e to vary with time.

Examined closely, the Allen size method is more complicated. The scattered radiation is both anisotropic and wavelength dependent, introducing two extra parameters. The phase function q is not in general known and has only been measured for a few bodies that can be observed over a very wide range of phase angles. Real surface materials will have (wavelength dependent) thermal emissivities <1, introducing another parameter. Most seriously of all, heat absorbed on the day-side of a rotating body can be carried by rotation onto the night-side, meaning that the emitted flux density depends on the heat-retaining capacity of the surface layers (traditionally characterized by the "thermal inertia parameter" $I = k\rho c_p$, where k is the thermal conductivity, ρ is the bulk density and c_p is the specific heat capacity, or by the "thermal diffusivity" defined by $\kappa = k/\rho c_p$. The magnitude of this interaction

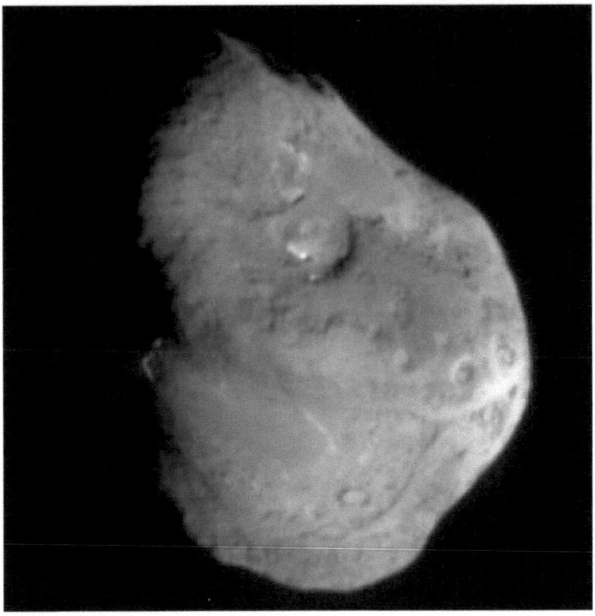

Fig. 17. Nucleus of P/Tempel 1 imaged from the NASA Deep Impact spacecraft. The effective radius is ∼3.1 km and surface albedo ∼0.05. Note the craters, the left–right gash across the nucleus, and the two regions of smooth terrain apparently occupying lowland positions. Courtesy Mike A'Hearn and NASA

between the rotation and the thermal emission introduces more parameters, for the thermal constants of the surface, and for the magnitude and orientation of the rotation vector relative to the line of sight. What looked like a conceptually simple method is in fact horribly complicated: the number of unknown parameters in the model generally exceeds the number of observational constraints.

What saves the Allen method is the empirical finding that assumed values for a great many of the unknown parameters can nevertheless give object cross-sections and albedos of useful accuracy. The "Standard Thermal Model" (STM) has arisen as a way to bundle many assumptions in such a way that they are not too visible to the user and so not too frightening! In STM, the thermal emission is assumed to emanate from a spherical body in which the surface temperature is set by instantaneous equilibrium with sunlight and where the effects of rotation are unimportant. This could mean that the surface heat retention is very small, so that heat is lost before rotation carries it away from the day-side, or it could mean that the rotation vector points exactly at the Sun, so that rotation does not change the surface heating pattern. Even with these and other assumptions for the emissivity (generally ∼0.9) and the angular dependence of the scattering, STM must include a fudge factor called η, the "beaming parameter," that is supposed to represent

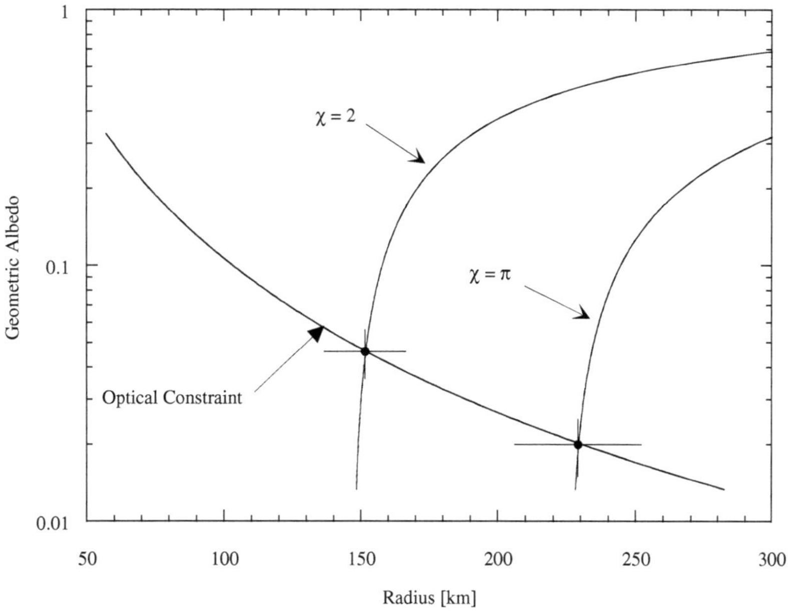

Fig. 18. Example of the thermal–optical method of determining the size and albedo of an object. The optical data place a constraint on the product $p_R r^2$, where p_R is the geometric albedo and r is the effective radius. The thermal data place, through a model of the surface temperature distribution, a constraint on $(1 - p_R)r^2$. The two curves labeled $\chi = 2$ and $\chi = \pi$ refer to the STM and ILM surface temperature approximations. The dots mark plausible solutions for these two models. Both yield low geometric albedos for this object. From [74]

the angular dependence of the emission from the surface caused by surface roughness and topographic effects. The value of η in STM is often taken to be $\eta = 0.756$ ([89]) but in fact it is very uncertain and recent work suggests that $\eta = 1$ may apply. For our purposes, the point is that the interpretation of thermal emission data in terms of object size and albedo depends on poorly specified parameters such as η.

A counterpart to the STM is the "Isothermal-Latitude Model" (ILM) which is best thought of as applying to a spherical body with the Sun in its equator and a rotation period so short that the temperature is independent of azimuth and a function only of latitude. The ILM model has lower mean surface temperatures than the STM and so requires a larger C_e (and smaller p) to generate a given thermal emission signal.

The strength of the Allen method is that it is widely applicable and seems mostly to give diameters accurate to ~5 or 10% when appropriately "tuned" by the selection of the uncertain parameters. It has been used to measure the cross-sections and albedos of about a dozen comets, as listed in Table 3.

Table 3. Well-Measured Cometary Nuclei

Object	r_e^a	p[b]	P[c]	b/a[d]
1P/Halley	5.5	0.04±0.02	52.8,177.6	2.0
2P/Encke	2.4	0.05±0.02	11?	2.6
9P/Tempel 1	3.1	0.05±0.02	41.0	1.4
10P/Tempel 2	5.3	0.022±0.005	9.0	1.7
19P/Borrelly	2.2	0.03	25.0	2.5
22P/Kopff	1.7	0.042±0.006	12.3	1.7
28P/Neujmin 1	10.7	0.03±0.01	12.75	1.5
49P/Arend-Rigaux	4.2	0.04±0.04	13.47	1.6
81P/Wild 2	2.1	0.03±0.01	12?	1.7
107P/Wilson-Harrington	1.7	0.05±0.01	6.1	1.2
C/1995 O1 (Hale-Bopp)	37	0.04±0.03	11.34	2.6
C/2001 OG108 (LONEOS)	8.9	0.030±0.005	57.19	1.3

[a]Effective radius [km]. [b]Visual albedo. [c]rotation period [hr]. [d]Axis ratio.

The size distribution of the cometary nuclei has been measured by different groups with different investigators reaching different conclusions. Usually the size distribution is represented as a power law with index q

$$n(r)\mathrm{d}r = \Gamma r^{-q}\mathrm{d}r \tag{11}$$

where Γ is a normalization constant. Reported values are $q = 3.6^{+0.3}_{-0.2}$ [46], $q = 2.6\pm0.03$ [157], $q = 2.6\pm0.3$ to 2.9 ± 0.3 [88], $q = 2.45\pm0.05$ for the radius range 1–10 km and $q = 1.91\pm0.06$ for radius between 2 and 5 km [108] and $q = 3.7\pm0.3$ also for radius between 2 and 5 km [144].

What do these values mean and why are they so different? The large scatter amongst the measurements has been attributed by Tancredi et al. [144] as the result of poor sample definition and, in some cases, the use of inaccurate nuclear magnitudes. When only JFC nuclei are considered, they obtain an index $q \sim 3.7$ regardless of which data set is used. While many of the measurements were taken apparently in the desire to make a comparison with the size distribution of Kuiper belt objects, this is a difficult comparison to make. First of all, the well-measured Kuiper belt objects (for which the size distribution index is $q = 4.0^{+0.6}_{-0.5}$ [151]) are one to two orders of magnitude larger than the measured cometary nuclei. There is no reason why a single power law should hold from the largest KBOs down to km-sized cometary nuclei, particularly if it is true that the smaller objects are collisional products while the larger KBOs are survivors from a primordial population [44]. Indeed, initial observations of fainter, smaller KBOs show that the size distribution flattens below ~100 km diameter, with an index q ~2.6 in this region [7].

More seriously, the measured cometary nuclei are a highly unrepresentative sample of cometary nuclei as a whole. The measured nuclei tend to be those of comets in which mass loss has almost certainly changed the nucleus shape (see below) and size. In fact, if the sublimation lifetime increases with the nucleus size (see (8) or (9)), then small nuclei should be destroyed faster than large ones, leading to a net flattening of the size distribution relative to the distribution in the initial (pre-heated) population. It is hard to see how the nucleus size distribution can tell us much of fundamental value about the comets so long as our measurements are confined to the relatively evolved nuclei of comets with perihelia in the terrestrial planet domain.

Nucleus Colors

Accurate determinations of the colors are available for a small number of cometary nuclei (Table 4). This value is consistent with the mean colors of various other inner- and middle-Solar system small-body populations, including the Jovian Trojans, the nuclei of dead JFCs, and the Damocloids (likely nuclei of dead Halley-family comets, see Fig. 19). However, the optical colors of comets are not consistent with those of KBOs or Centaurs, in the sense that the ultrared matter ($S' > 25\%/1000\,\text{Å}$) found on many of these objects is completely absent on the nuclei (Table 4). We will return to this observation in our discussion of the effects of a surface mantle.

Nucleus Shape and Rotation

The shapes and rotational states of cometary nuclei (and asteroids) can be determined from their rotational lightcurves (temporal variations in the scattered light). The main measurable parameters are the lightcurve period and the range. Two things are immediately worth mentioning. First, the relation between the lightcurve period and the underlying rotational period may not be obvious, a priori. If the lightcurve is caused by albedo spots, then the two are likely to be equal. If the lightcurve is caused by variations in the projected cross-section owing to aspherical shape, then the lightcurve period is likely to be half the rotational period. Where sufficient data exist to discriminate between these possibilities, the lightcurves are almost always found to be caused by aspherical shape more than by albedo spots. Second, the range of the lightcurve is routinely but inaccurately described in the literature as the amplitude (formally the amplitude is half the range). The measured range sets only

Table 4. Mean Optical Reflectivity Gradients [79]

Object	S'_{min}	S'_{max}	S'_{med}	$S' \pm \sigma$	Number
Damocloids	5	17	13	11.9 ± 0	12
Active JFC Nuclei	–5	22	11	11.6 ± 2.3	11
Inactive JFC Nuclei	–6	18	6	7.2 ± 2.0	12
Trojans	3	25	9	9.6 ± 0.9	32
Centaurs	0	43	19	20.3 ± 2.8	22
KBOs	–10	48	21	21.1 ± 1.4	83

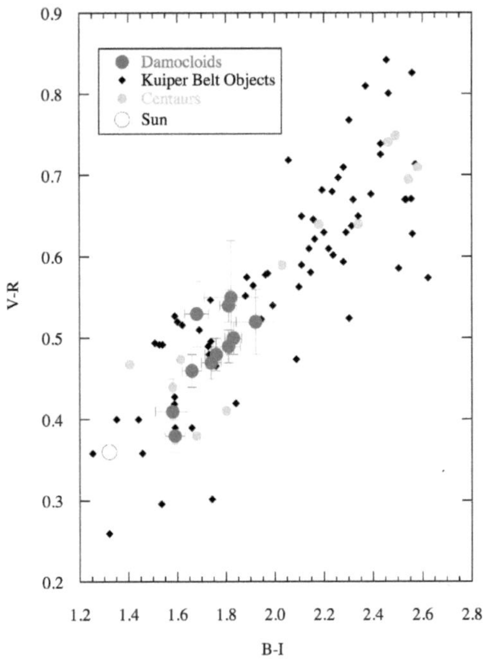

Fig. 19. Color-color plane showing the Damocloids, KBOs, and Centaurs. While the KBOs, and Centaurs show a wide range of surface colors (and, presumably, compositions) the Damocloid surfaces are entirely lacking in ultrared matter (spectral gradient >25%/1000 Å, corresponding to the upper right in this color-color diagram). From [79]

a lower limit to the nucleus axis ratio, because in general the rotational axis will not be aligned perpendicular to the line of sight. Repeated measurements under different geometries are needed to remove these effects of projection.

Figure 20 shows a range vs. period plot for those comets thought to be well-measured. Added to the plot are curves computed for two models. First, I show curves for prolate bodies in rotation about a minor axis, computed under the assumption that gravity at the tip of the spheroid exactly equals the centripetal acceleration there [68]. Second I assume that the nuclei are figures of rotational equilibrium and plot curves taken from Chandrasekhar's (in)famous book [17] in which the shapes of strengthless bodies are computed as a function of their density and angular momentum. The figure shows that the nuclei do not need to be very dense (in general, the critical densities are $<1000\,\mathrm{kg\,m^{-3}}$) in order to be stable against centripetal effects, regardless of which model is used. It is not known whether the nuclei behave at all like

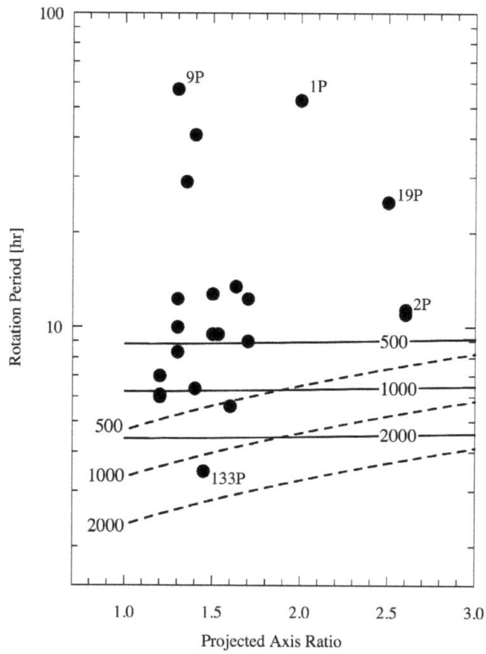

Fig. 20. rotation period vs. axis ratio (derived from lightcurve range) for cometary nuclei. Prolate spheroid curves (dashed lines) were computed as described in the text. The equilibrium spheroids (solid lines) were computed by Pedro Lacerda. The densities of the models are given in the figure and can be interpreted as limits to the nucleus densities under the assumption that the nuclei are strengthless

strengthless bodies, but the consensus view (influenced very strongly by the split comets, see [6]) is that this is likely to be a good approximation.

Nucleus Density

There are no good measurements of the densities of cometary nuclei, but there are many strong opinions held by planetary scientists about what those densities are! Perhaps because of preconceived ideas about the way in which comets formed, most planetary scientists believe that the nuclei are less dense than water. This might be true, but we do not know.

Several indirect methods have been invoked to measure the densities of the cometary nuclei.

- The range vs. period plot was first used to argue that the densities must be low [68]. A prolate ellipsoid nucleus model was used to estimate the density from the period and the lightcurve range. There is no particular reason to assume that the nuclei are well described by prolate ellipsoids and, as can be seen from Fig. 20, an alternate assumption gives substantially lower densities for a given period, range pair.

- D/Shoemaker-Levy 9 was disrupted while passing close to Jupiter (Fig. 21). Measurements of the spreading rate of the "string of pearls" comet after disruption, when interpreted as the product of tidal stresses acting on an aggregate body of negligible tensile strength, give a relatively robust estimate of the density $\rho = 600\,\mathrm{kg\,m^{-3}}$ [6].
- Asymmetrical outgassing exerts a "rocket" acceleration on the nucleus of magnitude

$$\alpha_{\mathrm{n}} = f_{\mathrm{r}} \frac{V}{M_{\mathrm{n}}} \frac{\mathrm{d}M_{\mathrm{n}}}{dt} \tag{12}$$

where M_{n} is the nucleus mass, V is the bulk speed of the material launched from the nucleus by sublimation and f_{r} is a dimensionless constant. The value of f_{r} depends on the angular distribution of the momentum flux in material launched from the nucleus. For a nucleus that ejects matter in a perfectly collimated beam $f_{\mathrm{r}} = 1$ while for isotropic ejection $f_{\mathrm{r}} = 0$. Consider a 1 km radius comet (mass $\sim 4 \times 10^{12}$ kg) ejecting mass at $10^3\,\mathrm{kg\,s^{-1}}$ in a collimated beam ($f_{\mathrm{r}} = 1$) while at 1 AU from the Sun. The rocket acceleration is $\alpha_{\mathrm{n}} \sim 3 \times 10^{-7}\,\mathrm{m\,s^{-1}}$, or about 10^{-5} times the Solar gravity at this distance. Although small, the long action time allows the rocket acceleration to produce measurable deviations from Keplerian motion. To use (12) to determine nucleus density, the acceleration α_{n} must first be measured from astrometry of the comet. Spectroscopy gives V from the Doppler shift of lines resonantly scattered from escaping gas and dM_{n}/dt can be estimated from the strengths of molecular

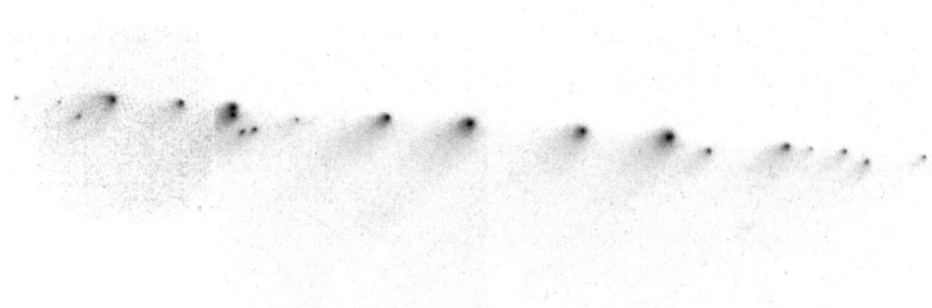

Fig. 21. Multiple components of the nucleus of D/Shoemaker-Levy 9 imaged from the Hubble Space Telescope. Each component sports a stubby tail, created by radiation pressure sweeping of emitted dust. Photometry shows that the emission was largely impulsive and occurred at the moment of break-up of the nucleus as it passed Jupiter (minimum distance 93,500 km or about 1.31 $\mathrm{R_J}$)

emission lines. Then, given a value of f_r, this equation gives the nucleus mass. Coupled with an estimate of the nucleus volume, the density can be determined.

This method has been used to estimate the densities of 81P/Wild 2 ($\rho <$ 600 to 800 kg m^{-3}; [30]), 67P/Churyumov-Gerasimenko ($\rho <$ 600 kg m^{-3}; [29]), 19P/Borrelly ($100 < \rho < 300$ kg m^{-3}; [28]). The low densities are interesting and in accord with the value obtained for D/Shoemaker-Levy 9 by a different method but, given the large amount of modeling needed to estimate f_r, I suspect that this method can give almost any density the user wants.

Still, accepting for the moment that the densities are <1000 kg m^{-3} and that the strengths are small, it is interesting to speculate about the possible internal structures of the nuclei. Most probably, the nuclei are porous dirt-ice mixtures with a broken internal structure consisting of blocks each much smaller than the aggregate size (middle panel of Fig. 22).

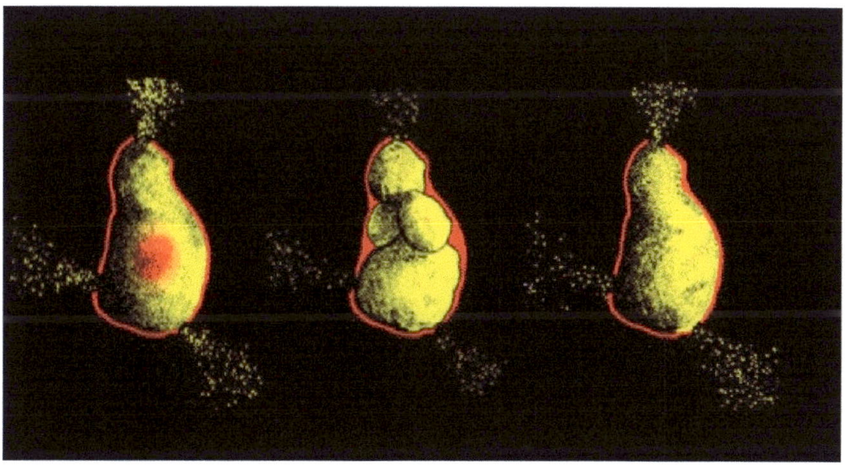

Fig. 22. Schematic of possible internal structure of the cometary nucleus. On the left, a differentiated nucleus in which the material properties (strength, composition) vary radially as a result of past heating, concentrated at the core. This model seems unlikely, given the high-volatile contents and low tensile strengths of comets. However, some of the larger nuclei could have experienced non-negligible internal heating from radioactive decays (enough to mobilize interior volatiles). In the middle, a multi-component (sometimes called "rubble pile") nucleus in which sub-elements in the body are loosely bound by gravity. This is probably closest to the real structure inside cometary nuclei. On the right, a monolithic nucleus with structural integrity over its whole diameter. Very small comets (like asteroids of <100 m scale), could be like this. The red skin on each object symbolizes the non-volatile mantle

3.1 Mantles

Observations show that the surfaces of cometary nuclei are largely non-volatile, consisting of refractory matter generally described as a "mantle" (crust might be a better word, and certainly less confusing given the stratigraphic relationship between the Earth's mantle and crust). Evidence for the existence of mantles includes

- Images from the ground and from space show that the mass loss from comets occurs from only a fraction of the total surface, suggesting that surface volatiles are not widely distributed (note: this says nothing about the distribution of volatiles inside the nucleus). Specifically, the mass loss occurs in jets and the total rate of production of water is less than would be expected if the whole nucleus were covered in water ice. The derived fractional "active areas" range from ~0.01 to ~10% [1].
- Spectral maps of comet 9P/Tempel 1 obtained from the NASA Deep Impact spacecraft show evidence for water only in a few locations occupying about 0.5% of the total surface [140].
- Temperatures of some nuclei are higher than can be sustained by a sublimating ice surface. Examples include 1P/Halley (peak temperature >360 K [41]), C/1996 B2 (Hyakutake) (320 K [98]) and 9P/Tempel 1 (peak temperatures ~330 K [140]).

The physical properties of the mantles remain poorly determined. This is a more serious problem for cometary science than it at first sounds, because almost everything we know about the comets is either controlled or at least strongly modulated by the mantles. Likewise, the physics behind mantle formation and destruction is not well known.

Figure 23 compares the colors of objects within each of several small-body populations. Color is parametrized by the normalized reflectivity gradient, S' [%/1000 Å], essentially the slope of the spectrum of the object after division by the spectrum of the Sun. Several features in Fig. 23 deserve comment.

(a) The nuclei of comets, both dead and alive, show a spread in color that matches that observed in the Trojans but which is distinct from the KBO color distribution. A few *blue* nuclei are known. We will argue below (Sect. 3.1) that these are most likely surfaces covered by rubble mantles.

(b) The Trojans (which are often but incorrectly described as consisting of very red D-type asteroids) in fact show a wide range of surface colors, down to neutral ($S' = 0$), and they are much less red than the majority of KBOs.

(c) Very red material is found only on the surfaces of the KBOs and the Centaurs. Specifically, if we define ultrared matter as having $S' \geq 25\%/1000$ Å [71], then the figure shows that ultrared matter is absent in the inner Solar system populations including the Jovian Trojans, the nuclei of active and inactive Jupiter family comets and the Damocloids (not shown here, but see Fig. 19; [79]). As the progression of objects from top to bottom in Fig. 23

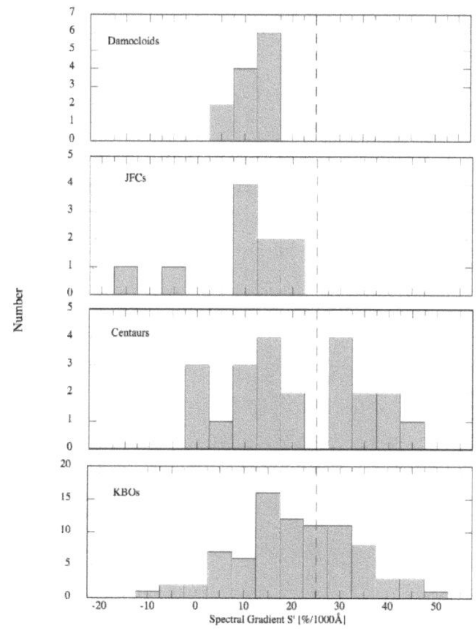

Fig. 23. Histogram showing the normalized reflectivity gradients measured in various small-body populations. Negative (positive) spectral gradients indicate blue (red) reflection spectra, relative to the Sun, which by definition has a spectral reflectivity gradient of zero. Material with $S' \geq 25\%/1000$ Å is defined as ultrared matter. Figure from [79]

represents (except for the Trojans) a dynamical progression from the Kuiper belt source inward, a plausible conclusion is that the ultrared matter cannot survive in the inner Solar system. One guess is that the ultrared objects are coated in organic matter that has been irradiated by long-term exposure to cosmic rays and other particles, creating an "irradiation mantle" (Sect. 3.1).

Rubble Mantles

A rubble mantle consists of refractory, particulate debris that is left behind on the surface of the nucleus by the sublimating gases. Particles bigger than a certain critical size, a_c, are too heavy to be launched against the gravitational attraction to the nucleus and remain behind. Assuming a spherical nucleus of radius r_n and density ρ_n, the surface gravitational force is just

$$g_\mathrm{n} = \frac{16}{9}\pi^2 G \rho_\mathrm{n} \rho_\mathrm{d} r_\mathrm{n} a^3 \tag{13}$$

where $G = 6.6 \times 10^{-11}\,\mathrm{N\,kg^{-2}\,m^2}$ is the gravitational constant and ρ_d and a are the density and radius of the dust grain. The gas drag force is a complicated function of the grain parameters (shape, roughness) and of the ratio of the grain size, a, compared to the mean free path in the gas, λ_mfp. In the case where the grain size, $a \ll \lambda_{\mathrm{mfp}}$, it is reasonable to consider the momentum of impacting gas molecules as being added one at a time, giving the classical drag force expression

$$F_{\mathrm{d}} = C_{\mathrm{d}} \pi a^2 \mu m_{\mathrm{H}} N_1 \Delta V^2 \tag{14}$$

in which C_{d} is the (dimensionless) drag coefficient, μ is the molecular weight of the sublimating gas ($\mu = 18$ for water), $m_{\mathrm{H}} = 1.67 \times 10^{-27}$ kg is the mass of the hydrogen atom, N_1 [m^{-3}] is the concentration of the gas at the nucleus surface and ΔV is the velocity of the gas relative to the grain. We calculate N_1 from the thermal equilibrium equation for sublimating ice. The velocity difference ΔV is roughly the bulk speed of the gas as it leaves the nucleus, which data, physics and models show is of order, V_{s}, the sound speed in the gas at the temperature of the sublimating surface. Balancing gravitational force on a spherical grain with the gas drag then gives

$$a_{\mathrm{c}} = \frac{\mu m_{\mathrm{H}} N_1 V_{\mathrm{s}}^2}{G \rho_{\mathrm{n}} \rho_{\mathrm{d}} r_{\mathrm{n}}} \tag{15}$$

for the critical size above which a grain cannot be accelerated to the escape speed from the nucleus and so which must fall back to the surface. We have ignored numerical constants in this expression and, given our state of ignorance, set $C_{\mathrm{d}} = 1$. Noting that

$$\frac{\mathrm{d}m}{\mathrm{d}t} = \mu m_{\mathrm{H}} N_1 V_{\mathrm{s}} \tag{16}$$

we can rewrite this expression as

$$a_c = \left[\frac{V_s}{G \rho_n \rho_d r_n}\right] \frac{dm}{dt} \tag{17}$$

where $\mathrm{d}m/\mathrm{d}t$ is obtained by solution of (3). Substitution gives us an immediate estimate of a_{c}. Consider a water ice nucleus 1 AU from the Sun and with radius $r_{\mathrm{n}} = 5$ km. The sublimation rate per unit area is $\mathrm{d}m/\mathrm{d}t \sim 10^{-4}\,\mathrm{kg\,m^{-2}\,s^{-1}}$ (Fig. 10). If we take $V_{\mathrm{s}} \sim 500\ \mathrm{R}_{\mathrm{AU}}^{-1/2}$ [m s^{-1}] as a first-order approximation to the gas speed at heliocentric distance R_{AU} [AU] and further take $\rho_{\mathrm{n}} = \rho_{\mathrm{d}} = 1000\,\mathrm{kg\,m^{-3}}$, we obtain $a_{\mathrm{c}} \sim 0.1$ m. Decimeter-sized bodies can be launched by gas drag against the gravitational attraction to the nucleus. This critical size decreases dramatically with increasing heliocentric distance owing to the rapid decline in the specific sublimation rate as the nucleus temperature drops. Beyond $R_{\mathrm{AU}} \sim 5$ or 6 AU, we find $a_{\mathrm{c}} < 0.1\,\mu$m, and the particles that can escape the gravity of the nucleus are those that are too small to efficiently scatter optical photons (with wavelengths $\lambda \sim 0.5\,\mu$m), rendering them unobservable. The magnitude of a_{c} is plotted in Fig. 24 as a function of nucleus size and heliocentric distance.

There are many weaknesses in this simple calculation and many papers have been written to refine it since Whipple's (1950) classic exposition. Still, the essential point is that very large particles, *if they exist in the nucleus*,

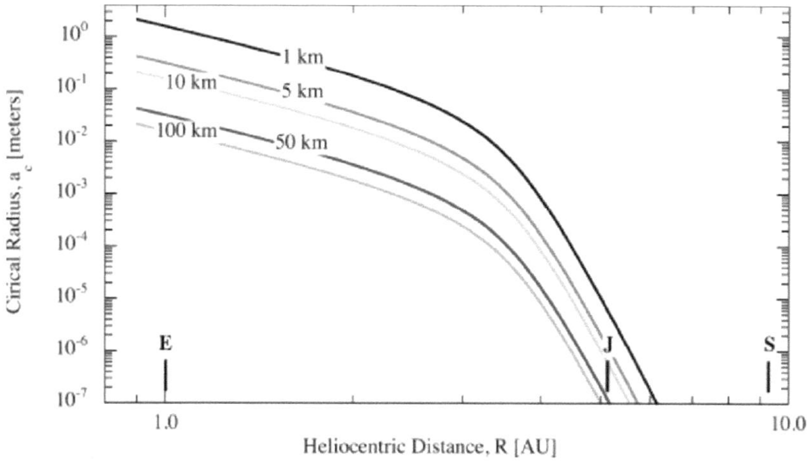

Fig. 24. Solution to (17) computed for dark (albedo 0.04) sublimating water ice nuclei as a function of heliocentric distance for nucleus radii from 1 to 100 km (as marked). The semimajor axes of the orbits of Earth, Jupiter, and Saturn are marked for reference. Particles larger than the wavelength of visible light, $\lambda \sim 0.5\,\mu\text{m}$, can be ejected all the way out to Jupiter's orbit but not much beyond

cannot be easily launched by gas drag into interplanetary space and will remain on the surface where they will impede the heating of surface ice and so diminish the sublimation gas flux. This is the rubble mantle (Fig. 1.25).

It is interesting to consider some consequences of this simple model. First, how thick must such a mantle be? The physical condition for the mantle to seriously impede the heating of ice is that the mantle thickness must rival or exceed the diurnal thermal skin depth. The latter is a measure of the depth to which heat can be carried from the surface by conduction, and is given by $L_{\rm D} \sim (\kappa P_{\rm rot})^{1/2}$, where κ is the thermal diffusivity and $P_{\rm rot}$ is the rotation period of the nucleus. With $\kappa = 10^{-7}\,\text{m}^2\,\text{s}^{-1}$ (appropriate for the porous dielectric materials likely to comprise the mantle matter) and $P_{\rm rot} =$ 10 h (typical of the well-observed cometary nuclei; see Table 3), the skin depth is only $L_{\rm D} \sim 0.06\,\text{m}$ (6 cm!) and the mantle need not be very thick to impede the gas production.

The timescale for such a mantle to form is

$$\tau_{\rm M} \sim \frac{\rho_{\rm n} L_{\rm D}}{f_{\rm M} {\rm d}m/{\rm d}t} \tag{18}$$

where $f_{\rm M}$ is the fraction of the solid mass that cannot be ejected by gas drag because it is contained in bodies with $a > a_{\rm c}$. For a power-law distribution in which the number of solid particles with sizes in the range a to $a + da$ is given by $n(a)da = \Gamma a^{-q} da$ (Γ and q are constants), the fraction $f_{\rm M}$ is easily calculated from

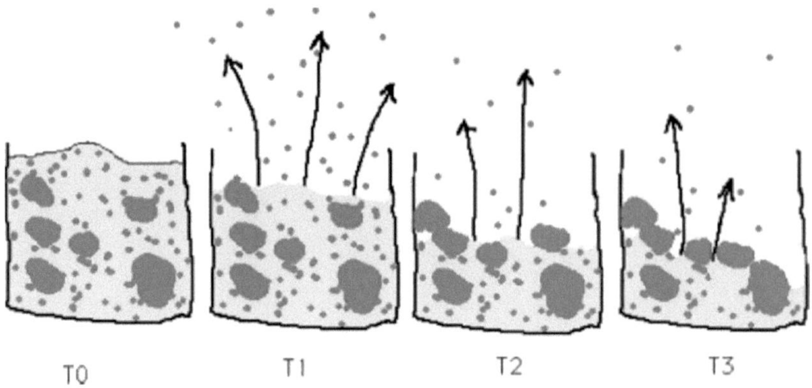

Fig. 25. Schematic cross-section in a cometary nucleus showing the formation of a rubble mantle. At initial time T0, the nucleus consists of a mix of "rocks" (dark) and ices (light). The nucleus is heated from above by sunlight, leading to the sublimation of the ices. Gas drag forces expel smaller rocks into the coma while larger solid particles are left behind. Movement of the sublimation surface into the nucleus exposes more rocks, including large ones that eventually clog the surface, creating a thermally insulating, non-volatile rubble mantle. Any mantle thicker than the diurnal skin depth (∼5 cm) can inhibit sublimation. The interval from T0 to T3 is a function of nucleus size and the pattern of insolation on the nucleus, but can be shorter than the orbit period for comets in the inner Solar system

$$f_{\mathrm{M}} = \frac{\int_{a_{\mathrm{c}}}^{a_+} a^{3-q} \mathrm{d}a}{\int_{a_-}^{a_+} a^{3-q} \mathrm{d}a} \tag{19}$$

where a_- and a_+ are the minimum and maximum sizes in the dust size distribution. This integral takes a particularly simple form when $q = 4$, and this happens to be not too different from the size distribution measured in the coma of 1P/Halley by the dust detectors of the Giotto spacecraft, at least for sizes near 100 µm [87]. Then

$$f_{\mathrm{M}} = \frac{ln(a_+/a_{\mathrm{c}})}{ln(a_+/a_-)} \tag{20}$$

provided $a_+ \geq a_{\mathrm{c}}$, and $f_{\mathrm{M}} = 0$ otherwise. The size of the largest "particle" in the cometary nucleus is unknown, but studies of bolides show that comets eject bodies of decimeter and larger sizes when near the sun. We take $a_+ = 0.1$ m and, based on observations of tiny dust particles in 1P/Halley, set $a_- = 10^{-8}$ m.

Combining (16, 17, 18, 19) and using (3) to calculate $\mathrm{d}m/\mathrm{d}t$, we obtain an estimate of the mantling time, τ_{M}, and the results are plotted in Fig. 26. Two volatiles have been used to estimate the timescales, water and carbon monoxide; the main difference being that the latent heats of sublimation of these materials are in the ratio of about 10:1. I further show curves computed

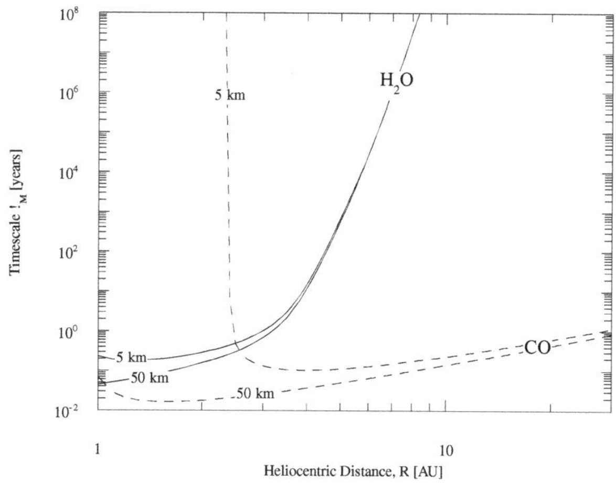

Fig. 26. Timescale for mantle formation from a simple model (18) as described in the text. Curves are shown for two volatiles (CO and H_2O) and two nucleus radii (5 and 50 km), with assumed density of 1000 kg m^{-3}. From [71]

for to values of the nucleus radius (at constant assumed density 1000 kg m^{-3}) to indicate the effect of size. Several features of Fig. 26 are worthy of note.

- The mantling timescales for water are less than 1 year for heliocentric distances $\leq$ 3 AU, for nuclei of both 5 and 50 km radius. This very short timescale means that rubble mantles can potentially grow within a single orbit. A patch of ice exposed to the Solar insolation would, in this model, seal itself against continued sublimation on a timescale of a year. If true, we should think of the mantle as a dynamic structure that can adapt to changes in the insolation.
- Mantling of the water nuclei slows with increasing heliocentric distance. At distances $R_{AU} \geq 6$, the mantling time exceeds the ~0.5 My dynamical lifetime of the Jupiter family comets [96]. Rubble mantles should not form at larger distances if formed only by the sublimation of water ice.
- Cometary activity powered by CO sublimation extends to much lower temperatures and larger heliocentric distances than for water. Indeed, CO is so volatile that it sublimates strongly across the entire planetary region of the Solar system. The mantling time because of CO is therefore very short even out to the orbits of the KBOs. One conclusion is that CO should not be found on the surfaces of the KBOs (unless held there by gravity on the largest objects). Another is that the past presence of CO in the Kuiper Belt would have led to rapid and complete encrustation of these bodies by rubble mantles.
- Figure 26 shows that the mantling times rise at the smallest heliocentric distances. This is most obvious for the CO, 5 km radius model, which rises toward infinity at about 2.5 AU. The physical reason for this is that when

sublimation is very strong, the gas drag forces are able to eject even the largest solid bodies in the distribution (i.e., $a_c > a_+$), and no mantles can form.

This simple model illustrates many of the key features of the rubble mantle. It needs to be only centimeters thick to protect nucleus ice from the heat of the Sun. It can form very quickly. A mantle formed at large heliocentric distance can be unstable to ejection at smaller distances. Mantles on large nuclei are more stable than on small nuclei. Depending on the size distributions in the refractory particles, very tiny nuclei might be unable to retain rubble mantles at all. Considerations like these have induced some researchers to consider models that couple mantle development with orbital evolution, particularly with the drop in perihelion distance that has occurred to most observed comets. The results are very interesting, and parallel to the qualitative ones presented here [130].

The given picture of rubble mantle development is highly simplistic, however. For example, the role of centripetal acceleration has been ignored. An elongated nucleus in rotation about its short axis will experience net reduction in gravity toward the tips that could render the rubble mantle unstable, producing bald spots. The existence of even a small tensile strength would overwhelm the significance of the tiny nuclear gravity and could give the mantle properties quite different from those inferred above. Lastly, and most importantly, what we have presented is no more than a hideous cartoon compared to the complex surface structures imaged by spacecraft on the nuclei of comets (Figs. 14, 15, 16 and 17). Making a deeper connection between the properties of the mantle and the surface morphology will require mechanical and other data from a surface lander. Perhaps ESA's Rosetta will do the job?

Irradiation Mantles

An entirely different type of mantle has long been postulated for the surfaces of cometary nuclei. This mantle is formed by the long-term bombardment of ices on the nucleus surface by energetic particles from the Sun, the Solar wind and galactic cosmic rays and is generally known as the "irradiation mantle" (see Fig. 27). Ironically, there is no specific evidence for irradiation mantles on the nuclei of comets. Instead, if they exist anywhere, they are most likely to be found on the exposed surfaces of the Kuiper belt objects. The reason for this is simple: rubble mantle formation timescales are *much* shorter than the timescales for radiation damage, given the known fluxes of energetic particles.

Energetic particles dissipate their energy in a complicated cascade of interactions that results in breaking the covalent bonds that hold common molecules together. New bonds can form, producing molecules that were not present in the initial mix. Hydrogen liberated from parent molecules in this way is small enough and sufficiently volatile to be able to escape, leaving behind C, N, and O to form complex molecules with whatever hydrogen remains.

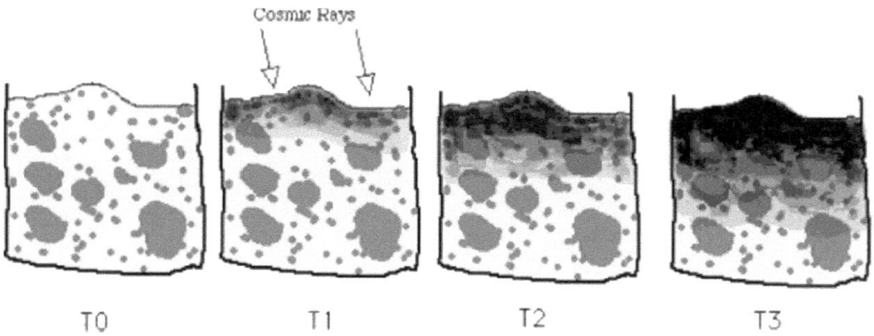

Fig. 27. Schematic cross-section in a cometary nucleus showing the formation of an irradiation mantle. At initial time T0, the nucleus consists of a mix of "rocks" (dark) and ices (light). Cosmic rays bombard the surface layers, breaking bonds in the ice molecules, allowing the formation of radicals, the preferential escape of hydrogen and the formation of a carbon-rich, low albedo "irradiation mantle." The thickness of the layer is of order 1 m (for bulk density 1000 kg m^{-3}). The interval from T0 to T3 is uncertain, but probably ∼100 My for complete processing. A thinner surface layer (affected only by low energy particles) could form on a shorter timescale

Experiments show that the result is a chemically complex mixture of organics, both aliphatic (carbon chain molecules) and aromatic (carbon ring molecules), in some cases polymerized to a very high molecular weight ($\mu > 100$s). High molecular weight corresponds to low volatility and the resulting irradiation mantle is stable against sublimation relative to the common ices. The mantle is also of low albedo, a reflection (pun intended) of the high carbon content. In fact, the molecular and chemical nature of this type of material is poorly defined. Related complex organic materials called "Tholins" are sometimes used as analogs, but these are produced by spark discharge in low pressure gases, and they may not be an appropriate analog for the mantle material. "Kerogens," high molecular weight hydrocarbons found in terrestrial oil shales, may be a good analog, although these are not produced by irradiation.

The depth to which material can be damaged by energetic particles is a function of the particle energy. In the planetary region, the largest fluxes are for low energy particles in the Solar wind (energy ∼1 to 10 keV), and these particles have very small penetration depths in ice. Much more energetic particles (MeV to GeV and beyond) are found in the cosmic rays but at relatively low fluxes. Damage occurs fastest at the surface but, given billions of years should extend to column densities ∼1000 kg m^{-2} (1 m in ice of density 1000 kg m^{-3}). Calculations of the timescale for delivery of 100 eV per oxygen atom are shown in Fig. 28, for heliocentric distances of 40 AU, 85 AU, and "∞ AU" (corresponding to the local interstellar medium). This energy dose is chosen because it corresponds to heavy damage to the exposed material. Major differences exist between these locations both because the flux of low energy particles from the Solar wind declines with the inverse square of the distance and because the magnetic interaction of the wind with the interstellar medium

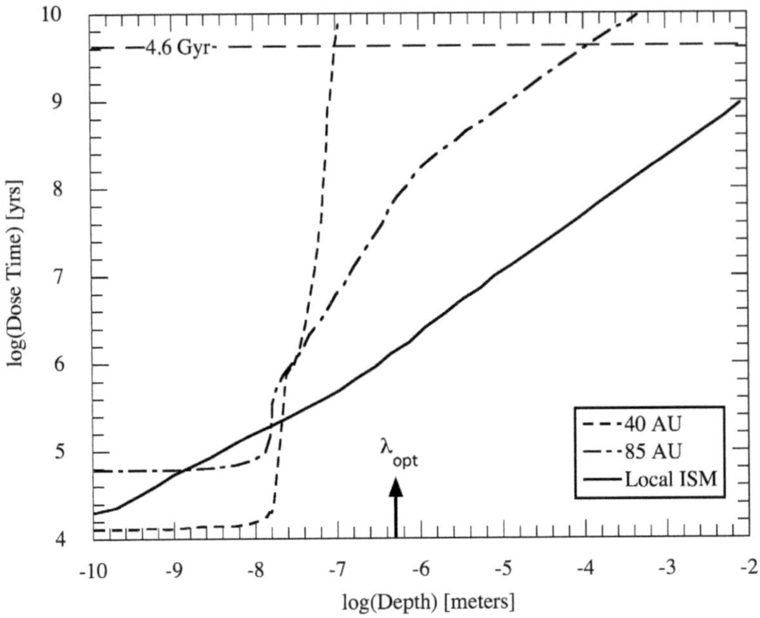

Fig. 28. Timescale for delivering 100 eV per $\mu = 16$ atom as a function of depth in water ice (density $1000\,\mathrm{kg\,m^{-3}}$) at three heliocentric distances. The dashed horizontal line at the top marks the age of the Solar system. The wavelength of visible photons is marked by λ_{opt} at the bottom. Damaged layers thicker than λ_{opt} are likely to have significant effect on the reflected light spectrum. Replotted from [23]

results in a gradient in the flux of energetic particles. The figure shows that the Kuiper belt objects at ∼40 AU in fact exist in a relatively benign radiation environment. Solar wind particles quickly irradiate a surface skin ∼100 Å thick (in 10^4 year) but damage to 0.1 μm takes a considerable fraction of the age of the Solar system. At 85 AU, close to the recently detected termination shock (where the Solar wind decelerates as it impacts the heliopause from the inside) the flux of energetic particles is increased and total damage occurs to depths of $\sim 10^{-4}$ m on billion-year timescales. In the open interstellar medium, the damage can reach depths in ice ∼1 m on the same timescale.

What does all this mean? First of all, the timescales for irradiation damage (Fig. 28) are vastly longer than those for the production of a rubble mantle (Fig. 26). I conclude that irradiation mantles should not be found on any object whose past life has allowed the possibility of mass loss and, so, of rubble mantle formation. Objects in the outer Solar system are too cold to sublimate water and so remain as candidates for irradiation mantling. Perhaps the ultrared matter ($S' \geq 25\%/1000\,\text{Å}$) that appears to be a unique feature of the KBOs and of some Centaurs, is irradiated mantle material. Consistent

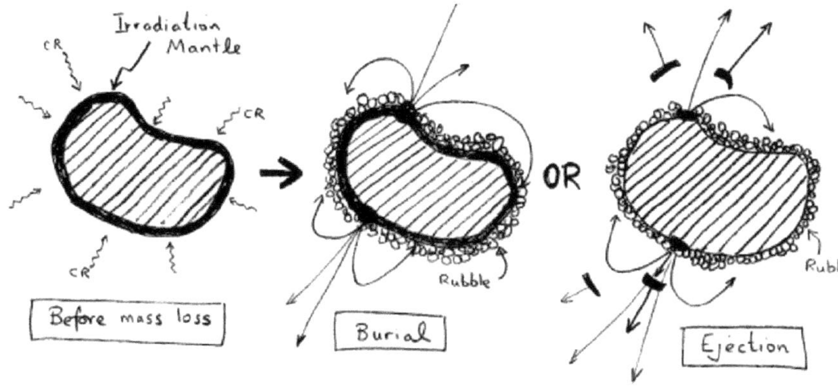

Fig. 29. Possible styles for the destruction of irradiation mantle. On the left, billions of years of exposure to energetic particles on a frigid surface has created an irradiation mantle (black) on a nucleus that is otherwise pristine (shaded). At the onset of sublimation-driven mass-loss, the irradiation mantle could be buried (middle) or cracked and ejected by gas drag (right), the exposed surface of the nucleus being replaced by a rubble mantle consisting of excavated, unirradiated matter in both cases

with this inference is the observation that ultrared matter does not survive approach to the sun within the orbit of Jupiter [71, 79], corresponding to the heliocentric distance inside which water begins to sublimate and the timescale for rubble mantle formation becomes short (Fig. 26). The mode of destruction of the irradiation mantle is not clear, however. The mantle could still be present but buried beneath a recently deposited rubble mantle consisting of (less red) debris excavated from beneath the ~1 m thick irradiated layer. Or it could be ejected by gas drag at the onset of strong sublimation inside ~5 AU (Fig. 29).

4 Kuiper Belt

Several of the important properties of the Kuiper Belt, established over the past 14 years by painstaking observational work around the world, have been summarized in the section of this book by Alessandro Morbidelli [112]. I will avoid duplication and instead focus on aspects of the Kuiper Belt that are less thoroughly covered elsewhere in this volume.

4.1 Kuiper Belt Physical Properties: Colors and Albedos

Ideally, we would use spectra to determine the surface compositions of KBOs and other Solar system bodies. The faintness of most such objects makes this ideal unreachable, and instead, broadband colors are often used as a

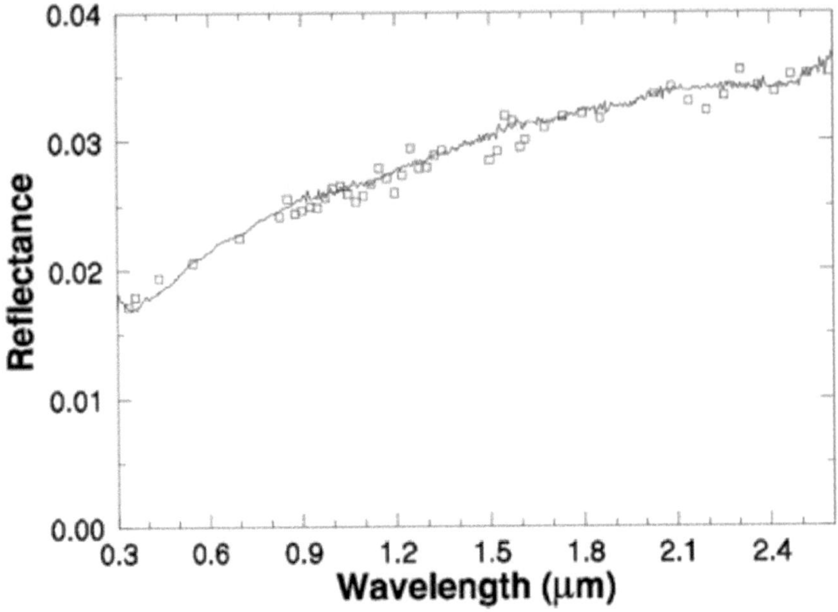

Fig. 30. Comparison of the optical near infrared reflection spectrum of D-type asteroid (368) Haidea (points) with the Tagish Lake meteorite (line), showing a nearly perfect match. Figure from [60]

proxy for the spectra and so for surface composition. Problems with this approach are numerous. Colors cannot, in general, be used to determine compositions Fig. 30. Colors are influenced by composition, but also by wavelength-dependent scattering effects in particulate regoliths, and by viewing geometry. On the other hand, colors can be used to classify objects into groups. The Holy Grail of colorimetric work on the KBOs has been for some years to find correlations between the colors and other properties such as size and orbital character [57, 75, 102, 106, 145–147]. Correlations like this might provide illuminating clues about the KBOs and their histories.

The use of color to learn about KBOs has been, to say the least, an up-hill battle. The first property to be measured was color diversity; the KBOs exhibit a range of surface optical colors that is large compared with the uncertainties of measurement. In fact, color diversity has emerged as the only physical property to be confirmed by every subsequent study. Later, color diversity at optical wavelengths was found to extend into the near infrared [27, 31, 33, 74, 106]. Moreover, the optical and infrared colors are correlated, which indicates that a single coloring agent is responsible for the wavelength dependence of the reflectivity across the wavelength range from B-band (0.45 μm) to J-band (1.2 μm) and perhaps beyond to K-band (2.2 μm).

The physical significance of color diversity is unclear. One possibility is that the different colors reflect intrinsically different compositions. This might be the case, but it is difficult to understand why the compositions of the measured KBOs would be so varied. After all, the measured objects are located in a comparatively narrow band between about 30 and 50 AU, where the radiation equilibrium temperatures vary from ∼40 to ∼50 K. This very small temperature range could scarcely effect the compositions of the KBOs enough to cause major color differences.

For this reason, a second model was proposed to explain the color dispersion. In this "resurfacing model," the hemispherically averaged color of a KBO is time-dependent and determined by a competition between collisional resurfacing and cosmic ray processing. For example, suppose that cosmic ray processing causes an exposed surface to become redder on timescale τ_{cr}. This process competes with impact-driven resurfacing, in which impacts excavate "fresh" material from beneath the irradiated layer. If the excavated matter has a different (neutral?) color, the instantaneous, hemispheric average color will vary stochastically between extremes set by fully radiation-processed matter and fresh, excavated material. Substantial color fluctuations are possible when the timescale for resurfacing, τ_{coll} is $\sim \tau_{cr}$.

Attractive though it at first seems, several predictions of the resurfacing model have not been confirmed by observations. The model predicts that rotational color variations on KBOs should be nearly as large as the color differences that exist between KBOs of a given size. This is not observed. The model also predicts that the range of colors observed should vary with KBO size, because the timescale for collisional resurfacing varies with object size while τ_{cr} does not. Again, this violates the observations. The model has been extended by the addition of color variations owing to possible outgassing effects [31] but the problems remain. Collisional resurfacing is unlikely to be responsible for the color dispersion of the KBOs, although it could conceivably be a contributing factor.

Tegler and Romanishin reported that the colors of KBOs were not just dispersed over a wide range but were *bimodally* distributed [145]. They continued to find bimodal color distributions with larger samples [146, 147] but failed to receive observational support for this finding from independent observers [31, 36, 57, 75]. The colors of the KBOs available at the time of writing (March 2006) are distinctly unimodal (see Figs. 31, 32 and 33). Recently, Peixinho et al. [121] reported that, while the KBO colors are indeed unimodally distributed, the Centaurs appear bimodal (see the next section). This is more than an academic distinction: a bimodal color distribution would have placed strong constraints on the nature of the KBOs, had it been real.

Few of the long-sought correlations between colors and other physical and dynamical properties have turned out to be observationally robust. The correlation that seems most likely to be real is between color and perihelion distance [146] or, equivalently, between color and inclination [152] amongst the classical KBOs. The perihelion vs. inclination ambiguity arises because these quantities are loosely related amongst the Classical objects. Doressoundiram [35] finds that the color vs. perihelion distance correlation is slightly

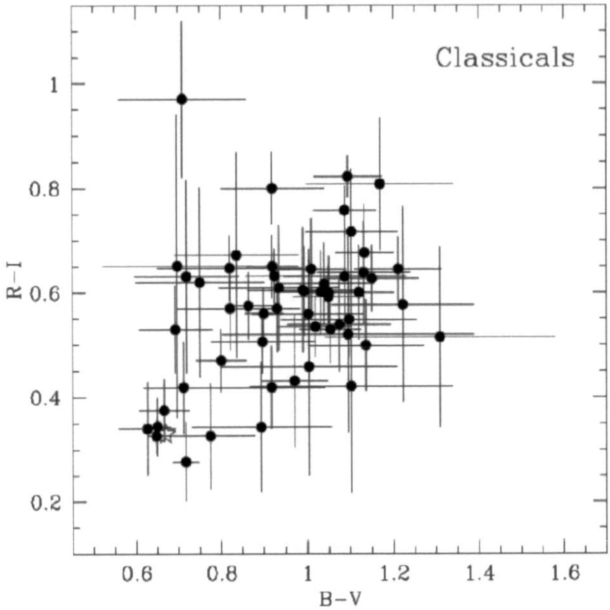

Fig. 31. Color-color diagram for classical KBOs. From [32]

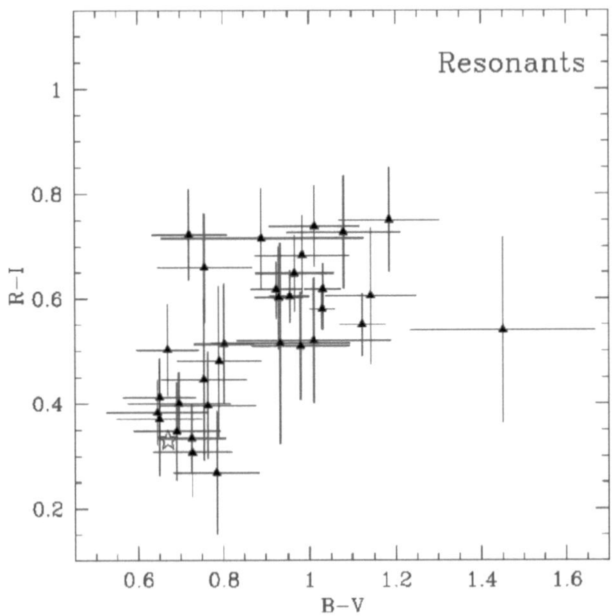

Fig. 32. Color-color diagram for resonant KBOs. From [32]

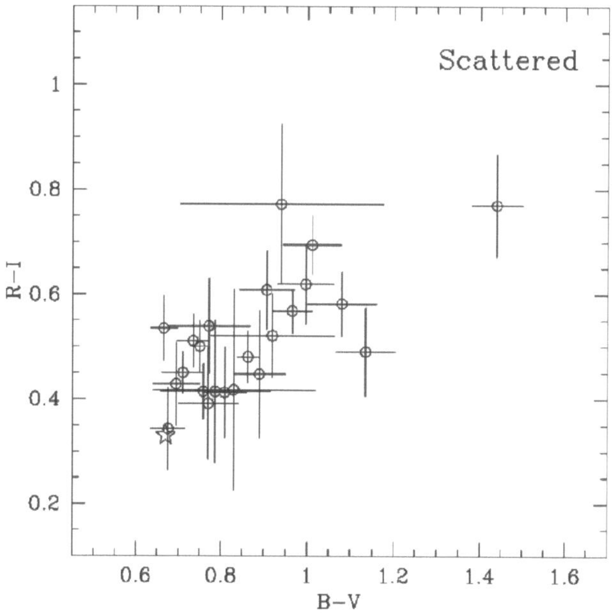

Fig. 33. Color-color diagram for scattered KBOs. From [32]

stronger than the color vs. inclination correlation. Trujillo and Brown [152] find that classical objects with small inclinations are redder, on average, than those with high inclinations. The latter observation has been factored into dynamical models by R. Gomes [53]. He asserts that the high inclination ("hot") Classical KBOs were scattered outward while the low inclination ("cold") Classical KBOs were formed exterior to Neptune, where they now reside [53]. Whether or not this is true, the central mystery that is unaddressed by dynamical models is why the cold and hot populations would have different colors. As measured by the B-I color index, the color vs. inclination (or color vs. perihelion) correlation appears secure at the 3σ or 4σ confidence level. However, the correlation is absent when V-R or V-I color indices are used [138]. One possibility is that the color correlation is forced by the B data (for example, there could be a B-band absorber whose distribution is correlated with inclination or perihelion distance but which would have no effect on color indices at wavelengths longer than B). As new observations are collected, it will be interesting to see whether or not the reported correlation will survive. No convincing explanation for the correlation, if real, has been suggested.

About a dozen KBOs possess both color and albedo determinations [25]. These are plotted in Fig. 34 together with corresponding data for the nuclei of comets, the Jovian Trojans, and Centaurs [80]. There it is seen that the wide dispersion of colors of the Centaurs and KBOs is matched by a wide dispersion in the albedos, with the large objects 2003 EL61 and Pluto defining one extreme. By comparison, the nuclei of the comets and the Jovian Trojans

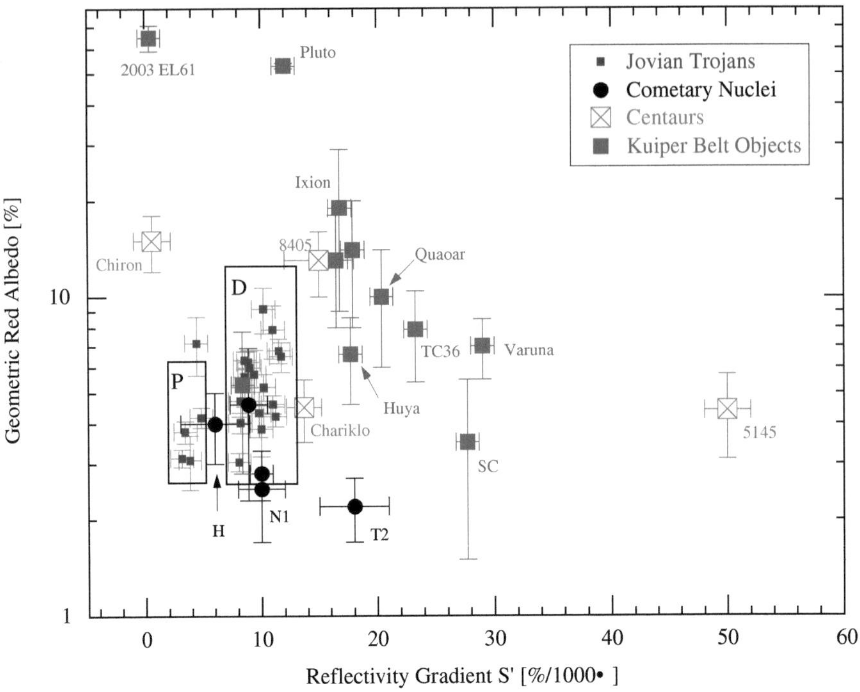

Fig. 34. Color-albedo plane for cometary nuclei, Jovian Trojans, Centaurs, and KBOs. Identities of particular objects are abbreviated for clarity: SC = 1993 SC, TC36 = 1999 TC36, H = 1P/Halley, N1 = 28P/Neujmin 1, and T2 = 10P/Tempel 2. Boxes mark the nominal positions of the P- and D-type asteroids. Data compiled from [25, 48, 80]

are confined to a small fraction of the color-albedo plane, with surfaces that are on average less red and darker than the KBOs and Centaurs. The diagram reinforces the conclusion that the surfaces of the comets and of the Trojans, while similar to each other, are not the same as the surfaces of the Centaurs and KBOs. If this difference reflects an evolutionary trend, then the fact that the Centaur and KBOs overlap in Fig. 34 shows that the modification occurs after the Centaur phase. Most likely it is associated with the onset of sublimation on bodies whose perihelia have approached or crossed the orbit of Jupiter (the rough boundary outside which water does not appreciably sublimate [71]). The very high albedos of EL61, Pluto, and perhaps some other objects are clearly associated with the presence of surface ice and the cleanliness of this ice suggests that it has been recently emplaced, probably by frost deposition from an atmosphere. None of the Trojans or cometary nuclei possess surface ice in quantities sufficient to influence the albedo, because they are too hot (surface ice would quickly sublimate). However, the simple removal of ice cannot explain why the surfaces of many low albedo KBOs and

Centaurs are so much redder than any seen in the comet or Trojan populations (cf. Fig. 23). Some form of instability of the ultrared matter is required.

4.2 Kuiper Belt Physical Properties: Spectra

Only ∼10 KBOs are bright enough for useful spectra to be obtained. The spectra fall into three basic classes.

The Water Worlds (Fig. 35). KBOs (50000) Quaoar [76], 2003 EL61 [143], and others show strong absorptions at 2.0 µm and 1.5 µm that are diagnostic of water ice. Water ice is stable against sublimation at Kuiper belt distances and temperatures, and it is appropriate to think of it as "bed rock" for other, more volatile species. The ice on Quaoar and 2003 EL61 is known to be crystalline as it shows a narrow band at 1.65 µm that is absent in the spectrum of amorphous ice. This is a puzzle, because ice at the ∼40 K to 50 K surface temperatures of the KBOs should be indefinitely stable in the amorphous form. Why should the ice instead be crystalline?

Crystallinity indicates that the ice has been raised above the critical temperature for transformation (roughly 100 or 110 K) at some point in its history. This heating could have occurred in the deep interiors of the KBOs provided that there is a way for heated ice at depth to be emplaced onto the surface.

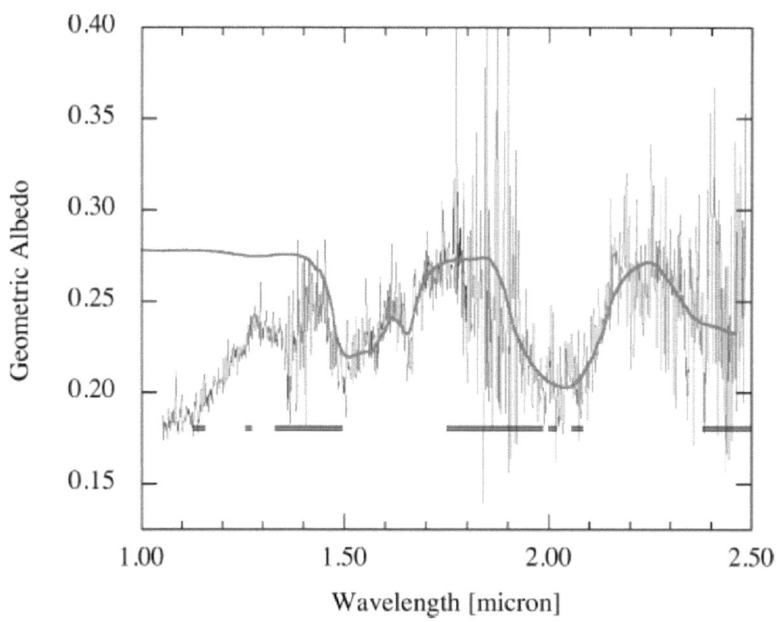

Fig. 35. Near infrared reflection spectrum of (50000) Quaoar. The solid line is a crystalline water ice spectrum over-plotted (not fitted) to the data. Note the feature at 1.65 µm that is diagnostic of crystalline ice. From [77]

One way for this to occur is through the past action of cryovolcanism; liquid water (or slush) might have erupted onto the surfaces of these KBOs when they were still internally hot from the decay of trapped radioactive nuclei. Conceivably, heating by micrometeorites is responsible, although this possibility is difficult to test given that we do not know the flux of impacting dust particles within the Kuiper belt. A more serious problem is that crystalline ice exposed to the unimpeded bombardment of energetic particles from the Solar wind and the cosmic rays should be transformed back toward the amorphous state, as the bonds in the crystalline lattice are systematically demolished. The timescale for this process is uncertain but probably short (~1–10 My). Hence, it appears that these KBOs must be resurfaced on a geologically very short timescale in order for the ice to have escaped back-conversion to the amorphous form. Again, the mechanisms for resurfacing are unknown. Comet-like outgassing (perhaps with CO playing the role of "volatile") is a possibility, but some effect related to micrometeorite "gardening" of the regolith, as is seen in the rocky fragmental layer on the surface of the Moon, seems more likely. The optically active surface layers may be continually churned together with buried crystalline ice that is protected from irradiation.

The issue of the crystalline state of water ice in small bodies deserves further exploration. Ice in comets is rarely directly detected, but in comets C/Hale-Bopp [26] and C/2002 T7 (LINEAR) [83], the absence of the 1.65 μm band shows that the ice is amorphous. Both objects are long-period comets, and it is possible that the amorphous nature of the ice is a result of energetic particle bombardment, rather than primordial in nature. The outgassing activity of some comets at heliocentric distances beyond the ~5 AU water sublimation zone (e.g. Fig. 24) is often interpreted as evidence for internal heating by the (exothermic) amorphous → crystalline phase transition [126]. An interesting question to be addressed observationally is the state of the ice in Jupiter family comets: is this ice crystalline as in the large KBOs or amorphous, as in the two measured long-period comets?

The Methanoids (Fig. 36). KBOs Pluto, 2003 UB313 [153], and 2005 FY9 [99], show evidence for surface methane, with distinct bands in the near infrared spectral region. (Triton, likely to be a large KBO captured by Neptune, also shows a methane-rich spectrum).

Methane is interesting from two perspectives. First of all, methane is unstable to sublimation on long timescales at the distances and temperatures of most Kuiper belt objects. This instability has been explored in detail for Pluto, where it is found that the escape of methane is limited by the flux of energetic (EUV) Solar radiation [65], but can still exceed several kilometers equivalent thickness over the age of the Solar system. The escape from smaller bodies will be dramatically faster, perhaps explaining why the known Methanoids are large (but not explaining why ~1200 km diameter Quaoar is methane-free). Second, the origin of the methane is problematic. Low temperatures and pressures in the solar nebula are thought to

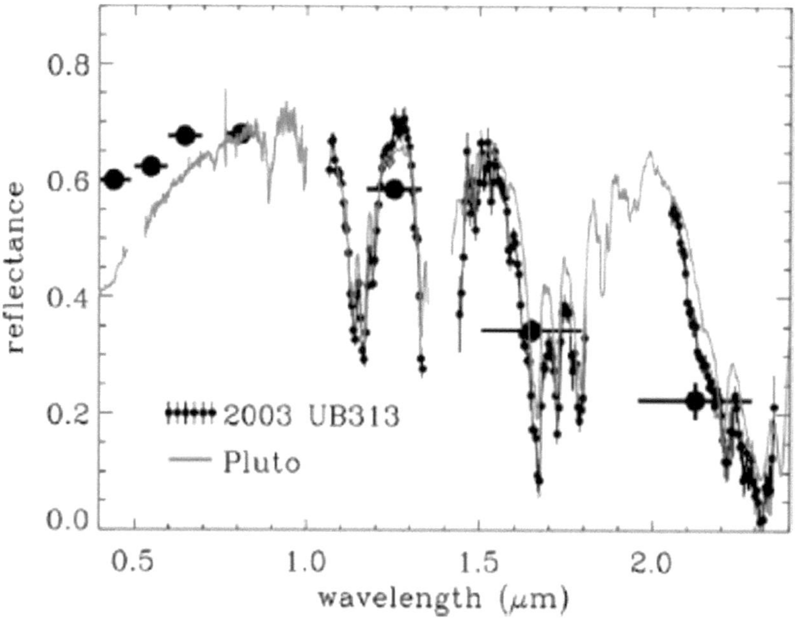

Fig. 36. Optical and near infrared reflection spectra of large KBOs Pluto (line) and 2003 UB313 (points). The principal absorptions in both spectra are due to methane. From [13]

favor the incorporation of carbon atoms in the oxidized form as CO and CO_2, rather than in the reduced form of CH_4 [125]. Therefore, it seems unlikely that the methane was delivered to these bodies from the nebula. One possibility is that CH_4 arrived as a clathrate (a physical cage in crystalline water ice in which sufficiently small "guest molecules" can be trapped). In my mind, it seems more likely that the CH_4 is produced in the interiors of these bodies, probably from hydrogen released by serpentinization followed by Fischer–Tropsch reactions and then outgassed on to the surface. The lack of methane on small KBOs could then reflect a lack of production, because only bodies large and hot enough to sustain liquid water can experience serpentinization.

Featureless Class. Objects in this class have sloped but otherwise featureless near infrared spectra. Obviously, all spectra are featureless when observed at sufficiently low signal-to-noise ratio, so here "featureless" is probably a relative term and many objects labeled as such will resolve into the other classes once better spectra are secured. By analogy with the featureless spectra of many mantled objects already observed at decent signal-to-noise ratios, including the nuclei of dead comets and the Jovian Trojans (e.g., [37,42,101]), however, it is likely that a subset of the featureless objects will remain so even under more intense scrutiny.

4.3 Kuiper Belt Physical Properties: Shapes, Spins

The shapes and spins of Kuiper belt objects are studied from their rotational lightcurves [86, 134]. The most informative way to present these data is in a plot showing the photometric range as a function of the rotational frequency (rotations per day), as here in Fig. 37, from [134]. The range–frequency plane is divided into three regions, based on the original prescription of Leone et al. [93].

Region A shows lightcurves of small range and any period, for which the lightcurve could be affected by surface albedo variations and for which, in any case, the interpretation is likely to be highly ambiguous. Strictly, *any* lightcurve can be produced by a surface albedo distribution of arbitrary complexity. However, studies of the lightcurves of hundreds of asteroids show few examples where albedo variations are important, perhaps because regolith transport is efficient and albedo differences are quickly smeared out by the redistribution of dust. Those examples are confined to rotational ranges $\Delta m \sim$ 0.1–0.2 mag. To be conservative, in Fig. 37, we have marked Region A as extending up to $\Delta m = 0.3$ mag. The most notable exception to this empirical rule is Saturn's 1460 km diameter satellite Iapetus, which shows a hemispherical albedo asymmetry, with the leading hemisphere being ~6 times darker

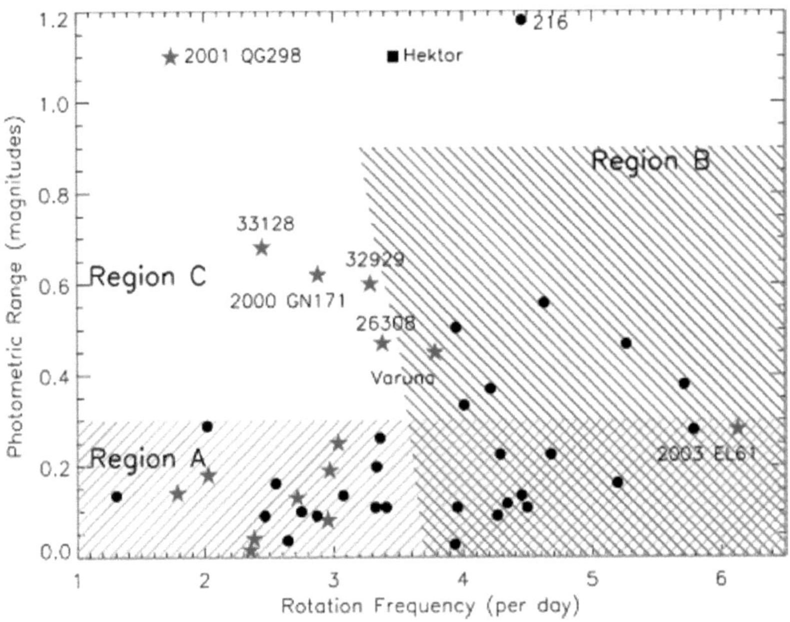

Fig. 37. Rotational range vs. frequency (rotations per day), modified by Scott Sheppard from [134]. Black dots denote large main-belt asteroids (diameters >200 km) while KBOs are marked as stars. Note that Hektor is a Jovian Trojan while 33128 is a Centaur

than the trailing. However, the Iapetus albedo asymmetry is a consequence of its synchronous rotation about the planet (which leads to hemispherically asymmetric fluxes of incident charged particles from Saturn's magnetosphere and of Saturn-orbiting dust particles), a circumstance which is not replicated in the KBOs.

Region B shows objects rotating sufficiently rapidly that centripetal distortion of the shape constitutes a likely explanation of the lightcurve. The region is marked for an assumed density $\rho = 1000\,\mathrm{kg\,m^{-3}}$ and calculated from the figures of equilibrium by Chandrasekhar [17]. Higher (lower) densities would push the left boundary of Region B to the right (left). The implicit assumption is that the tensile strengths are zero and, while this is unlikely to be exactly correct, it is a reasonable approximation for bodies that have been internally fractured by past collisions. Two KBOs fall in Region B; (20000) Varuna ($\rho \sim 1000\,\mathrm{kg\,m^{-3}}$ [72] and 2003 EL61 ($\rho \sim$ 2600–3340 $\mathrm{kg\,m^{-3}}$ [127]).

Region C shows locations in the range vs. frequency plot where close and contact binaries would plot. A binary consisting of two spheres viewed equatorially would have $\Delta m = 2.5\log(2) = 0.7$ mag. Mutual gravitational deformation would elongate the components, raising Δm to 0.9 mag [93]. Objects with $\Delta m > 0.9$ mag are not explainable as rotationally deformed single bodies and contact binaries are preferred. In the whole Solar system, very few objects have been found with such large photometric range. The main examples are Trojan (624) Hektor, which is believed to be a 150 km scale binary, 200 km main – belt asteroid (216) Kleopatra and ~260 km KBO 2001 QG298 [134]. The inferred abundance (admittedly from a single detection) of contact or very close binaries in the Kuiper belt is at least 10–20% [134].

To give a short summary, rotational studies of KBOs have revealed a number of interesting cases for rotational deformation (Varuna and 2003 EL61) and close or contact binaries (the best case remains 2001 QG298 but other KBOs in Region C of Fig. 37, like 2000 GN171, are candidates for contact binaries observed non-equatorially). The appearance of these examples in a still-small (N~40) observational sample is evidence that rotationally deformed and contact-binary structures must be common in the Kuiper belt. Preliminary evidence suggests that the shape distributions of KBOs larger and smaller than 400 km diameter are not the same [86]. If confirmed by future work, this observation might find a natural explanation in terms of collisional effects at small sizes and self-gravity at larger sizes.

4.4 Kuiper Belt Physical Properties: Multiple Objects

About 20 examples of multiple KBOs have been reported as of early 2006 (many are not yet properly published, appearing only in electronic circulars). Multiple KBOs in Table 5 have been collected from [32] and [118] and from a few recent electronic publications. The objects are binaries except for Pluto (three satellites known) and 2003 EL61 (two satellites known), but this is no doubt an effect of observational selection against small, faint companions

Table 5. Multiple KBOs

Object	a [km][a]	e[b]	i[deg][c]	Type[d]	θ[arcsec][e]	P[days][f]	Δmag
Pluto				3:2			
Charon	19,400	0.00	96	—	0.9	6.4	1.3
S/2005 P1	64,700?	—	—	—	2.2	38.3?	9.0
S/2005 P2	49,400?	—	—	—	1.7	25.5?	9.4
1995 TL_8	—	—	—	Sca	—	—	—
(58534) 1997 CQ_{29}	8,010(80)	0.45	—	Cla	0.2	312(3)	0.3
(26308) 1998 SM_{165}	11,310(11)	—	—	2:1	0.2	130	1.9
1998 WW_{31}	22,300	0.82	42	Cla	1.2	574	0.4
(79360) 1999 CS_{29}	—	—	—	Cla	—	—	—
1999 OJ_4	—	—	—	Cla	—	—	—
1999 RZ_{253}	4,660(170)	0.46	—	Cla	–	46	—
(47171) 1999 TC_{36}	7,640(460)	—	—	3:2	0.4	50.5	1.9
2000 CF_{105}	—	—	—	Cla	0.8	—	0.9
2000 CQ_{114}	—	—	—	Cla	—	—	—
2000 CM_{105}	—	—	—	Cla	—	—	—
2000 CM_{114}	—	—	—	Cla	0.07	—	0.5
2000 OJ_{67}	—	—	—	Cla	—	—	—
2000 YW_{134}	—	—	—	Sca	—	—	—
2001 QC_{298}	3,690(70)	—	—	Cla	0.17	19.2	N/A
(88611) 2001 QT_{297}	27,300(340)	0.24	—	Cla	0.6	—	0.5
2001 QW_{322}	—	—	—	Cla	4.0	—	0.4
2002 CR_{46}	—	—	—	Sca	0.11	—	1.2
2003 EL_{61}				Sca			
S/2005 (2003 EL_{61}) 1	49,500(400)	0.050(0.003)	234.8(0.3)	—	1.3	49.12 ± 0.03	3.3
S/2005 (2003 EL_{61}) 2	39,300?	—	—	—	1.0	34.1?	4.5
2003 QY_{90}	—	—	—	Cla	—	—	—
2003 UB_{313}	36,000	—	—	Sca	0.5	14	4.2
2003 UN_{284}	—	—	—	Cla	—	—	—
2005 EO_{304}	—	—	—	Cla	—	—	—

[a] semi-major axis of the binary system.
[b] eccentricity.
[c] inclination.
[d] Dynamical type: 3:2, 2:1 = resonant, Clas = Classical, Sca = Scattered.
[e] Angular separation.
[f] Orbital period.

and a larger fraction of the KBOs must have multiple satellites. The largest satellite of the first-known (but mis-labeled) KBO Pluto has been known for decades, but it still surprising to see how many KBOs observed at high angular resolution are double. What can we learn from the binaries?

First, binaries are present in the classical, scattered, and resonant KBO populations. Systematic observations of 81 KBOs spread across these classes reveal 9 binaries at the resolution (and magnitude difference) accessible to the Hubble Space Telescope and its NICMOS camera, giving an average binary fraction of $11^{+5}_{-2}\%$ [139]. Given that binaries of very small separation and those having a large magnitude difference between the components cannot be detected, this must be taken as a strong lower limit to the binary fraction.

Second, low inclination ($i < 5°$) Classical KBOs have a binary fraction $22^{+10}_{-5}\%$ [139], which is different from the average value at the $\sim 2\sigma$ level. The mean value for all KBOs other than the $i < 5°$ Classicals is $5.5^{+4}_{-2}\%$, which is different enough from $22^{+10}_{-5}\%$ to be interesting. The difference, if real, could be a hint that the diverse dynamical histories of the bodies have had an effect on the survival of binaries. For example, perhaps whatever excited, the orbital inclinations and eccentricities of KBOs also acted to split a fraction of the binaries.

Third, the binaries appear to be of different types. Pluto (and probably 2003 UB313 and others) have short orbital periods and orbital eccentricities $e \sim 0$. Together these strongly suggest the effects of tidal damping. Close binaries like these might be produced by glancing impacts between large precursors [15]. The number density of large KBOs is presently far too low to account for such collisions. If this is the correct explanation, the collisionally produced binaries must be relics from an earlier time at which the number density in the belt was much (probably two to three orders of magnitude) higher than now [15, 71].

Where measured, most KBOs have periods from months to years and the eccentricities of the orbits are in the range $0.2 \leq e \leq 0.8$ (see Table 5). These wider, more eccentric binaries are unlike the binaries expected to be produced by glancing, massive impacts, and other explanations must be sought. Several have already been proposed, including binary formation through dynamical friction [52], three-body interactions [52, 156], and exchange reactions [50]. These models are all good in the sense that they make observationally testable predictions. The exchange model predicts binary eccentricities larger than observed and can probably be ruled out, at least in its simplest form. Three-body interactions should produce mainly weakly bound binaries. It is not yet clear if the distribution of semimajor axes of the known binaries is incompatible with three-body captures, but this seems likely (Table 5). Capture by dynamical friction (exerted on large, growing bodies by the "sea" of smaller bodies surrounding them and now dissipated) is expected to produce a large binary fraction (as observed) with a high abundance of tight binaries (maybe consistent with the data). Continued action of dynamical friction should lead the binary components to spiral together, making contact binaries (one, 2001 QG298, is already suspected), but it is not clear that observed eccentricities $0.2 \leq e \leq 0.8$ can be explained. At this early stage, I do not know if the proposed models fail because they are completely wrong, or because they tell only part of the story. Binaries could form by dynamical friction, for example, and then be excited by external agents after the source of dynamical friction had dissipated. Long-term (4 Gy) survival of the KBO binaries appears to be possible, but the existing pairs may constitute only a fraction of those initially present, with the softest binaries having all been disrupted [122].

4.5 Kuiper Belt Physical Properties: Densities

Densities have been discussed here and there throughout this chapter. For convenience, I have summarized them graphically in Fig. 38, where they are plotted as a function of the object diameters. The densities of cometary nuclei plotted in the figure have been estimated from various techniques as discussed in Sect. 3. Densities of KBOs are estimated from binary motions and size estimates (Pluto, Charon, and 1999 TC36, [137]), from lightcurves interpreted as rotational deformation of the shape ((20000) Varuna [71] and 2003 EL61 [127]) and from a contact binary model (2001 QG298, [134]). The densities of the planetary satellites are obtained nearly directly from gravitational perturbations on the motions of spacecraft, except that the densities of small Saturnian satellites including Pandora and Prometheus are estimated from a more complicated model of these satellites' interaction with nearby rings.

What does Fig. 38 show? The most obvious feature is a general trend toward higher densities at larger diameters, adequately described by the power law relation $\rho = 340\ \mathrm{D}^{0.2}$ (with ρ in kg m^{-3} and D in km). This trend is

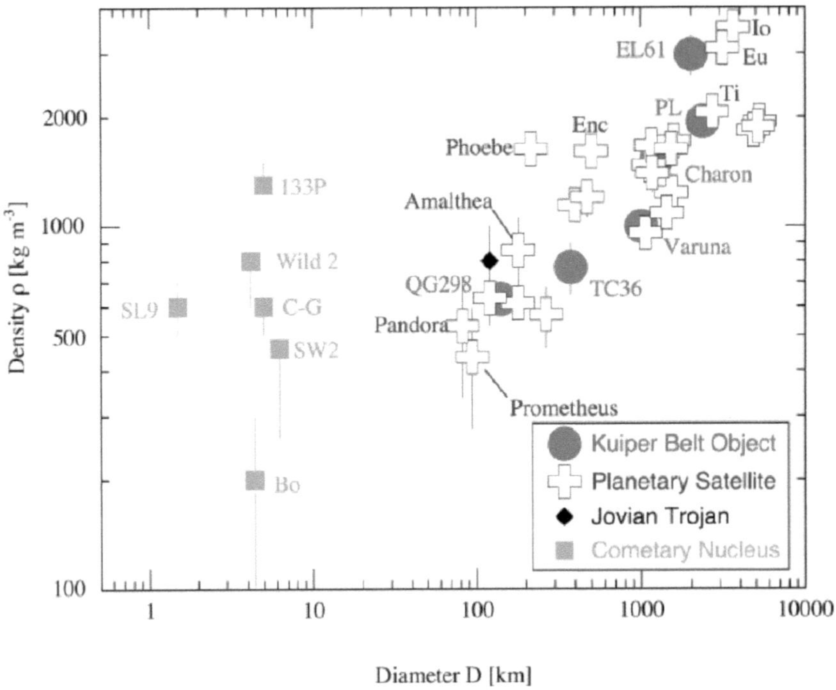

Fig. 38. Densities of KBOs, cometary nuclei, planetary satellites, and Jovian Trojan Patroclus. Abbreviations in the plot are **Comets** Bo = 19P/Borrelly [28], C-G = 67P/Churyumov-Gerasimenko [29], SL9 = D/Shoemaker-Levy 9 [6], SW2 = 31P/ Schwassmann-Wachmann 2, Wild 2 = 81P/Wild 2 [30], 133P = 133P/Elst-Pizarro [62] **Kuiper Belt Objects** EL61 = 2003 EL61 [127], Pl = Pluto, TC36 = 1999 TC36 [137], QG298 = 2001 QG298 [134, 141] **Planetary Satellites** Enc = Enceladus, Ti = Titan, Eu = Europa. These densities are culled from the NASA-JPL site at http://ssd.jpl.nasa.gov/, mostly based on data from the Voyager, Galileo, and Cassini missions. The single Trojan is (617) Patroclus [105]. Plotted error bars are 1σ uncertainties. Single-sided errors below or above the points indicate either upper limits or lower limits to the density, respectively

apparent within the various populations (i.e., the planetary satellites and the KBOs independently show this trend) and, although there is considerable scatter in the densities of bodies at any particular diameter, the trend appears to be real.

The mean density of a composite body consisting of rock and ice is

$$\overline{\rho} = \rho_{\mathrm{i}} f_{\mathrm{i}} + \rho_{\mathrm{r}} f_{\mathrm{r}} \tag{21}$$

where ρ_{i} and ρ_{r} are the densities of ice and rock and f_{i} and f_{r} are the fractional volumes occupied by ice and rock, respectively. The latter are related by

$$f_{\mathrm{i}} + f_{\mathrm{r}} + f_{\mathrm{v}} \equiv 1 \tag{22}$$

in which f_{v} is the fractional void space, also known as "porosity." In the context of Fig. 38, much of the trend in the bulk density is likely to be related to size-dependent variations in f_{v}. This is because self-compression of ice and rock is not very important across most of the plotted diameter range [the central hydrostatic pressure in a body of radius r and average density ρ is $P_{\mathrm{c}} \sim G\rho^2 r^2$. With $\rho = 1000\,\mathrm{kg\,m^{-3}}$ and $r = 500\,\mathrm{km}$, $P_{\mathrm{c}} \sim 20\,\mathrm{MPa}$ (Mpa $= 10^6\,\mathrm{N\,m^{-2}}$)], or roughly 200 bars, but densification through collapse of void space is likely. Laboratory experiments with ice at 77 K show brittle failure at comparable pressures [39] and suggest that part of the density-radius correlation may result from self-compression, particularly by the closing of void-space in porous bodies [71, 107].

Any object less dense than pure water ice ($\rho \sim 1000\,\mathrm{kg\,m^{-3}}$) must be porous. This includes most of the comets in Fig. 38 (but not 133P, the one MBC for which we possess a density constraint) and several of the co-orbital satellites of Saturn (Pandora and Prometheus both have $\rho \sim 500\,\mathrm{kg\,m^{-3}}$). More surprisingly, Jupiter's innermost satellite Amalthea ($\sim$160 km in diameter) has $\rho = 800 \pm 200\,\mathrm{kg\,m^{-3}}$ [5] and so is likely porous and ice-rich. This is a big surprise, given that before the density determination, Amalthea was always described as one of the most refractory, high-temperature products of Jupiter's long-gone accretion disk. The evidence for porosity is strong and independent infrared spectral observations [142] show a deep hydration feature that supports a watery constitution.

Porosity can be due to large, empty spaces ("macroporosity") or to open structure on a small scale "microporosity" and everything in between. Microporosity in stony meteorites averages 10% and can reach 30% in some samples [12]. Macroporosity can be produced by past impacts that have cracked and even dissociated bodies leading to their re-assembly as a collection of irregularly shaped blocks with considerable internal void space. Evidence for this is seen even in the main-asteroid belt (e.g., rocky asteroid (253) Mathilde has $\rho = 1300 \pm 200\,\mathrm{kg\,m^{-3}}$ [11,159]). Porosity caused by collisional shattering and reassembly should become less important at larger diameters both because sufficiently energetic impacts are rare and because of closure of pore space at the higher hydrostatic pressures in large objects. I suspect that most of the slope in the ρ vs. D relation seen in Fig. 38 is caused by systematic decrease in the porosity as D grows larger. The equation of state for self-gravitating ice bodies [100] is too flat to fit the trend apparent in Fig. 38.

4.6 Centaurs

The Centaurs are bodies strongly interacting with the giant planets. Several definitions exist. When defined as non-Trojan bodies having both perihelia and semimajor axes between the orbits of Jupiter, $a_J = 5.2$ AU, and Neptune, $a_N = 30$ AU, there are about 87 known examples of Centaurs as of early 2006. Of these, five or six display comae and so are double-designated as comets [the most famous and prototypical example is (2060) Chiron; Table 4.6]. A detailed classification scheme has recently been proposed [61].

The differential size distribution of the Centaurs is consistent with a power law having an index $q \sim 4$, and this is compatible with the size distribution measured for KBOs of similar size [132]. The known Centaurs tend to be intermediate in size between the nuclei of well-studied comets (typically a few to 10 km diameter) and the well-studied KBOs (mostly ~100 to ~1000 km diameter). The latter is simply an effect of selection: the Centaurs are intermediate in distance between the perihelia of the well-studied comets and the Kuiper belt.

In terms of their albedos and surface colors, the Centaurs resemble the KBOs more than any other Solar system population (Fig. 34). This is consistent with the recent extraction of the Centaurs from the Kuiper Belt. Dynamical models (see the parallel Saas Fee review by Alessandro Morbidelli [112]) show that the Centaurs have dynamical lifetimes limited by scattering from (and occasional impacts into) the giant planets. Their most usual fate is to be ejected to the interstellar medium, on a median timescale $\sim 10^8$ y, but some survive entanglement with Jupiter and are kicked into orbits with perihelia $q < 5$ AU, where they begin to sublimate strongly in the heat of the sun, and are thereafter labeled as comets.

One property that has been reported to differ between the Centaurs and KBOs is the distribution of optical colors [121,148]. The available data suggest that the Centaur colors may be bimodally distributed (Fig. 39) whereas the

Table 6. The known cometary centaurs

Object	Perihelion [AU]	Semimajor axis [AU]	Eccentricity	Inclination [deg]	T_J
C/2001 M10	5.30	26.66	0.80	28.0	2.59
29P/SW1	5.72	5.99	0.04	9.4	2.98
39P/Oterma	6.83	7.25	0.24	1.9	3.01
2060 Chiron	8.45	13.62	0.38	6.9	3.36
C/2001 T4	8.56	13.92	0.38	15.4	3.29
(60558) 2000 EC98	5.83	10.73	0.46	4.3	3.03

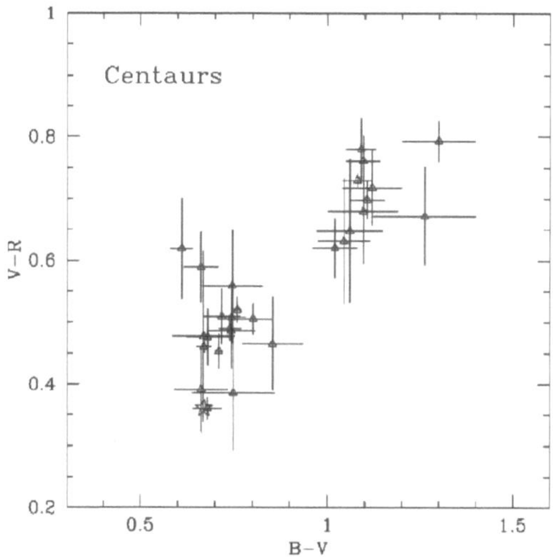

Fig. 39. Color-color diagram for Centaurs showing evidence for bimodality. From [32]

KBOs, as noted in a previous section (Figs. 31–33), are not. It is tempting to imagine that this effect (which is formally statistically significant) could be caused by past or present activity on the Centaurs. However, a search for correlations between Centaur color and such likely indicators as perihelion distance, semimajor axis, nucleus size, or current outgassing activity has revealed nothing of importance. The Centaur bimodality, if it is real, is unexplained.

4.7 Irregular Satellites

There are two, largely distinct types of planetary satellite, based on dynamical characteristics. The most familiar satellites have small eccentricities and inclinations, and orbit from a few to a few dozen planetary radii from their parent planets. These are the regular satellites, most thought to have formed by accretion within circumplanetary disks that were present around the planets during the formation epoch (the details of satellite formation in disks remain obscure and are the subject of interesting speculation and ongoing research). Other satellites, out-numbering the regulars by a considerable margin, follow eccentric and highly inclined orbits with large semimajor axes. These "irregular satellites" in fact sweep-out a considerable fraction of the Hill spheres of their planets. The Hill sphere is the region in which planetary gravity is dominant over Solar gravity and has radius (roughly the distance from the planet to the innermost Lagrange point) of

$$R_{\mathrm{H}} = a\left(\frac{m_{\mathrm{p}}}{3M_{\odot}}\right)^{1/3} \tag{23}$$

Table 7. Planetary hill spheres

Object	Mass/$M_\oplus$ [a]	a [AU] [b]	R_H [AU] [c]	θ_H [deg] [d]
Jupiter	316	5	0.35	5
Saturn	95	10	0.43	2.8
Uranus	15	20	0.47	1.4
Neptune	17	30	0.77	1.5

[a] Planetary mass. [b] semimajor axis. [c] Hill sphere radius from Equation (23). [d] apparent angular radius of the Hill sphere from Earth.

where a is the planet semimajor axis, m_p is the planet mass, and $M_\odot$ is the Solar mass (Table 7). The irregular satellites are interesting in the context of the Saas Fee workshop from the point of view of their possible origin. They cannot have been formed like the regular satellites within accretion disks surrounding the planets. Instead, they must have been captured. It is not known from where they were captured, but there are two broad possibilities. First, they might have had a local source. The irregular satellites could be planetesimals that were initially in heliocentric orbits and were captured by the planets as a result of their sudden growth (we will discuss the "standard models" for satellite capture in a moment). In this case, the irregulars are interesting because they are surviving samples of the kinds of solid bodies most of which were accreted into the cores of the giant planets, or which were scattered out of the Solar system soon after the giant planets attained their final masses. A second possibility is that the irregular satellites are captured comets or, equivalently, captured KBOs. In this case, the irregulars would take on new significance as (relatively) local examples of objects from the much more distant Kuiper Belt.

Research into the irregular satellites is in the midst of a sudden burst of new work, driven by the application of large-format CCD detectors to the problem of their detection. Less than a dozen irregular satellites were discovered in the entire twentieth Century. Most of these were chance detections made by observers using photographic plates and long exposures on large telescopes. Within the past ~half-decade, nearly 100 new irregulars have been identified, most as the result of surveys conducted using various telescopes and large cameras on Mauna Kea [133,135]; an updated summary of the data may be found at http://www.ifa.hawaii.edu/~jewitt/irregulars.html. These surveys continue, and more irregular satellites discoveries are anticipated, but we already are beginning to see new patterns in the distribution of the satellites that raise problems concerning the mechanisms of capture.

The central problem of permanent capture is that a body that follows an orbit initially unbound to a planet must lose or otherwise redistribute some of its kinetic energy to become bound to the planet. For a long time, the standard model for the capture of the irregular satellites has been through the action of gas drag forces on heliocentric planetesimals passing through the bloated gaseous envelopes of the young giant planets. This model, which was developed in parallel with models for the formation of gas giant planets like Jupiter and Saturn, implies that the irregular satellites observed today are those objects that were neither too small (ablated and absorbed in the gaseous envelopes like meteors in the Earth's upper atmosphere) nor too large (passed through the envelopes with negligible deceleration to continue in heliocentric orbits). It also relies upon the sudden collapse of the extended envelopes to leave the satellites behind: continued friction would lead to all trapped bodies spiraling into the planets.

A problem with this gas-drag capture model is that the new surveys show that Uranus and Neptune possess irregular satellite systems of their own. In fact, when corrected for the magnitude-limited nature of the observational surveys to the best of our ability, the new surveys show that the gas giants and the ice giants possess about the same number of irregular satellites, measured down to a given satellite absolute magnitude or size. This is seen by comparing Fig. 40 (the apparent magnitude distributions of the satellites of all four giant planets) with Fig. 41 (same as Fig. 40 but corrected for the varying distances of the planets using the inverse-square law [78]). Within the errors, the irregular satellite absolute brightness (size) distributions are the same. This is a remarkable and unexpected observational result. It is difficult to see how Uranus and Neptune, which are relatively gas-free ice giants, formed by processes quite different from those that produced the gas giants Jupiter and Saturn, could capture the irregular satellites by gas drag. At least, gas-drag capture has never been demonstrated for the ice giants in any publication

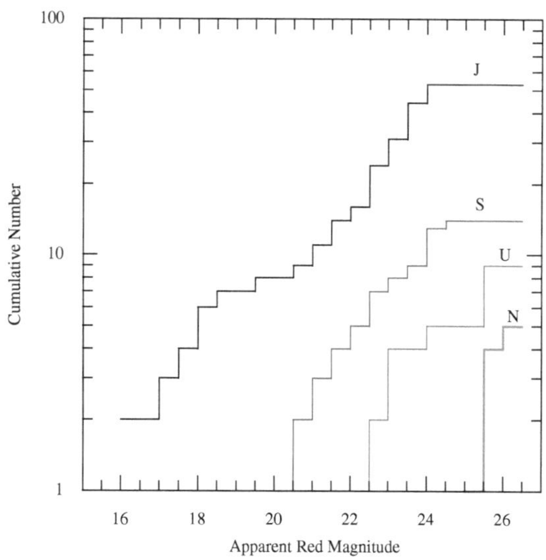

Fig. 40. Number of irregular satellites of each planet as a function of apparent magnitude. From [78]

of which I am aware. Taken as a whole, the uniform abundance of irregular satellites around the gas and ice giants argues against gas-drag capture.

What about other capture processes? A separate mechanism has been proposed in which runaway mass-growth of a planet leads to the permanent trapping of objects initially moving within the Hill sphere. This is called "pull-down capture" [59]. Like gas-drag, pull-down capture works best for the gas giants, which had a runaway growth of mass as they attracted gas from the protoplanetary nebula in a hydrodynamic in-flow (Sect. 2). The ice giants, instead, grew slowly by successive collisions with solid bodies in the disk, and they did not experience a runaway growth in mass. Therefore, it seems unlikely that pull-down capture can explain the irregular satellite systematics revealed in Fig. 41.

This leaves the generic class of "three-body interactions" as possible explanation of the capture of the irregular satellites. Three-body capture is appealing because it separates the capture mechanism from the details of planet formation. All that is needed is a sufficient density of objects for three-body interactions (two small bodies within the Hill-sphere of a large one) to occur with high enough frequency to be relevant. Although suggested long ago [22], three-body captures have rarely been discussed in the context of the irregular satellites precisely because the densities of small bodies in the Solar system are so low that the frequency of interaction is negligible. Our changing perspective, in which the density of small bodies may have been

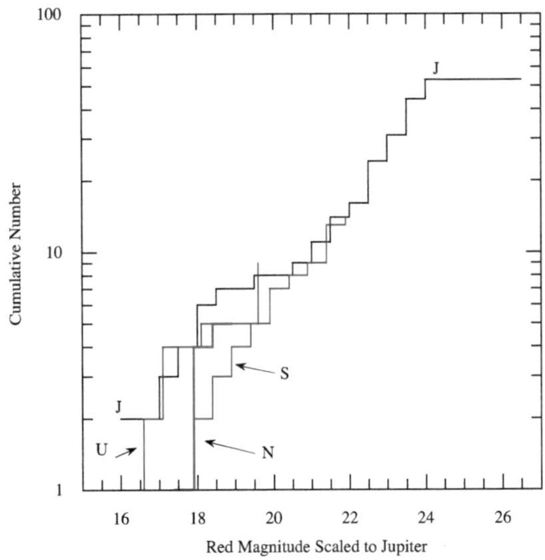

Fig. 41. Number of irregular satellites of each planet as a function of reduced magnitude (i.e., corrected for their differing heliocentric and geocentric distances using the inverse square law). From [78]

hundreds or thousands of times larger than now, makes three-body processes more attractive.

Is there any evidence that the irregular satellites were captured from a local source as opposed to a Kuiper belt source, or vice versa? The color distribution of the irregular satellites is different from the color distribution in the Kuiper belt [54, 55] with the main difference being that the ultrared matter is absent on the satellites but common on both KBOs and Centaurs. This could indicate that the Kuiper belt is not the source of the irregular satellites, suggesting that sources local to each planet are more likely. Alternatively, there could be a delivery mechanism from the Kuiper belt that operates selectively to exclude the ultrared objects. At Jupiter, it is possible that the colors of the satellites have been modified by rubble mantle formation or by another process, as is inferred for the Trojans at the same heliocentric distance. The authors of the Nice, France model [115] are careful to note that objects captured by Jupiter as Trojans have mostly spent time at smaller heliocentric distances (by which they mean to say that the color differences between Trojans and KBOs may be explained by past outgassing). The same argument could be made for the irregular satellites of Jupiter. Modification by mantling seems unlikely at Saturn, Uranus, and Neptune, however, because of the lower temperatures at 10, 20, and 30 AU and the expected lack of sublimation driven activity at these distances.

The size distribution of the irregular satellites ($q \sim 2$; [78]) is flatter than the corresponding distribution of the KBOs ($q \sim 4$; [151]). This does not rule out an origin by the capture of KBOs, however, because the satellite size distribution could have been strongly modified either by the capture process or by size-dependent evolutionary effects [117].

Measurements of the density ($1630 \pm 33\,\mathrm{kg\,m^{-3}}$) of Saturn's large irregular satellite Phoebe (Fig. 42) have been claimed as evidence for Kuiper belt origin [81]. The argument is that Phoebe is denser than most other Saturnian satellites and that the higher density more closely resembles the densities of Kuiper belt objects such as Pluto and Triton ($\rho \sim 1900\,\mathrm{kg\,m^{-3}}$). This is a difficult argument to sustain, however, given that the densities of KBOs seem to vary over a wide range and that the Saturnian regular satellite Enceladus has a density ($1606 \pm 12\,\mathrm{kg\,m^{-3}}$) essentially identical to that of Phoebe (but there is no suggestion that Enceladus is captured). I note without further comment that the *low* density of Jovian Trojan (617) Patroclus ($\rho = 800^{+200}_{-100}\,\mathrm{kg\,m^{-3}}$) has been asserted as evidence for its origin by capture from the Kuiper Belt [105]. The bottom line is that there is no simple link between density and formation location, and it seems impossible to me to use one to predict the other.

Measurements of diverse surface composition on Phoebe, including ices of water, trapped CO_2, and organics and cyanide compounds, suggest to some that this body was formed at a remote location and then captured [21]. Again, the argument is an indirect one, and, as the authors note, it is possible that

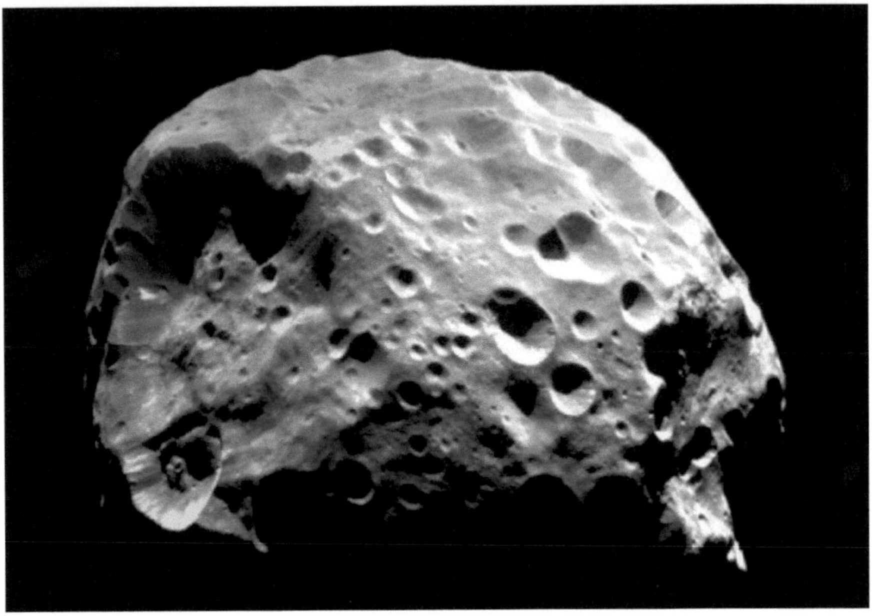

Fig. 42. Saturnian irregular satellite Phoebe, roughly 220 km in diameter and in possession of a magnificent impact crater almost half its size. Courtesy Cassini Imaging Team and NASA/JPL/SSI

the surface ice on Phoebe is in part a coating from the impact of a comet itself from distant regions.

4.8 Trojans

The origin of the Trojans has long been a source of mystery. Objects colliding near the Lagrangian L4 and L5 resonances have a small but finite probability of being captured there, particularly if they were already nearly co-moving with Jupiter [19, 104, 160]. Icy asteroids near the growing Jupiter could also be pulled into trapped orbits by the mass growth of Jupiter [49,104]. It has also been suggested that the Trojans might have originated at remote locations in the Solar system and were captured through the action of outgassing forces [160] or a chaotic disturbance that would have resulted if Jupiter and Saturn were once in 2:1 mean-motion resonance with each other [115].

In terms of what we know from observations, the Trojans may have no connection at all to the Kuiper belt or they may be genetically closely related. The observational constraints are presently too weak for us to determine the origin of these intriguing bodies at any level above the conjectural. One reason for this sorry state of affairs is that most Trojans are twice as distant and so $2^4 = 16$ times fainter than main-belt asteroids of corresponding size. By comparison, the main-belt asteroids represent "low hanging fruit" to most observers, and so, they have received the lion's share of the attention. This situation has only recently started to change. Indeed, until recently *only* Jovian Trojans were known. Now we are also aware of Trojans of Mars and of Neptune. Planned all-sky surveys should greatly improve our knowledge of the populations and size distributions of these bodies. In this section, we briefly review the known properties of the Trojans and compare them with the KBOs and other bodies.

Surveys show that the number of Jovian Trojans rivals the number of main-belt asteroids when measured down to a common limiting diameter [70, 136]. There are about 1.5×10^5 Trojans larger than 1 km in radius. They occupy two banana-shaped clouds in Jupiter's orbit, leading and trailing the planet by $\pm 60°$. Objects in the clouds librate around the L4 and L5 Lagrangian points in response to the combined gravitational attractions to the Sun and Jupiter (see [49] for a nice discussion of Trojan dynamics, from which the following is taken). In the idealized planar, restricted three-body (Sun-Jupiter-Trojan) approximation their equation of motion is

$$\frac{\mathrm{d}^2\phi}{\mathrm{d}t^2} = \left(\frac{27}{4}\right)\mu n_{\mathrm{J}}^2\phi = 0 \tag{24}$$

where ϕ is the angular separation between the Trojan and its Lagrangian point, t is time, μ is approximately the ratio of the mass of Jupiter to the mass of the Sun, and n_{J} is the mean motion of Jupiter in its orbit. The solution to Equation 24 is

$$\phi = \frac{A}{2}\cos\left[\left(\frac{27}{4}\mu n_{\mathrm{J}}^2\right)^{1/2} t + B\right] \tag{25}$$

where A and B are constants representing the amplitude of the libration and the phase, respectively. With $\mu \sim 0.001$, $n_{\mathrm{J}} \sim 0.52\,\mathrm{year}^{-1}$, the characteristic frequency is $\omega = (27\mu n_{\mathrm{J}}^2/4)^{1/2} \sim 0.043\,\mathrm{year}^{-1}$, corresponding to a period $2\pi/\omega \sim 150$ years, or about 10 times Jupiter's orbital period. The distribution of libration amplitudes, A, is very broad, with a mean near 30° [110, 136]. In addition to wide excursions about the Lagrangian points in the orbital plane of the planet, the Trojans also occupy a broad distribution of inclinations, with a bias-corrected mean of 14° [69] to 17° [136]. As a consequence, the velocity dispersion amongst the Trojans ($\sim 5\,\mathrm{km\,s^{-1}}$) rivals that amongst the main-belt asteroids. Collisions between Jovian Trojans are expected to be highly erosive.

Limited work on the long-term stability of Trojans at Jupiter suggests two loss mechanisms. There is a slow leak because dynamical chaos [97], with a timescale that depends on A (larger A being less stable). The more significant loss process is due to ejection from the Lagrangian clouds following collisions or near-miss interactions. Kilometer-sized and larger bodies are lost this way at a rate $\sim 10^3\,\mathrm{My}^{-1}$ [103], meaning that the observed population of small objects should vanish in a few $\times 10^2$ My. That small Trojans remain is presumably a result of a collisional cascade, with the small objects being both lost and continually supplied from the break-up of larger bodies. Ejected Trojans pursue orbits that are scattered by the planets, quickly becoming indistinguishable from the orbits of Jupiter family comets. Up to ∼10% of the latter could be escaped Jovian Trojans [70, 103]: the contributions from the Trojan swarms of other planets are unknown.

Several physical properties of the Trojans have been measured. The size distribution is a broken power law [70,136]. Objects with absolute magnitudes V(1,1,0) <9.5 (corresponding to diameters >84 km, for albedo 0.04) are well described by a differential power law index $q = 5.5 \pm 0.9$ (Fig. 43). Those with $11 \leq$ V(1,1,0) ≤ 14 (diameters $4.4 \leq D \leq 40$ km, for the same albedo) instead have $q = 3.0 \pm 0.3$ [70]. The index for the smaller objects is close to the nominal value expected for a system in collisional equilibrium [34], consistent with the idea that these smaller bodies are part of a collisionally-produced cascade. The steep slope of the large Trojans presumably reflects a "production function": at least, these big bodies seem unlikely to have been molded

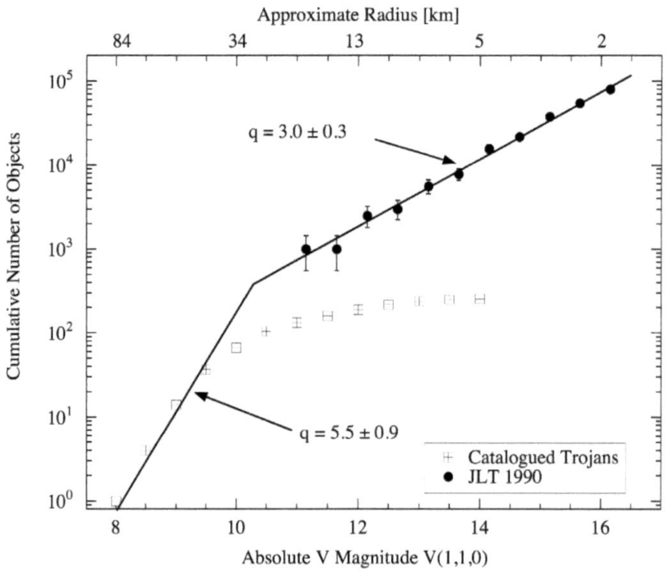

Fig. 43. Brightness distribution of the Jovian Trojans , showing the break in the size distribution. Red points show the numbered Trojans. The roll-over above V(1,1,0) ∼ 10 is due to observational incompleteness. The blue points are from [70], scaled to correct for the small area of the Trojan swarms observed in that survey. The difference in slope between the large and small objects is independent of the scaling. The radius scale at the top is computed on the assumption that the Trojans all have albedo 0.04. From [70]

much by energetic collisions. For comparison, the $D \geq 100$ km KBOs occupy a distribution with $q \sim 4$ [151]. Within the errors ($\sim 2\sigma$), this is compatible with the size distribution of the larger Trojans, as might be expected if the latter were captured from the Kuiper belt [115].

As already noted, the optical color distribution of the Trojans is different from that of the KBOs and Centaurs because the Trojans lack ultrared matter. This could mean that there is no relation between the Trojans and the KBOs or Centaurs, or it could mean that the surfaces of the Trojans have been modified in some way by their exposure to sunlight (as have the surfaces of the Jupiter family comet nuclei, which very likely do come from the Kuiper belt). We prefer the latter explanation, but it does not tell us anything about the source of the Trojans, because the surface modification process could operate regardless of the origin of the bodies. Any object formed beyond the snow-line (perhaps originally at $\sim$3 AU or slightly closer) is expected to be icy and should evolve when heated to develop a surface mantle. In the same vein, the albedo distribution of the Trojans is very narrow compared to that of the KBOs and Centaurs [48] but more similar to the nuclei of Jupiter family comets (Fig. 34). This is probably also a result of surface modification on bodies that have been heated strongly by the sun but, again, we cannot use this information to specify the source of the Trojans. In terms of their spectra, the Trojans have steadfastly resisted every attempt to assess surface composition from observations taken in the optical and near infrared [37, 42, 101]. The absence of features is consistent with the very dark surfaces of these bodies and suggests (but does not require) an organic-rich compositional nature [24]. Observations at thermal wavelengths have revealed features consistent with emission from silicates in three Trojans (624 Hektor, 911 Agamemnon and 1172 Aneas; [25]).

Lastly, the density of Trojan (617) Patroclus has been estimated from infrared observations [48] and from its dynamical system mass as $\rho = 800^{+200}_{-100}\,\mathrm{kg\,m^{-3}}$ [105]. Although the authors of [105] cite this low density as evidence that Patroclus is a captured KBO, in fact low density is only evidence for a high mass fraction of ice and/or vacuum ("porosity") and cannot be diagnostic of the Trojan source. In fact, any object formed at any distance beyond the snow-line would be expected to have a high ice content and correspondingly low density. Simply put, "density is not destiny."

An accurate summary is that the available physical data on Trojans, from their surface colors [69] to their albedos [48] to the one measured density [105] are similar to the corresponding quantities reported for the nuclei of comets but not similar to those of the KBOs. The measured Trojan properties very probably reflect refractory surface mantles left behind following ancient mass-loss, but we cannot uniquely determine the source of the Trojans from the physical data. An interesting exercise for the readers of this article is to think of observations that could be taken to uniquely determine the source of the Trojans. I, for my part, will be trying to do exactly the same.

References

1. M. A'Hearn, R. Millis, D. Schleicher, D. Osip and P. Birch. The Ensemble Properties of Comets. Icarus 118, 223–270 (1995).
2. F. Adams and G. Laughlin. Constraints on the Birth Aggregate of the Solar System. Icarus, **150**, 151–162 (2001).
3. D. A. Allen. Infrared Diameter of Vesta. Nature, **227**, 158–159 (1970).
4. S. Andrews and J. Williams. Circumstellar Dust Disks in Taurus-Auriga. Ap. J. **631**, 1134–1160 (2005).
5. J. Anderson et al. Amalthea's Density is Less than that of Water. Science 308, 1291–1293 (2005).
6. E. Asphaug and W. Benz. Structure of Comet Shoemaker-Levy 9 Inferred from the Physics of Tidal Breakup. Icarus, 121, 225–248 (1996).
7. G. Bernstein et al. The Size Distribution of Trans-Neptunian Objects. Astron. J. 128, 1364–1390 (2004).
8. A. Boss. Gas Giant Protoplanet Formation: Disk Instability Models with Thermodynamics and Radiative Transfer. Ap. J. 563, 367–373 (2001).
9. A. Boss, G. Wetherill and N. Haghighipour. Rapid Formation of Ice Giant Planets. Icarus 156, 291–295 (2002).
10. C. Briceno et al. The CIDA-QUEST Large Scale Survey of Orion OB1: Evidence for Rapid Disk Dissipation in a Dispersed Stellar Population. Science 291, 93–97 (2001).
11. D. Britt and G. Consolmagno. The Structure of High Porosity Asteroids. Icarus 152, 134–139 (2001).
12. D. Britt and G. Consolmagno. Stony Meteorite Porosities and Densities. Meteor. Plan. Sci 38, 1161–1180 (2003).
13. M. Brown, C. Trujillo and D. Rabinowitz. Discovery of a Planetary-Sized Object in the Scattered Kuiper Belt. Ap. J. 635, L97–100 (2005).
14. K. Cai et al. Effects of Metallicity and Grain Size on Gravitational Instabilities in Protoplanetary Disks. Ap. J. 636, L149–152 (2006).
15. R. Canup. A Giant Impact Origin of Pluto-Charon. Science 307, 546–550 (2005).
16. J. Carvano, T. Mothe-Diniz and D. Lazzaro. Search for Relations Among a Sample of 460 Asteroids with Featureless Spectra. Icarus 161, 356–382.
17. S. Chandrasekhar. Figures of Equilibrium (New York: Dover).
18. J. Chambers. Making More Terrestrial Planets. Icarus 152, 205–224 (2001).
19. E. Chiang and Y. Lithwick. Neptune Trojans as a Test Bed for Planet Formation. Ap. J., 628, 520–532 (2005).
20. F. Ciesla and J. Cuzzi. The Evolution of the Water Distribution in a Viscous Protoplanetary Disk. Icarus 181, 178–204 (2006).
21. R. Clark et al. Compositional Maps of Saturn's Moon Phoebe from Imaging Spectroscopy. Nature 435, 66–69 (2005).
22. G. Columbo and F. Frankin. On the Formation of the Outer Satellite Groups of Jupiter. Icarus 15, 186–189 (1971).
23. J. Cooper, E. Christian, J. Richardson and C. Wang. Proton Irradiation of Centaur, Kuiper Belt and Oort Cloud Objects. Earth, Moon and Planets 92, 261–177 (2003).
24. D. Cruikshank et al. Constraints on the Composition of Trojan Asteroid 624 Hektor. Icarus 153, 348–360 (2001).

25. D. Cruikshank, M. Barucci, J. Emery, Y. Fernandez, W. Grundy, K. Noll and J. Stansberry. Physical Properties of Trans-Neptunian Objects. Protostars and Planets V (eds. B. Reipurth, D. Jewitt and K. Keil), University Az. Press, Tucson, p. 879–892. (2006).
26. J. Davies et al. The Detection of Water Ice in Comet Hale-Bopp. Icarus 127, 238–245 (1997).
27. J. Davies et al. Visible and Infrared Photometry of Fourteen KBOs. Icarus 146, 252–262 (2000).
28. B. Daviddson and P. Gutierrez. Estimating the Nucleus Density of 19P/Borrelly. Icarus 168, 392–408 (2004).
29. B. Daviddson and P. Gutierrez. Nucleus Properties of 67P/Churyumov Gerasimenko. Icarus 176, 453–477 (2005).
30. B. Daviddson and P. Gutierrez. Non-gravitational force modeling of Comet 81P/Wild 2. Icarus 180, 224–242 (2006).
31. A. Delsanti, O. Hainaut, E. Jourdeuil, K. Meech, H. Boehnhardt and L. Barrera. Simultaneous visible-near IR Photometric Study of Kuiper Belt Objects. Astron. Ap. 417, 1145–1158 (2004).
32. A. Delsanti and D. Jewitt. The Solar System beyond the Planets. In Solar System Update (eds. Ph. Blondel and J. Mason), Springer-Praxis, Berlin, p. 267–294 (2006).
33. Delsanti, A., Peixinho, N., Boehnhardt, H., Barucci, A., Merlin, F., Doressoundiram, A., and Davies, J. K. Astron. J. 131, 1851–1863 (2006).
34. J. Dohnanyi. Collisional Models of Asteroids and Their Debris. J. Geophys. Res. 74, 2531–2554 (1969).
35. A. Doressoundiram. Color Properties and Trends in Trans-Neptunian Objects. Earth, Moon and Planets 92, 131–144 (2003).
36. A. Doressoundiram et al. Meudon Multicolor Survey (2MS) of Centaurs and Trans-Neptunian Objects. Icarus 174, 90–104 (2005).
37. C. Dumas, T. Owen and M. Barucci. Near-Infrared Spectroscopy of Low-Albedo Surfaces in the Solar System. Icarus 133, 221–232 (1998).
38. M. Duncan, Quinn and S. Tremaine. The Origin of Short-Period Comets. Ap. J. 328, L69–73 (1988).
39. W. Durham, H. Heard and S. Kirby. Experimental Deformation of Polycrystalline H_2O Ice at High Pressure and Low Temperature. J. geo. res. 88, B377–392 (1983).
40. K. Edgeworth. J. Br. Astron. Soc., 53, 181–188 (1943).
41. C. Emerich et al. Temperature and Size of the Nucleus of P/Halley Deduced from IKS Infrared Vega 1 Measurements. Astron. Ap., 187, 839–842 (1987).
42. J. Emery and R. Brown. Constraints on the Surface Composition of Trojan Asteroids from Near Infrared (0.8–4.0μm) Spectroscopy. Icarus 164, 104–121 (2003).
43. E. Everhart. Intrinsic Distributions of Cometary Perihelia and Magnitudes. Astron. J., 72, 1002 (1967).
44. P. Farinella and D. Davis. Short-period Comets: Primordial Bodies or Collisional Fragments? Science 273, 938–941 (1996).
45. J. Fernandez. On the Existence of a Comet Belt Beyond Neptune. MNRAS 192, 481–491 (1980).
46. J. Fernandez, G. Tancredi, H. Rickman and J. Licandro. The Population, Magnitudes, and Sizes of Jupiter Family Comets. Astron. Ap., 352, 327–340 (1999).

47. J. Fernandez, T. Gallardo and A. Brunini. Icarus 159, 358 (2002).
48. Y. Fernandez, S. Sheppard and D. Jewitt. Albedo Distribution of Jovian Trojan Asteroids. Astron. J. 126, 1563–1574 (2003).
49. H. Fleming and D. Hamilton. On the Origin of the Trojan Asteroids: Effects of Jupiter's Mass Accretion and Radial Migration. Icarus, 148, 479–493 (2000).
50. Y. Funato, J. Makino, P. Hut, E. Kokubo and D. Kinoshita. The Formation of Kuiper Belt Binaries Through Exchange Reactions. Nature 427, 518–520 (2004).
51. P. Francis. The Demographics of Long–Period Comets. Ap. J., 635, 1348–1361 (2005).
52. P. Goldreich, Y. Lithwick and R. Sari. Formation of Kuiper Belt Binaries by Dynamical Friction and Three–Body Encounters. Nature 420, 643–646 (2002).
53. R. Gomes. The Origin of the Kuiper Belt High Inclination Population. Icarus 161, 404–418 (2003).
54. T. Grav, M. Holman, B. Gladman and K. Aksnes. Photometric Survey of the Irregular Satellites. Icarus 166, 33–45 (2003).
55. T. Grav, M. Holman and W. Fraser. Photometry of Irregular Satellites of Uranus and Neptune. Ap. J. 613, L77–80 (2004).
56. N. Haghighipour and A. Boss. On Gas Drag-Induced Rapid Migration of Solids in a Nonuniform Solar Nebula. Ap. J. 598, 1301–1311 (2003).
57. O. Hainaut and A. Delsanti. Colors of Minor Bodies in the Outer Solar System: A Statistical Analysis. Astron. Ap. 389, 641–664 (2002).
58. J. Heisler and S. Tremaine. The Influence of the Galactic Tidal Field on the Oort Comet Cloud. Icarus, 65, 13–26 (1986).
59. T. Heppenheimer and C. Porco. New Contributions to the Problem of Capture. Icarus, 30, 385–401 (1977).
60. T. Hiroi, M. Zolensky and C. Pieters. The Tagish Lake Meteorite: Possible Sample from a D-Type Asteroid. Science 293, 2234–2236 (2001).
61. J. Horner, N. Evans, M. Bailey and D. Asher. The Populations of Comet-Like Bodies in the Solar System. MNRAS 343, 1057–1066 (2003).
62. H. Hsieh, D. Jewitt and Y. Fernandez. The Strange Case of 133P/Elst-Pizarro. Astron. J. 127, 2997–3017 (2004).
63. H. Hsieh and D. Jewitt. A Population of Comets in the Main Asteroid Belt. Science, 312, 561–563 (2006).
64. D. Hughes. The Magnitude Distribution, Perihelion Distribution and Flux of Long-period Comets. MNRAS, 326, 515–523 (2001).
65. D. Hunten and A. Watson. Stability of Pluto's Atmosphere. Icarus 51, 665–667 (1982).
66. S. Ida, J, Larwood and A. Burkert. Evidence for Early Stellar Encounters in the Orbital Distribution of Edgeworth-Kuiper Belt Objects. Ap. J. **528**, 351–356 (2000).
67. S. Jacobsen. The Hf-W Isotopic System and the Origin of the Earth and Moon. Ann. Rev. Earth and Planet Sci. 33, 531–570 (2005).
68. D. Jewitt and K. Meech. Optical Properties of Cometary Nuclei and a Preliminary Comparison with Asteroids. Ap. J. 328, 974–986 (1988).
69. D. Jewitt and J. Luu. CCD Spectra of Asteroids II. The Trojans as Spectral Analogs of the Cometary Nuclei. Astron. J. 100, 933–944 (1990).
70. D. Jewitt, C. Trujillo and J. Luu. Population and Size Distribution of Small Jovian Trojans. Astron. J. 120, 1140–1147 (2000).

71. D. Jewitt. From Kuiper Belt Object to Cometary Nucleus: The Missing Ultra-Red Matter. Astron. J., 123, 1039–1049 (2002).
72. D. Jewitt and S. Sheppard. Physical Properties of Trans-Neptunian Object (20000) Varuna. Astron. J. 123, 2110–2120 (2002).
73. D. Jewitt and J. Luu. Discovery of the Candidate Kuiper Belt Object 1992 QB1. Nature, 362, 730–732 (1992).
74. D. Jewitt and J. Luu. Optical-Infrared Spectral Diversity in the Kuiper Belt. Astron. J. 115, 1667–1670 (1998).
75. D. Jewitt and J. Luu. Colors and Spectra of Kuiper Belt Objects. Astron. J., 122, 2099–2114 (2001).
76. D. Jewitt and J. Luu. Crystalline Water Ice on Kuiper Belt Object (50000) Quaoar. Nature 432, 731–733 (2004).
77. D. Jewitt, S. Sheppard and C. Porco. In Jupiter (eds. F. Bagenal et al), Cambridge University Press. (2004).
78. D. Jewitt and S. Sheppard. Irregular Satellites in the Context of Giant Planet Formation. Space Sci. Rev. 116, 441–456 (2005).
79. D. Jewitt. A First Look at the Damocloids. Astron. J., 129, 530–538 (2005).
80. D. Jewitt. From the Cradle to the Grave: The Rise and Demise of the Comets. In Comets II (eds. M. Festou, H. Weaver and H. U. Keller), University Az. Press, Tucson, pp. 659–676 (2005).
81. T. Johnson and J. Lunine. Saturn's Moon Phoebe as a Captured Body from the Outer Solar System. Nature, 435, 69–71 (2005).
82. T. Jones, L. Lebofsky, J. Lewis and M. Marley. The Composition and Origin of the C-, P- and D-Asteroids. Icarus 88, 172–192 (1990).
83. H. Kawakita et al. Evidence of Icy Grains in Comet C/2002 T7 (LINEAR at 3.52 AU. Ap. J. 601, L191–194 (2004).
84. K. Keil. Thermal Alteration of Asteroids: Evidence from Meteorites. Planet. Space Sci. 48, 887–903 (2000).
85. G. Kuiper. On the Origin of the Solar System. In Astrophysics (ed. J. Hynek), McGraw-Hill, New-York, pp. 357–424 (1951).
86. P. Lacerda and J. Luu. Analysis of the Rotational Properties of Kuiper Belt Objects. Astron. J. 131, 2314–2326. (2006).
87. P. Lamy, E. Grun and J. Perrin. Astron. Ap., 187, 767.
88. P. Lamy, I. Toth, Y. Fernandez and H. Weaver. In Comets II, (eds. M. Festou, H. Weaver and H. Keller), Univesity Az. Press, Tucson, pp. 223–264 (2005).
89. L. Lebofsky et al. A Refined Standard Model for Asteroids Based on Observations of 1 Ceres and 2 Pallas. Icarus 68, 239–251 (1986).
90. T. Lee, F. Shu, H. Shang, A. Glassgold and K. Rehm. Ap. J., **506**, 898–912 (1976).
91. T. Lee, D. Papanastassiou and G. Wasserburg. Geophys. Res. Lett. **3**, 41–44 (1998).
92. F. Leonard. The New Planet Pluto. Leaflet Astron. Soc. Pac. No 30, Aug, 121–124 (1930).
93. G. Leone, P. Farinella, P. Paolicchi and V. Zappala. Equilibrium Models of Binary Asteroids. Astron. Ap. 140, 265–272 (1984).
94. H. Levison et al. The Mass Disruption of Oort Cloud Comets. Science, 296, 2212–2215 (2002).
95. H. Levison, D. Terrell, P. Wiegert, L. Dones and M. Duncan. On the Origin of the Unusual Orbit of Comets 2P/Encke. Icarus, 182, 161–168 (2006).

96. H. Levison and M. Duncan. Long Term Dynamical Behavior of Short-Period Comets. Icarus, 108, 18–36 (1994).
97. H. Levison, G. Shoemaker and C. Shoemaker. The Dispersal of the Trojan Asteroid Swarm. Nature 385, 42–44 (1997).
98. C. Lisse et al. The Nucleus of Comet Hyakutake (C/1996 B2). Icarus 140, 189–204 (1999).
99. J. Licandro, N. Pinilla-Alona, M. Pedani, E. Oliva, G. Tozzi and W. Grundy. Methane Ice Rich Surface of Large TNO 2005 FY9. Astron. Ap. 445, L35–38 (2006).
100. M. Lupo and J. Lewis. Mass-Radius Relationships in Icy Satellites. Icarus 40, 157–170 (1979).
101. J. Luu, D. Jewitt and E. Cloutis. Near-Infrared Spectroscopy of Primitive Solar System Objects. Icarus 109, 133–144 (1994).
102. J. Luu and D. Jewitt. Color Diversity Among the Centaurs and Kuiper Belt Objects. Astron. J. 112, 2310- (1996).
103. F. Marzari, P. Farinella and V. Vanzani. Are Trojan Collisional Families a Source for Short-Period Comets? Astron. Ap. 299, 267 (1995).
104. F. Marzari, P. Tricarico and H. Scholl. Clues to the Origin of Jupiter's Trojans: The Libration Amplitude Distribution. Icarus 162, 453–459 (2003).
105. F. Marchis et al. Nature 439, 565–567 (2006).
106. N. McBride et al. Visible and Infrared Photometry of Kuiper Belt Objects: Searching for Evidence of Trends. Icarus 161, 501–510 (2003).
107. W. McKinnon. On the Initial Thermal Evolution of Kuiper Belt Objects. Proceeding of Asteroids, Comets, Meteor, ESA SP-500, Noordwijk, Netherlands, pp. 29–38 (2002).
108. K. Meech, O. Hainaut and B. Marsden. Comet nucleus size distributions from HST and Keck telescopes. Icarus 170, 463–491 (2004).
109. R. Meier and T. Owen. Cometary Deuterium. Space Sci. Rev. 90, 33–43 (1999).
110. A. Milani. Trojan Asteroids Belt: Proper Elements, Stability, Chaos and Families. Cel. Mech. Dyn. Astron. 57, 59–94 (1993).
111. H. Mizuno, K. Nakazawa and C. Hayashi. Instability of a Gaseous Envelope Surrounding a Planetary Core and Formation of Giant Planets. Prog. Theor. Physics 60, 699–710 (1978).
112. A. Morbidelli. Comets and their Reservoirs: Current Dynamics and Primordial Evolution. In: Trans-Neptunian Objects and Comets (eds. K. Altwegg, W. Benz and N. Thomas), Vol. 35, pp. 79–164 (2008).
113. A. Morbidelli et al. Source Regions and Time Scales for the Delivery of Water to Earth. Meteoritics and Planet. Sci. 35, 1309–1320 (2000).
114. A. Morbidelli, and H. Levison. Scenarios for the Origin of the Orbits of the Trans-Neptunian Objects 2000 CR105 and 2003 VB12 (Sedna). Astron. J., **128**, 2564–2576 (2004).
115. A. Morbidelli, H. Levison, K. Tsiganis and R. Gomes. Chaotic Capture of Jupiter's Trojan Asteroids in the Early Solar System. Nature 435, 462–465 (2005).
116. S. Mostefaoui, G. Lugmair, P. Hoppe and A. El-Goresy. New Astronomy Review, **48**,155–159 (2004).
117. D. Nesvorny, J. Alvarellos, L. Dones and H. Levison. Orbital and Collisional Evolution of the Irregular Satellites. Astron. J., 126, 398–429 (2003).

118. K. Noll. Solar System Binaries. Proceedings of Asteroids, Comets, Meteors 2005 (eds. S. Ferraz-Mello and D. Lazzaro), IAU Symp. 229, Cambridge University Press, Cambridge (2006).
119. J. Oort. The Structure of the Cloud of Comets Surrounding the Solar System and a Hypothesis Concerning its Origin. Bull. Astron. Inst. Neth., 11, 91–110 (1950).
120. T. Owen, P. Mahaffy, H. Niemann, S. Atreya, T. donahue, A. Bar-Nun and I. de pater. A Low Temperature Origin for the Planetesimals that Formed Jupiter. Nature 402, 269–270 (1999).
121. N. Peixinho, A. Dorresoundiram, A. Delsanti, H. Boehnhardt, M. Barucci and I. Belskaya. Reopening the TNOs Color Controversy: Centaurs Bimodailty and TNOs Unimodality. Astron. Ap. 410, L29–32 (2003).
122. J. Petit and O. Mousis. KBO Binaries: How Numerous were they? Icarus 168, 409–419 (2004).
123. J. Pollack, O. Hubickyj, P. Bodenheimer, J. Lissauer, M. Podolak and Y. Greenzweig. Formation of the Giant Planets by Concurrent Accretion of Solids and Gas. Icarus 124, 62–85 (1996).
124. K. R. Popper. The Logic of Scientific Discovery. Hutchinson and Co., London.
125. R. Prinn and B. Fegley. Kinetic Inhibition of CO and N_2 Reduction in Circumplanetary Nebulae. Ap. J. 249, 308–317 (1981).
126. D. Prialnik. In Comets II (eds. M. Festou, H. Weaver and H. U. Keller), University Az. Press, Tucson, pp. 359–387 (2005).
127. D. Rabinowitz et al. Photometric Observations of 2003 EL61. Ap. J. 639, 1238–1251 (2006).
128. H. Rauer. Comets. In: Trans-Neptunian Objects and Comets (eds. K. Altwegg, W. Beuz and N. Thomas), Vol. 35, pp. 165–254 (2008).
129. S. Raymond, T. Quinn and J. Lunine. Making Other Earths: Dynamical Simulations of Terrestrial Planet Formation and Water Delivery. Icarus 168, 1–17 (2004).
130. H. Rickman, J. Fernandez and B. Gustafson. Astron. Ap., 237, 524 (1990).
131. G. Rieke et al. Decay of Planetary Debris DIsks. Ap. J., 620, 1010–1026 (2005).
132. S. Sheppard et al. A Wide Field CCD Survey for Centaurs and Kuiper Belt Objects. Astron. J. 120, 2687–2694 (2000).
133. S. Sheppard and D. Jewitt. An Abundant Population of Irregular Satellites Around Jupiter. Nature 413m 261–263 (2003).
134. S. Sheppard and D. Jewitt. Extreme Kuiper Belt Objects 2001 QG298 and the Fraction of Contact Binaries. Astron. J., 127, 3023–3033 (2004).
135. S. Sheppard, D. Jewitt and J. Kleyna. Ultradeep Survey for Irregular Satellites of Uranus: Limits to Completeness. Astron. J., 129, 518–525 (2005).
136. G. Shoemaker, C. Shoemaker and R. Wolfe. In Asteroids II (eds. T. Gehrels and M. Matthews), University Az. Press, pp. 487–523 (1989).
137. J. Stansberry et al. Albedos, Diameters and Density of Kuiper Belt Object (47171) 1999 TC36. Ap. J, 643, 556–566 (2006).
138. D. Stephens and K. Noll. HST Photometry of Trans-Neptunian Objects. Earth, Moon and Planets 92, 251–260 (2004).
139. D. Stephens and K. Noll. Detection of Six Trans-Neptunian Binaries with NICMOS. Astron. J., 131, 1142–1148 (2006).
140. J. Sunshine et al. Exposed Water Ice Deposits on the Surface of Comet Tempel 1. Science 311, 1453–1455 (2006).

141. S. Takahashi and W. Ip. A Shape and Density Model of the Putative Binary EKBO 2001 QG298. PASJ 56, 1099–1103 (2004).
142. N. Takato, S. Bus, H. Terada, T-S. Pyo and N. Kobayashi. Detection of a Deep 3 μm Absorption Feature in the Spectrum of Amalthea. Science 306, 2224–2227 (2004).
143. N. Takato, H. Terada and T-S. Poo. Crystalline Water Ice on the Satellite of 2003 EL61. Preprint (2006).
144. G. Tancredi, J. Fernandez, H. Rickman and J. Licandro. Nuclear magnitudes and the Size Distribution of Jupiter Family Comets. Icarus, 182, 527–549 (2006).
145. S. Tegler and W. Romanishin. Two distinct populations of Kuiper-belt objects. Icarus, 392, 49–51 (1998).
146. S. Tegler and W. Romanishin. Extremely Red Kuiper-belt Objects in Near-circular Orbits Beyond 40 AU. Nature 407, 979–981 (2000).
147. S. Tegler and W. Romanishin. Resolution of the Kuiper Belt Object Color Controversy: Two Distinct Color Populations. Icarus 161, 181–191 (2003).
148. S. Tegler, W. Romanishin, and S. Consolmagno. Color Patterns in the Kuiper Belt: A Possible Primordial Origin. Ap. J. 599, L49–52 (2003).
149. E. Thommes, M. Duncan and H. Levison. The Formation of Uranus and Neptune Between Jupiter and Saturn. Astron. J., 123, 2862–2883 (2002).
150. I. Toth. Impact-Generated Activity Period of the Asteroid 7968 Elst-Pizarro in 1996. Astron. Ap. 360, 375–380 (2000).
151. C. Trujillo, D. Jewitt and J. Luu. Properties of the Trans-Neptunian Belt: Statistics from the Canada-France-Hawaii Telescope Survey. Astron. J., 122, 457–473 (2001).
152. C. Trujillo and M. Brown. A Correlation between Inclination and Color in the Classical Kuiper Belt. Ap. J. 566, L125–128 (2002).
153. C. Trujillo, M. Brown, D. Rabinowitz and T. Geballe. Near-Infrared Surface Properties of the Two Intrinsically Brightest Minor Planets. Ap. J., 627, 1057–1065 (2005).
154. K. Tryka, R. Brown and V. Anicich. Near-Infrared Absorption Coefficients of Solid Nitrogen as a Function of Temperature. Icarus 116, 409–414 (1995).
155. W. Ward and J. Hahn. Dynamics of the Trans-Neptune Region: Apsidal Waves in the Kuiper Belt. Astron. J., 116, 489–498 (1998).
156. S. Weidenschilling. On the Origin of Binary TransNeptunian Objects. Icarus 160, 212–215 (2002).
157. P. Weissman and S. Lowry. The Size Distribution of Jupiter-Family Cometary Nuclei. Lunar Planet. Sci. Conf. Abstract No. 2003. (2003).
158. P. Wiegert and S. Tremaine. The Evolution of Long-Period Comets. Icarus, 137, 84–121 (1999).
159. D. Yeomans et al. Estimating the Mass of Asteroid 253 Mathilde from Tracking Data During the NEAR Flyby. Science 287, 2106–2109 (1997).
160. C. Yoder. Notes on the Origin of the Trojan Asteroids. Icarus 40, 341–344 (1979).
161. B. Zuckerman, T. Forveille and J. Kastner. Inhibition of Giant-Planet Formation by Rapid Gas Depletion Around Young Stars. Nature 373, 494–496 (1995).

Comets and Their Reservoirs: Current Dynamics and Primordial Evolution

A. Morbidelli

Comets are often considered to be the gateway for understanding Solar System formation. In fact, they are probably the most primitive objects of the Solar System because they formed in distant regions where the relatively cold temperature preserved the pristine chemical conditions. For this reason, they have been the target of very sophisticated and expensive space missions (such as *Giotto*, *Stardust*, and *Rosetta*) for in-situ analysis or sample return. To best exploit the information collected by ground-based and space-based observations, however, it is necessary to know where comets come from, where they are formed, and how they evolved in the distant past. For instance, did they form at 5, 30 or at 100 AU? Are they chunks of larger objects that presumably underwent significant thermal and collisional alteration or are they pristine planetesimals that failed to grow larger?

In addition, the orbital structure of the comet reservoirs records information of the dynamical processes that occurred when the Solar System was taking shape. For example, it carries evidence of the migration of the giant planets and/or of close encounters between our Sun and other stars. Modeling these dynamical processes, and comparing their outcomes with the observed structures, gives us a unique opportunity to reconstruct the history of the formation of the planets and of their primordial evolution.

The purpose of this chapter is to review our current understanding of comets from the dynamical point of view and to underline the open issues which still need more investigation. The first part is devoted to the current Solar System. In Sect. 1, I describe the orbital and dynamical properties of the trans-Neptunian population: the Kuiper belt and the Scattered disk. Section 2 is devoted to the evolution of comets from their parent reservoirs – the trans-Neptunian population or the Oort cloud – to the inner Solar System. As we know the current Solar System quite well – the orbits of the planets and the galactic environment – the results discussed in these sections are quite secure. In contrast, the second part of the chapter focusses on more controversial topics, as it is devoted to the origin of the Solar System, how the comet reservoirs formed and acquired their current shapes. More precisely, Sect. 3 is

A. Morbidelli, *Comets and Their Reservoirs*. In: K. Altwegg et al., Trans-Neptunian Objects and Comets, pp. 79–163 (2008)
DOI 10.1007/978-3-540-71958-8_2

devoted to the formation of the Oort cloud, Sect. 4 to the primordial sculpting of the trans-Neptunian population and Sect. 5 discusses a recently proposed connection between these events and the Late Heavy Bombardment of the terrestrial planets. In the final section, I will speculate on a scenario for the primordial evolution of the Solar System that would put all these aspects together in a coherent scheme.

1 The Trans-Neptunian Population

Our observational knowledge of the trans-Neptunian population[1] is very recent. The first object, Pluto, was discovered in 1930, but unfortunately this discovery was not quickly followed by the detection of other trans-Neptunian objects. Thus, Pluto was thought to be an exceptional body – a planet – rather than a member of a numerous small body population, of which it is not even the largest in size. It was only in 1992, with the advent of CCD cameras and a lot of perseverance, that another trans-Neptunian object – 1992 QB_1 – was found [86]. Now, 13 years later, we know more than 1,000 trans-Neptunian objects. Of them, about 500 have been observed for at least 3 years. A time-span of 3 years of observations is required in order to compute their orbital elements with some confidence. In fact, the trans-Neptunian objects move very slowly, and most of their apparent motion is simply a parallactic effect. Our knowledge of the orbital structure of the trans-Neptunian population is therefore built on these $\sim$500objects.

Before moving to discuss the orbital structure of the trans-Neptunian population, in the next subsection, a brief overview of the basic facts of orbital dynamics is given. The expert reader can move directly to Sect. 1.2.

1.1 Brief Tutorial on Orbital Dynamics

Neglecting mutual perturbations, all bodies in the Solar System move on an elliptical orbit relative to the Sun, with the Sun at one of the two foci of the ellipse. Therefore, it is convenient for astronomers to characterize the relative motion of a body by quantities that describe the geometrical properties of its orbital ellipse and its instantaneous position on the ellipse. These quantities are usually called *orbital elements*.

The shape of the ellipse can be completely determined by two orbital elements: the semi-major axis a and the eccentricity e (Fig. 1). The name *eccentricity* comes from e being a measure of the distance of the focus from the center of the ellipse, in units of semi-major axis' length ("eccentric" means

[1]There is no general consensus on nomenclature, yet. In this work, I use "trans-Neptunian population" to describe the collection of small bodies with semi-major axis (or equivalently orbital period) larger than that of Neptune, with the exception of the Oort cloud (semi-major axis larger than 10,000 AU).

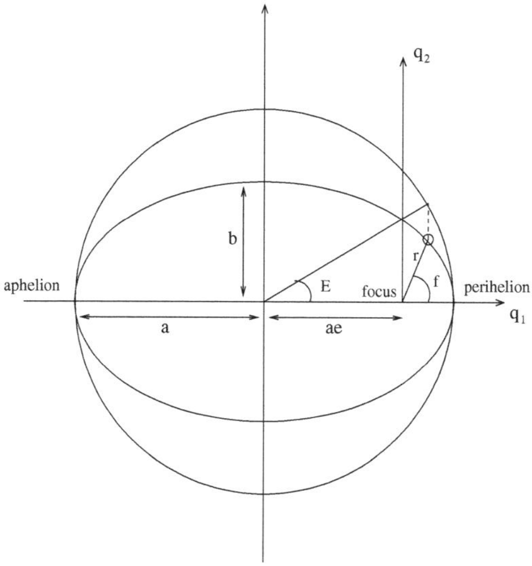

Fig. 1. Keplerian motion: definition of a, e and E

"away from the center"). The eccentricity is therefore an indicator of how much the orbit differs from a circular one: $e = 0$ means that the orbit is circular, whereas $e = 1$ means that the orbit is a segment of length $2a$, the Sun being at one of the extremes. Among all "elliptical" trajectories, the latter is the only collisional one, if the physical radii of the bodies are neglected. A semi-major axis of $a = \infty$ and $e = 1$ denotes parabolic motion, while the convention $a < 0$ and $e > 1$ is adopted for hyperbolic motion. I will not deal with these kinds of unbounded motion in this chapter and will therefore concentrate, hereafter, on the elliptic case. On an elliptic orbit, the closest point to the Sun is called the *perihelion*, and its heliocentric distance q is equal to $a(1-e)$; the farthest point is called the *aphelion*, and its distance Q is equal to $a(1+e)$.

To denote the position of a body on its orbit, it is convenient to use an orthogonal reference frame q_1, q_2 with origin at the focus of the ellipse occupied by the Sun and q_1 axis oriented towards the perihelion of the orbit. Alternatively, polar coordinates r, f can be used. The angle f is usually called the *true anomaly* of the body. From Fig. 1, using elementary geometrical relationships, it can be seen that

$$q_1 = a(\cos E - e) \ , \quad q_2 = a\sqrt{1-e^2}\sin E \tag{1}$$

and

$$r = a(1 - e\cos E) \ , \quad \cos f = \frac{\cos E - e}{1 - e\cos E} \tag{2}$$

where E, as Fig. 1 shows, is the angle subtended at the center of the ellipse by the projection – parallel to the q_2 axis – of the position of the body on the circle that is tangent to the ellipse at perihelion and aphelion. This angle is called the *eccentric anomaly*. The quantities a, e and E are enough to characterize the position of a body in its orbit.

From Newton equations, it is possible to derive [28] the evolution law of E with respect to time, usually called the *Kepler equation*:

$$E - e \sin E = n(t - t_0) \tag{3}$$

where

$$n = \sqrt{\mathcal{G}(m_0 + m_1)}a^{-3/2} \tag{4}$$

is the orbital frequency, or *mean motion*, of the body, m_0 and m_1 are the masses of the Sun and of the body, respectively, and $\mathcal{G}$ is the gravitational constant; t is the time and t_0 is the time of perihelion passage.

Astronomers like to introduce a new angle

$$M = n(t - t_0) \tag{5}$$

called the *mean anomaly*, as an orbital element that changes linearly with time. M also denotes the position of the body in its orbit, through equations (3) and (2).

To characterize the orientation of the ellipse in space, with respect to an arbitrary orthogonal reference frame (x, y, z) centered on the Sun, one has to introduce three additional angles (see Fig. 2). The first one is the inclination, i, of the orbital plane (the plane that contains the ellipse) with respect to the (x, y) reference plane. If the orbit has a nonzero inclination, it intersects the (x, y) plane in two points, called the *nodes* of the orbit. Astronomers

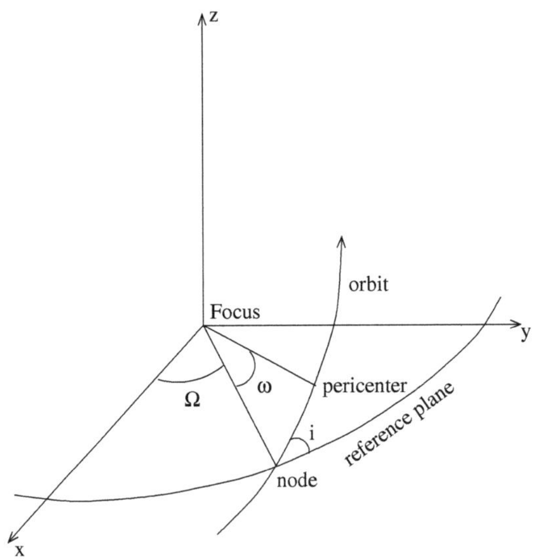

Fig. 2. Keplerian motion: definition of i, Ω and ω

distinguish between an *ascending* node, where the body passes from negative to positive z, and a *descending* node, where the body plunges towards negative z. The orientation of the orbital plane in space is then completely determined when one gives the angular position of the ascending node from the x axis. This angle is traditionally called the *longitude of ascending node* and is usually denoted by Ω. The last angle that needs to be introduced is the one characterizing the orientation of the ellipse in its plane. The *argument of perihelion*, ω, is defined as the angular position of the perihelion, measured in the orbital plane relative to the line connecting the central body to the ascending node.

In the definition of the orbital elements above, note that when the inclination is zero, ω and M are not defined, because the position of the ascending node is not determined. Moreover, M is also not defined when the eccentricity is zero, because the position of the perihelion is not determined. Therefore, it is convenient to introduce the *longitude of perihelion* $\varpi = \omega + \Omega$ and the *mean longitude* $\lambda = M + \omega + \Omega$. The first angle is well defined when $i = 0$, whereas the second one is well defined when $i = 0$ and/or $e = 0$.

In the absence of external perturbations, the orbital motion is perfectly elliptic. Thus, the orbital elements a, e, i, ϖ, Ω are fixed, and λ moves linearly with time, with frequency given by (4). When a small perturbation is introduced (for instance the presence of an additional planet), two effects are produced. First, the motion of λ is no longer perfectly linear. Correspondingly, the other orbital elements have short periodic oscillations with frequencies in the order of the orbital frequencies. Second, the angles ϖ and Ω start to rotate slowly. This motion is called *precession*. Typical precession periods in the Solar System are of the order of 10,000–100,000 years. Correspondingly, e and i have long periodic oscillations, with periods of the order of the precession periods.

The regularity of these short and long periodic oscillations is broken when one of the following two situations occur: (i) the perturbation becomes large, for instance when there are close approaches between the body and the perturbing planet, or when the mass of the perturber is comparable to that of the Sun (as in the case of encounters of the Solar System with other stars) or (ii) the perturbation becomes resonant. In either of these cases, the orbital elements a, e, i can have large nonperiodic, irregular variations.

A resonance occurs when the frequencies of λ, ϖ or Ω of the body, or an integer combination of them, are in an integer ratio with one of the time frequencies of the perturbation. If the perturber is a planet, the perturbation is modulated by the planet orbital frequency and precession frequencies. We speak of *mean-motion resonance* when $k\mathrm{d}\lambda/\mathrm{d}t = k'\mathrm{d}\lambda'/\mathrm{d}t$, with k and k' integer numbers and λ' denoting the mean longitude of the planet. We speak of *linear secular resonance* when $\mathrm{d}\varpi/\mathrm{d}t = \mathrm{d}\varpi'/\mathrm{d}t$ or $\mathrm{d}\Omega/\mathrm{d}t = \mathrm{d}\Omega'/\mathrm{d}t$, prime variables referring again to the planet. Other types of resonances exist in more complicated systems (non-linear secular resonances, three-body resonances, Kozai resonance etc.). Resonant motion will be discussed more specifically in

Sect. 1.3, when reviewing the dynamical properties of some *trans*-Neptunian sub-populations.

1.2 The Structure of the *Trans*-Neptunian Population

The trans-Neptunian population is "traditionally" divided into two sub-populations: the *Scattered disk* and the *Kuiper belt.* The definition of these sub-populations is not unique, with the Minor Planet center and various authors often using slightly different criteria. Here I propose a partition based on the dynamics of the objects and their relevance for the reconstruction of the primordial evolution of the outer Solar System, keeping in mind that all bodies in the Solar System must have been formed on orbits typical of an accretion disk (e.g. with very small eccentricities and inclinations).

I call the *Scattered disk* the region of the orbital space that can be visited by bodies that have encountered Neptune within a Hill's radius,[2] at least once during the age of the Solar System, assuming no substantial modification of the planetary orbits. The bodies that belong to the Scattered disk in this classification do not provide us any relevant clue to uncover the primordial architecture of the Solar System. In fact their current eccentric orbits might have been achieved starting from quasi-circular ones in Neptune's zone by pure dynamical evolution, in the framework of the current architecture of the planetary system.

I call the *Kuiper belt* the trans-Neptunian region that cannot be visited by bodies encountering Neptune. Therefore, the non-negligible eccentricities and/or inclinations of the Kuiper belt bodies cannot be explained by the scattering action of the planet on its current orbit, but they reveal that some excitation mechanism, which is no longer at work, occurred in the past (see Sect. 4).

To categorize the observed trans-Neptunian bodies into the Scattered disk and Kuiper belt, one can refer to previous works on the dynamics of trans-Neptunian bodies in the framework of the current architecture of the planetary system. For the $a < 50$ AU region, one can use the results by [38] and [103], who numerically mapped the regions of the (a, e, i) space with $32 < a < 50$ AU, which can lead to a Neptune encountering orbit within 4 Gy. Because dynamics are reversible, these are also the regions that can be visited by a body after having encountered the planet. Therefore, according to the definition above, they constitute the Scattered disk. For the $a > 50$ AU region, one can use the results in [107] and [39], where the the evolutions of the particles that encountered Neptune in [38] have been followed for another 4 Gy time-span. Although the initial conditions did not cover all possible configurations, one

[2]The Hill's radius is given by the formula $R_{\rm H} = a_{\rm P}(m_{\rm P}/3)^{1/3}$, where $m_{\rm P}$ is the mass of the planet relative to the mass of the Sun and $a_{\rm P}$ is the planet's semi-major axis. It corresponds approximately to the distance from the planet of the Lagrange equilibrium points L_1 and L_2.

can reasonably assume that these integrations cumulatively show the regions of orbital space that can be visited by bodies transported to $a > 50$ AU by Neptune encounters. Again, according to my definition, these regions constitute the Scattered disk.

Figure 3 shows the (a, e, i) distribution of the trans-Neptunian bodies, which have been observed during at least three oppositions. The bodies that belong to the Scattered disk according to my criterion are represented as red dots. The Kuiper belt population is in turn subdivided into two subpopulations: the *resonant population* (green dots) and the *classical belt* (blue dots). The former is made of objects located in the major mean-motion resonances with Neptune (essentially the 3:4, 2:3 and 1:2 resonances, but also the 2:5 – see [23]), while the classical belt objects are not in any noticeable resonant configuration. Mean-motion resonances offer a protection mechanism against close encounters with the resonant planet (see Sect. 1.3). For this

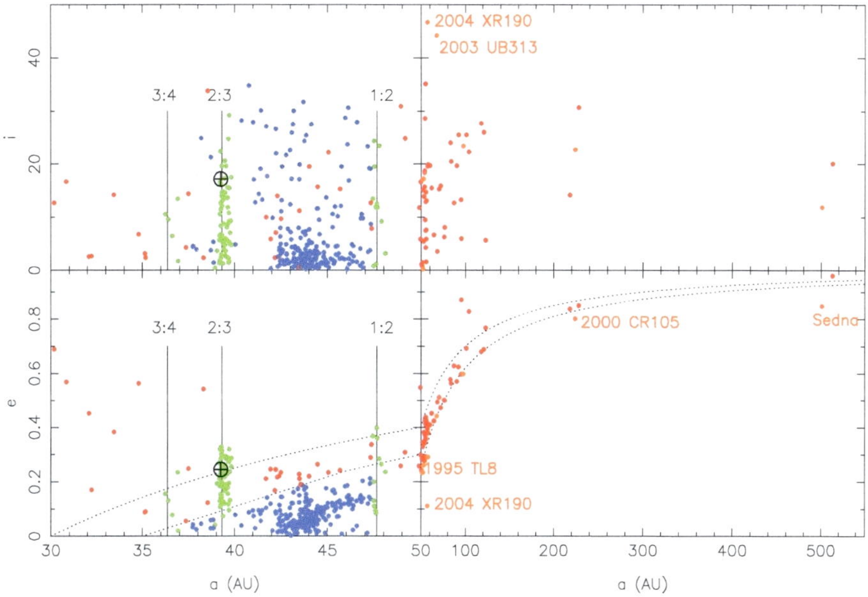

Fig. 3. The orbital distribution of multi-opposition trans-Neptunian bodies, as of Aug. 26, 2005. Scattered disk bodies are represented in red, Extended Scattered disk bodies in orange, classical Kuiper belt bodies in blue and resonant bodies in green. We qualify that, in the absence of long-term numerical integrations of the evolution of all the objects, and because of the uncertainties in the orbital elements, some bodies could have been mis-classified. Thus, the figure should be considered as an indicative representation of the various subgroups that compose the trans-Neptunian population. The dotted curves in the bottom left panel denote $q = 30$ AU and $q = 35$ AU; those in the bottom right panel $q = 30$ AU and $q = 38$ AU. The vertical solid lines mark the locations of the 3:4, 2:3 and 1:2 mean-motion resonances with Neptune. The orbit of Pluto is represented by a crossed circle

reason, the resonant population can have perihelion distances much smaller than those of the classical belt objects. Stable resonant objects can even have Neptune-crossing orbits ($q < 30$ AU) as in the case of Pluto (see Sect. 1.3). The bodies in the 2:3 resonance are often called *Plutinos*, because of the similarity of their orbit with that of Pluto. According to [168], the Scattered disk and the Kuiper belt constitute roughly equal populations, while the resonant objects, altogether, make up about 10% of the classical objects.

In Fig. 3, the existence of bodies on highly eccentric orbits with $a > 50$ AU can be seen. These objects do not belong to the Scattered disk according to my definition (orange dots). Among them are 2000 CR_{105} ($a = 230$ AU, perihelion distance $q = 44.17$ AU and inclination $i = 22.7°$), Sedna ($a = 495$ AU, $q = 76$ AU), 2004 XR_{190} ($a = 57.4$ AU, $q = 51$ AU) and the current size-record holder 2003 UB_{313} ($a = 67.7$ AU, $q = 37.7$ AU but $i = 44.2°$; diameter equal to 2400 ± 100 km [18]), although for some objects the classification is uncertain for the reasons explained in the figure caption. Following [56], I call these objects *Extended Scattered disk* objects for three reasons. (i) They are very close to the Scattered disk boundary. (ii) Bodies of the sizes of the three objects quoted above (300–2000 km) presumably formed much closer to the Sun, where the accretion timescale was sufficiently short [153]. This implies that they have been transported in semi-major axis space (e.g. scattered), to reach their current locations. (iii) The lack of objects with $q > 41$ AU and $50 < a < 200$ AU should not be due to observational biases, given that many classical belt objects with $q > 41$ AU and $a < 50$ AU have been discovered (see Fig. 6). This suggests that the Extended Scattered disk objects are *not* the highest eccentricity members of an excited belt beyond 50 AU. These three considerations indicate that in the past the true Scattered disk extended well beyond its present boundary in perihelion distance. The reason for this is still under debate. Some ideas will be presented in Sect. 4.

Given that observational biases become more severe with increasing perihelion distance and semi-major axis, the currently known Extended Scattered disk objects may be like the tip of an iceberg, the emerging representatives of a conspicuous population, possibly outnumbering the Scattered disk population [56].

The Excitation of the Kuiper Belt

An important clue to the history of the early outer Solar System is the dynamical excitation of the Kuiper belt. While the eccentricities and inclinations of resonant and scattered objects are expected to have been affected by interactions with Neptune, those of the classical objects should have suffered no such excitation. Nonetheless, the confirmed classical belt objects have an inclination range up to at least 32° and an eccentricity range up to 0.2, significantly higher than expected from a primordial disk, even accounting for mutual gravitational stirring.

The observed distributions of eccentricity and inclination in the Kuiper belt are highly biased. High eccentricity objects have closer approaches to the Sun, and thus, they become brighter and are more easily detected. Consequently, the detection bias roughly follows curves of constant q. At first sight, this bias might explain why, in the classical belt beyond $a = 44$ AU, the eccentricity tends to increase with semi-major axis. However, the resulting (a, e) distribution is significantly steeper than a curve q =constant. Thus, the apparent relative under-density of objects at low eccentricity in the region $44 < a < 48$ AU is likely to be a real feature of the Kuiper belt distribution.

High-inclination objects spend little time at the low latitudes[3] at which most surveys take place, while low-inclination objects spend no time at the high latitudes where some searches have occurred. Using this fact, [16] computed a de-biased inclination distribution for classical belt objects (Fig. 4).

A clear feature of this de-biased distribution is its bi-modality, with a sharp drop around 4° and an extended, almost flat distribution in the 4–30° range. This plateau is required to fit the presence of objects with large inclinations. The bi-modality can be modeled with two Gaussian functions and suggests the presence of two distinct classical Kuiper belt populations, called hot ($i > 4$) and cold ($i < 4$) after [16].

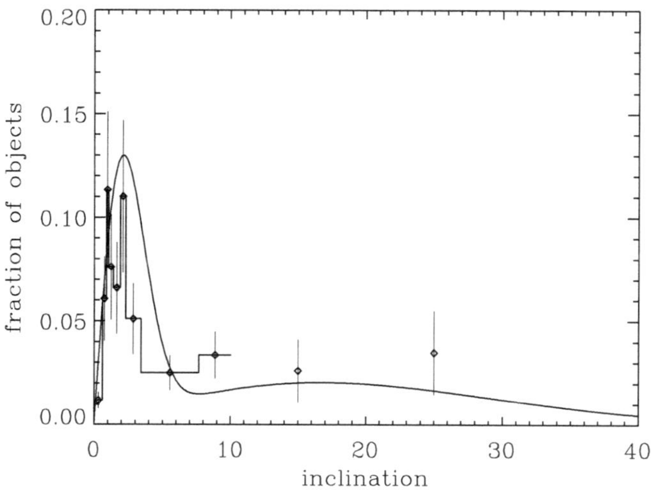

Fig. 4. The inclination distribution (in degree) of the classical Kuiper belt, from [131]. The points with error bars show the model-independent estimate constructed from a limited subset of confirmed classical belt bodies, while the smooth line shows the best fit two-population model $f(i)di = \sin(i)[96.4\exp(-i^2/6.48) + 3.6\exp(-i^2/288)]di$ [16]. In this model, ~60% of the objects have $i > 4°$

[3]Latitude (angular height over a reference curve in the sky) and inclination should be defined with respect to the local Laplace plane (the plane normal to the orbital precession pole), which is a better representation for the plane of the Kuiper belt than is the ecliptic [41].

Physical Evidence for Two Populations in the Classical Belt

The co-existence of a hot and a cold population in the classical belt could be caused in one of two general manners. Either a subset of an initially dynamically cold population was excited, leading to the creation of the hot classical population, or the populations are truly distinct and formed separately. One manner in which one can attempt to determine which of these scenarios is more likely is to examine the physical properties of the two classical populations. If the objects in the hot and cold populations are physically different, it is less likely that they were initially part of the same population.

The first suggestion of a physical difference between the hot and the cold classical objects came from [108] who noted that the intrinsically brightest classical belt objects (those with lowest absolute magnitudes) are preferentially found on high-inclination orbits. This conclusion has been recently verified in a bias-independent manner in [171], with a survey for bright objects which covered ~70% of the ecliptic and found many hot classical objects but few cold classical ones.

The second possible physical difference between hot and cold classical Kuiper belt objects is their colors, which relate in an unknown manner to surface composition and physical properties. Several possible correlations between orbital parameters and color were suggested by [163] and further investigated by [34]. The issue was clarified by [170] who quantitatively showed that for the classical belt, the inclination is correlated with color. In essence, the low-inclination classical objects tend to be redder than higher inclination objects (see Fig. 5). This correlation has since been confirmed by several other authors [35,41]. Whether or not there is also a correlation between color and perihelion distance is still a matter of debate [35].

More interestingly, we see that the colors naturally divide into distinct low-inclination and high-inclination populations at precisely the location of the divide between the hot and cold classical objects. These populations differ at a 99.9% confidence level. In addition, the cold classical population also differs in color from the Plutinos and the scattered objects at the 99.8% and 99.9% confidence level, respectively, while the hot classical population appears identical in color to these other populations [170]. The possibility remains, however, that the colors of the objects, rather than being markers of different populations, are actually *caused* by the different inclinations. For example [157] has suggested that the higher average impact velocities of the high-inclination objects will cause large-scale resurfacing by fresh water ice, which could be blue to neutral in color. However, careful analysis has shown that there is no clear correlation between average impact velocity and color [165].

In summary, the significant color and size differences between the hot and cold classical objects imply that these two populations are physically different, in addition to being dynamically distinct.

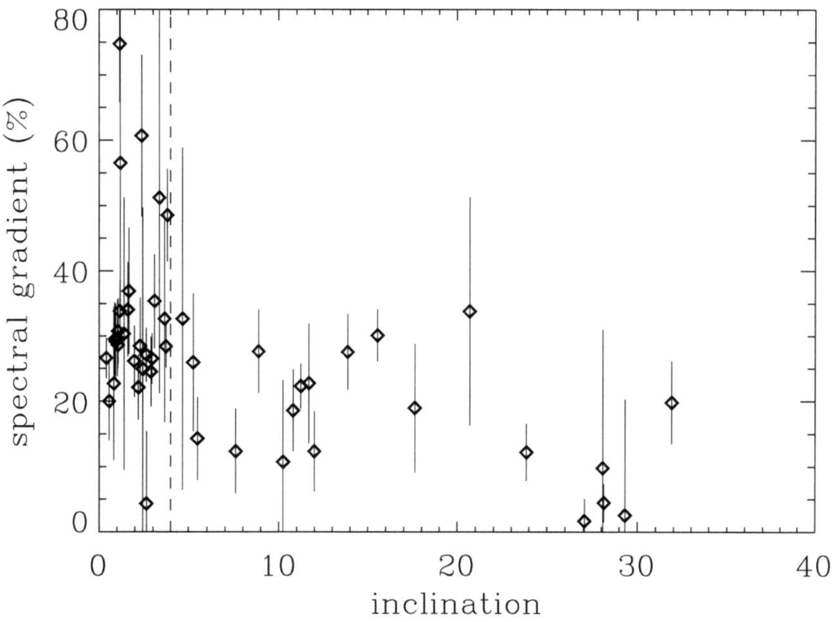

Fig. 5. Color gradient versus inclination in the classical Kuiper belt (from [131], using the database in [68]). Color gradient is the slope of the spectrum, in % per 100 nm, with 0% being neutral and large numbers being red. The hot and cold classical objects have significantly different distributions of color

The Radial Extent of the Kuiper Belt

Another important property of interest for understanding the primordial evolution of the Kuiper belt is its radial extent. While the initial expectations were that the mass of the Kuiper belt should smoothly decrease with heliocentric distance – or perhaps even increase in number density by a factor of ~100 [153], back to the level given by the extrapolation of the minimum mass solar nebula [69] beyond the region of Neptune's influence – the lack of objects detected beyond about 50 AU soon began to suggest a drop off in number density [22, 87, 168]. It was often argued that this lack of detections was the consequence of a simple observational bias caused by the extreme faintness of objects at greater distances from the Sun, but [4] and [5] showed convincingly that for a fixed absolute magnitude distribution, the number of objects with semi-major axis less than 50 AU was larger than the number greater than 50 AU, and thus, some density decrease is present.

The characterization of the density drop beyond 50 AU was hampered by the weak statistics characterising each individual survey, because of the small number of objects that each of them found. A method using the objects detected by all surveys to estimate a radial distribution of the Kuiper belt and to test hypothetical distributions against the known observations was

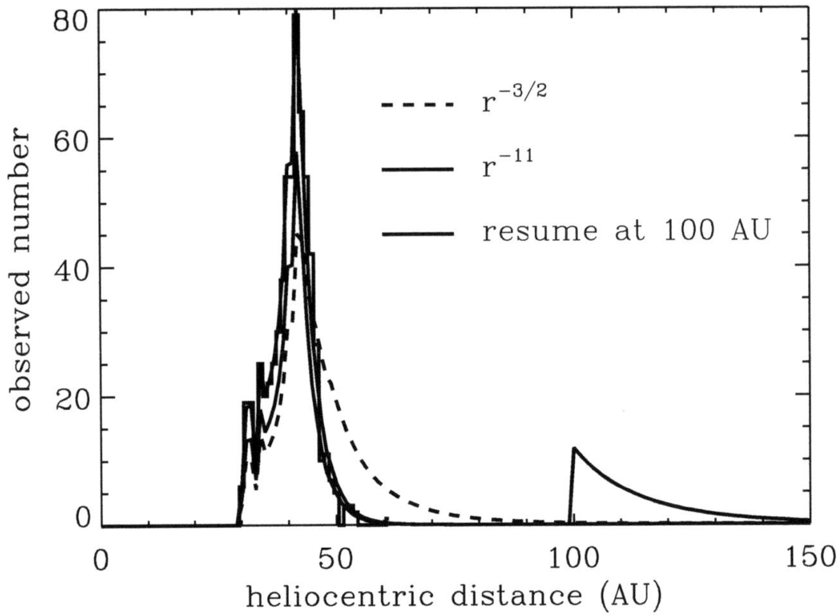

Fig. 6. The observed radial distribution of Kuiper belt objects (solid histogram) compared to radial distributions that would be observed for models where the surface density of Kuiper belt objects decreases by $r^{-3/2}$ beyond 42 AU (dashed curve), by r^{-11} beyond 42 AU (solid curve), and where the surface density at 100 AU increases by a factor of 100 to the value expected from an extrapolation of the minimum mass solar nebula (dashed-dotted curve). From [131]

developed in [169]. The analysis reported in that work is reproduced in Fig. 6. The drop off of the discovery heliocentric distance distribution of Kuiper belt objects beyond 42 AU is clearly inconsistent with a smooth decline of the surface density distribution proportional to $r^{-3/2}$. Instead, it can be fitted with a surface density distribution with a much sharper decay, $r^{-11\pm4}$ (error bars are 3σ), i.e. by assuming the existence of an effective edge in the radial Kuiper belt distribution. This steep radial decay should presumably hold up to ~ 60 AU, beyond which a much flatter distribution because of the Scattered disk objects should be found.

It has been conjectured [153] that, beyond some range of Neptune's influence, the number density of Kuiper belt objects could increase back up to the level expected for the minimum mass solar nebula [69]. Such an increase can be ruled out at the 3σ level within 115 AU from the Sun. Beyond this distance, the biases because of the slow motion of the objects also become important; so no definite conclusion can be drawn from the current data about objects beyond this threshold. If the model is slightly modified to make the maximum object mass proportional to the surface density at a particular distance, a 100 times resumption of the Kuiper belt can still be ruled out inside 94 AU.

Although the drop off in the heliocentric distance distribution starts at 42 AU, a visual inspection of Fig. 3 shows that the edge of the Kuiper belt in semi-major axis space is precisely at the location of the 1:2 mean-motion resonance with Neptune. This is a very important feature, which points to a role for Neptune in the final positioning of the edge. I will come back to this in Sect. 4.

The Missing Mass of the Kuiper Belt

The absolute magnitude[4] distribution of the Kuiper belt objects can be determined from the so-called *cumulative luminosity function*, which is given by the number of detections that surveys report as a function of their limiting magnitude, weighted by the inverse area of sky that they covered. If one assumes that the albedo distribution of the Kuiper belt objects is size independent, the slope of the absolute magnitude distribution can be readily converted into the slope of the cumulative size distribution.

The size distribution turns out to be very steep, with exponent of the cumulative power law falling between -3.5 and -3 for bodies larger in diameter than ~200 km [55]. Actually, the size distribution is slightly shallower for the hot population than for the cold population, as shown in a recent analysis [10] (see Fig. 7). This is not surprising, given that – as we have seen above – the hot and the cold populations contain roughly the same total number of objects, but the former hosts the largest members of the Kuiper belt.

The HST survey in [10] also reported the detection of a change in the size distribution for objects fainter than $H = 9$–10, corresponding to about 100 km in diameter, assuming a standard albedo of ~4%. The slopes of the size distribution below this limit, however, remain very uncertain because of small number statistics. Some researchers still dispute the validity of the detection of any turnover in the size distribution [144]. Given these uncertainties, as well as uncertainties on the mean albedo of the Kuiper belt objects (required to convert a given absolute magnitude into a size) and their bulk density, the total mass of the Kuiper belt is uncertain to at least an order of magnitude, estimates ranging from $0.01\,M_\oplus$ [10] to $0.1\,M_\oplus$ [55].

Whatever the real value (in this range, or slightly beyond), it nevertheless seems certain that the total mass of the Kuiper belt is now very small, in particular, compared with the mass of the disk of solids from which the Kuiper belt objects had to form. There are two lines of argument to estimate this primordial mass.

A first argument follows the reasoning that led Kuiper to conjecture the existence of a band of small planetesimals beyond Neptune. [104] The minimum mass solar nebula inferred from the total planetary mass (plus lost volatiles; [69]) smoothly declines from the orbit of Jupiter until the orbit of

[4] The absolute magnitude, H, is a measure of the intrinsic brightness of an object. It corresponds to the visual magnitude that an object would have in the paradoxical situation of being simultaneously at 1 AU from the Sun and the Earth, at opposition!

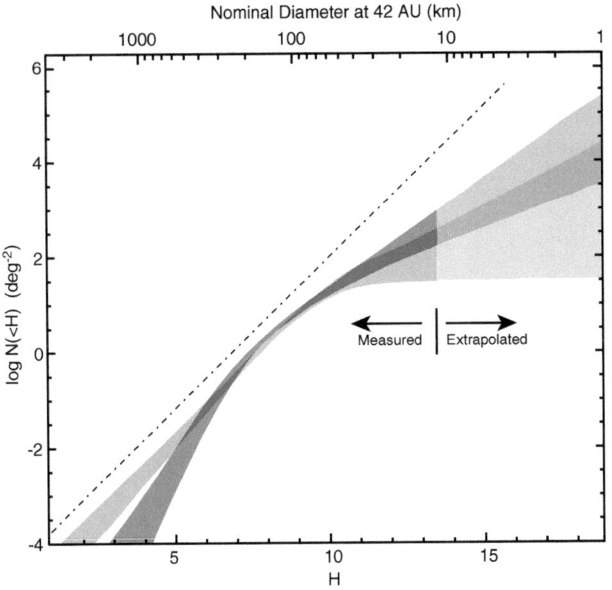

Fig. 7. The H or size distribution in the Kuiper belt (adapted from [10] with the permission of Bernstein). The red and green bands show the uncertainties for the cold and the hot population, respectively (although the definition for hot and cold used in that work do not exactly match those adopted in this paper). Absolute magnitudes have been computed assuming that all detections occurred at 42 AU (the maximum of the radial surface density distribution of the Kuiper belt), and the conversion to diameters uses the assumption that the mean albedo is 4%

Neptune; why should it drop abruptly beyond the last planet? The extrapolation and integration of this surface density distribution predicts that the original total mass of solids in the 30–50 AU range should have been $\sim 30\,M_\oplus$.

The second argument for a massive primordial Kuiper belt was first raised in [152], where it was found that the objects currently in the Kuiper belt could not have formed in the present environment: collisions are sufficiently infrequent that 100 km objects cannot be built by pairwise accretion within the current population over the age of the Solar System. Moreover, owing to the large eccentricities and inclinations of Kuiper belt objects – and consequently to their high encounter velocities – the collisions that do occur tend to be erosive rather than accretional, making bodies smaller rather than larger. Stern suggested that the solution of this dilemma is that the primordial Kuiper belt was both more massive and dynamically colder, so that more collisions occurred, and they were gentler and therefore generally accretional.

Following this idea, detailed modeling of accretion in a massive primordial Kuiper belt was performed [92, 93, 94, 153, 154, 155]. While each model includes different aspects of the relevant physics of accretion, fragmentation, and velocity evolution, the basic results are in approximate agreement. First, with $\sim 10\,M_{\oplus}$ or more of solid material in an annulus from about 35 to 50 AU on very low eccentricity orbits ($e \leq 0.001$), all models naturally produce a few objects of the size of Pluto and approximately the right number of ~ 100 km objects, on a timescale ranging from several 10^7 to several 10^8 years. The models suggest that the majority of mass in the disk was in bodies approximately 10 km and smaller. The accretion stopped when the formation of Neptune (or other dynamical phenomena; see Sect. 4) began to excite eccentricities and inclinations in the population that were high enough to move the collisional evolution from the accretional to the erosive regime.

A massive and dynamically cold primordial Kuiper belt is also required by the models that attempt to explain the formation of the numerous observed binary Kuiper belt objects [6, 54, 57, 175].

Therefore, the general formation picture of an initially massive Kuiper belt appears to be secure, and understanding the ultimate fate of the 99% of the initial mass that appears no longer to be in the Kuiper belt is a crucial step in reconstructing the history of the outer Solar System.

1.3 Dynamics in the Kuiper Belt

In this section, I will give an overview of the dynamical properties of the Kuiper belt. Without any pretension of being exhaustive, the goal is to understand which properties of the Kuiper belt orbital structure can be explained from the evolution of the objects in the framework of the current architecture of the Solar System and which, conversely, require an explanation built on a scenario of primordial sculpting (as in Sect. 4).

Figure 8 shows a map of the dynamical lifetime of trans-Neptunian bodies as a function of their initial semi-major axis and eccentricity, for an inclination of 1° and a random choice of the orbital angles λ, ϖ, and Ω [38]. Similar maps, referring to different choices of the initial inclination or different projections on orbital element space can be found in [103] and [38]. These maps have been computed numerically, by simulating the evolution of massless particles from their initial conditions, under the gravitational perturbations of the giant planets. The latter were assumed to be initially on their current orbits. Each particle was followed until it suffered a close encounter with Neptune. Objects encountering Neptune would then evolve in the Scattered disk for a typical time of order $\sim 10^8$ years (but much longer residence times in the Scattered disk occur for a minority of objects), until they are transported by planetary encounters into the inner Solar System or to the Oort cloud, or are ejected to the interstellar space. This issue is described in more detail in Sect. 2.

In Fig. 8, the colored strips indicate the timespan required for a particle to encounter Neptune, as a function of its initial semi-major axis and eccentricity. Strips that are colored yellow represent objects that survive for the length of the simulation, 4×10^9 years (the approximate age of the Solar System) without encountering the planet. The figure also reports the orbital

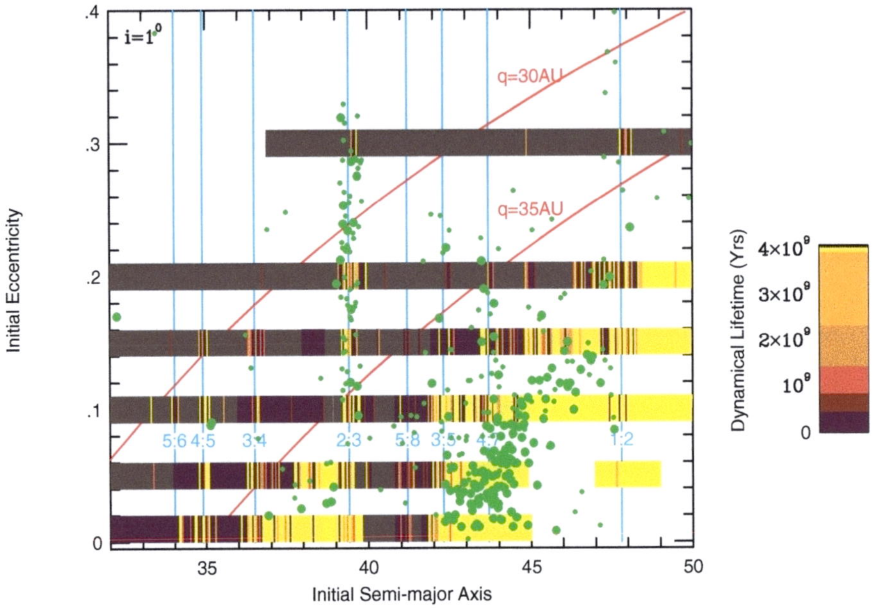

Fig. 8. The dynamical lifetime for small particles in the Kuiper belt derived from 4 billion year integrations [38]. Each particle is represented by a narrow vertical strip of color, the center of which is located at the particle's initial eccentricity and semi-major axis (the initial orbital inclination for all objects was 1°). The color of each strip represents the dynamical lifetime of the particle. Strips colored yellow represent objects that survive for the length of the integration, 4×10^9 years. Dark regions are particularly unstable on short timescales. For reference, the locations of the important Neptune mean-motion resonances are shown in blue and two curves of constant perihelion distance, q, are shown in red. The (a, e) elements of the Kuiper belt objects with well-determined orbits are also shown as green dots. Large dots are for $i < 4°$, small dots otherwise

elements of the known Kuiper belt objects. Big dots refer to bodies with $i < 4°$, consistent with the low inclination at which the stability map has been computed. Small dots refer to objects with larger inclination and are plotted only for completeness.

As can be seen in the figure, the Kuiper belt has a complex dynamical structure, although some general trends can be easily explained.

Stability Limits Imposed by Close Encounters with Neptune

Most objects with perihelion distances less than ~ 35 AU are unstable. This is because they pass sufficiently close to Neptune that they are destabilized during the encounters. In fact, in these cases, Neptune's gravity is no longer a "small perturbation" relative to that of the Sun. The regularity of the oscillation of the orbital elements is broken. The semi-major axis suffers jumps at each encounter with the planet, and the eccentricity has correlated jumps to keep the perihelion distance roughly constant (more precisely, to conserve the *Tisserand parameter,* see Sect. 2). Through one encounter after another, the object wanders over the (a, e) plane: the object is effectively a member of the Scattered disk. Consequently, the $q = 35$ AU curve can be considered as the approximate border between the Kuiper belt and the Scattered disk, in the 30–50 AU semi-major axis range. The real border, however, has a more complicated, fractal structure, illustrated by the boundary between the black and the yellow regions in Fig. 8.

Not all bodies with $q < 35$ AU are unstable. The exception is those objects in mean-motion resonances with Neptune. These objects, despite approaching (or even intersecting) the orbit of Neptune at perihelion, never approach the planet to short distance. This happens because the resonance plays a role protecting against close encounters.

The stabilizing role of a mean-motion resonance can be understood in simple, qualitative terms. For instance, Fig. 9 illustrates the mechanism for the case of Pluto (2:3 mean-motion resonance). The trajectory of Pluto is shown in the figure in a frame that rotates with Neptune. Pluto moves in a clockwise direction when further from the Sun than Neptune and moves in a counter-clockwise direction when closer to the Sun. In the figure, an object with Pluto's eccentricity and exactly at Neptune 2:3 mean-motion resonance would have a trajectory that is a double-lobed structure oriented as in Fig. 9a. The configuration shown in the figure will remain fixed only if the object's semi-major axis is *exactly* equal to that characterizing the location of the resonance. For an object with semi-major axis slightly displaced, the double-lobed structure will slowly precess in the rotating frame.

If the semi-major axis of the object is slightly larger than that corresponding to the exact location of the resonance, the double-lobed trajectory will slowly precess toward that shown in Fig. 9b. If the precession continued indefinitely, eventually the trajectory would pass over the location of Neptune and a close encounter or a physical collision would occur. However, because the new trajectory is no longer symmetric with respect to Neptune, the object receives its largest acceleration (a_{m}) from Neptune when in or near the upper lobe. At this point, the object is leading Neptune in its orbit, and thus, it is slowed down in its heliocentric motion. Consequently, its semi-major axis decreases.

When the semi-major axis of the object becomes smaller than that corresponding to the exact location of the resonance, the situation reverses. Now the double-lobed trajectory slowly precesses in the opposite direction. The configuration of Fig. 9a is restored, and then the trajectory continues to precess toward the configuration of Fig. 9c. In this case, the object gets its largest acceleration when it is near perihelion and is trailing Neptune in their orbits (near the lower lobe of the trajectory). Thus, the object's orbital velocity is increased, increasing its semi-major axis.

When the semi-major axis of the object again becomes larger than the exact resonant value, the precession of the double-lobed trajectory reverses

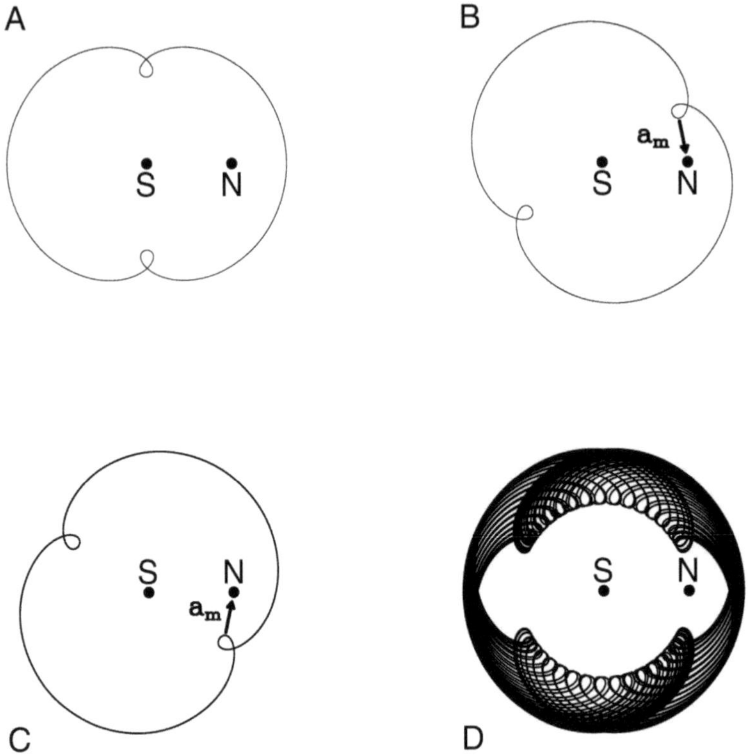

Fig. 9. The dynamics of an object in the 2:3 mean-motion resonance with Neptune. The double-lobed curve represents the orbit of an object with the eccentricity of Pluto. The coordinate frame rotates counterclockwise at the average speed of Neptune. Thus, Neptune (dot labeled "N") is stationary in this figure. The location of the Sun is labeled "S". A) The orbit of an object whose semi-major axis is equal to that characterizing the exact location of the resonance. The gravitational perturbations of Neptune cancel out because of the symmetry in the geometry. Thus, this orbit does not precess in the rotating frame. B) If the symmetry is broken, there is a net acceleration because of Neptune. Here, the strongest perturbation ($a_{\rm m}$) is at the upper lobe. The object is leading Neptune at this lobe, so the net acceleration will decrease its semi-major axis. C) The strongest perturbation is in the lower lobe. Consequently, the object's semi-major axis has to increase. D) The orbit of an object that *librates* in the resonance. Courtesy of H. Levison

once more. The trajectory goes back to the configuration of Fig. 9a and then to that of Fig. 9b, and the cycle repeats indefinitely. Each cycle is called a *libration*. Over a full libration cycle, the pattern drawn by the object's dynamics in the frame co-rotating with Neptune is that illustrated in Fig. 9d.

Therefore, the mean-motion resonance exerts on the object a *restoring torque* that reverses the precession of its double-lobed trajectory before a close encounter can occur. This of course happens only if the object is not too far from the exact resonance location, otherwise the precession is too fast, and the magnitude of the restoring torque is not sufficient. The limiting distance from the exact resonance location within which the restoring torque is effective defines the *resonance width.*

The analytic computation of resonance widths is detailed in [128]. This calculation, however, overestimates the width of the region where resonant objects are stable over the Solar System's age. In fact, the situation is complicated by the interaction between the libration motion inside the resonance and the precession motion of the orbits of the object and of the perturbing planet. A detailed exploration of the stability region inside the two main mean-motion resonances of the Kuiper belt, the 2:3 and 1:2 resonances with Neptune, has been done in [136, 137]. Its results are beyond the scope of this chapter.

Secular Resonance Instabilities

In Fig. 8, one can see that the dark region extends significantly below the $q = 35\,\mathrm{AU}$ line for $40 < a < 42\,\mathrm{AU}$ (and also for $35 < a < 36\,\mathrm{AU}$). The instability in these regions is due to the presence of a secular resonance, such that $\mathrm{d}\varpi/\mathrm{d}t \sim \mathrm{d}\varpi_{\mathrm{N}}/\mathrm{d}t$, where ϖ is the perihelion longitude of the object and ϖ_{N} that of Neptune.

This resonance forces large variations in the eccentricity of the trans-Neptunian object, so that – even if the initial eccentricity is zero – the perihelion distance eventually decreases below 35 AU, and the object enters the Scattered disk [82, 126].

The destabilizing effect of a secular resonance between the longitude of perihelia can be understood in easy qualitative terms. Consider a simple case where the orbits of the object and of *two* planets are in the same plane. The presence of two planets is necessary, otherwise the planetary orbit would be a fixed, non-precessing ellipse. The orbit of the small body also precesses under the planets' perturbations. The left plot in Fig. 10 shows the long-term trajectories of these objects in a fixed frame. The middle plot shows the same system in a frame that rotates with the precession rate of the small body. Note that the orbit of the small body (the outermost orbit) is, in this frame, a fixed ellipse. If the precession rates of the planetary orbits are different from that of the small body, the trajectories of the two planets in the rotating frame are still, on average, axisymmetric, and thus, the small body experiences no long-term torques. However, if one of the planets precesses at the same rate as the small body, as in the right plot in Fig. 10, its long-term trajectory is also a fixed ellipse in the rotating frame, and it is no longer axisymmetric. Thus, the small body feels a significant long-term torque, which can lead to a significant change in its eccentricity (which is related to the angular momentum).

The location of secular resonances in the Kuiper belt has been computed in [98]. This work showed that this secular resonance is present only at small inclination. Large inclination orbits with $q > 35\,\mathrm{AU}$ and $40 < a < 42\,\mathrm{AU}$ are

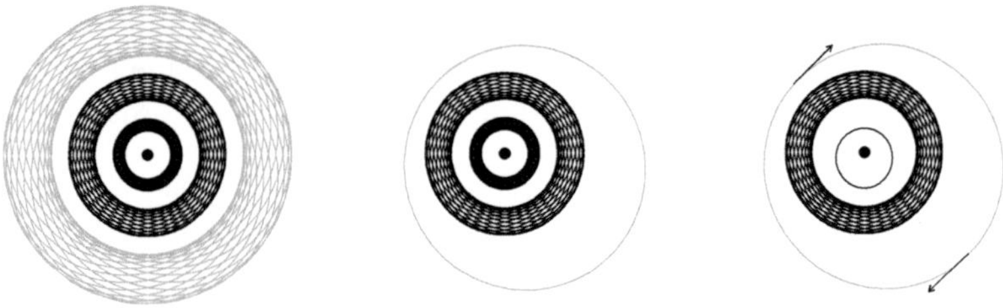

Fig. 10. The dynamics of a secular resonance. Three orbits are shown in each panel. The inner two are planets, which are shown as black lines. The outer orbit (gray line) is for a small object. The orbits of each object are ellipses, and the ellipses are precessing due to the mutual gravitational effects of the planets. Left: The orbits of the objects over a period of time that is long compared to the precession time of the orbits. Here, we are looking in a fixed, non-rotating reference frame. Each orbit sweeps out a torus of possible positions. Center: The same as in the left plot, except that we are looking in a frame that rotates at the precession rate of the small outer body. Thus, its orbit is again an ellipse. This panel shows the geometry if no secular resonance exists. Note that the trajectories of the planets look axisymmetric. Therefore, there is no net torque on the outer small object. Right: Same as the middle plot, except that the outer object is in a secular resonance with the inner planet, i.e. both orbits precess at the same rate. As a result, the outer object no longer sees an axisymmetric gravitational perturbation from the inner planet. Indeed, it feels a significant torque. Courtesy of H. Levison

therefore stable. Indeed, Fig. 8 shows that many objects with $i > 4°$ (small dots) are present in this region. Only large dots, representing low-inclination objects, are absent.

Chaotic Diffusion in the Kuiper Belt

Figure 8 also shows the presence of narrow bands with brown colors, crossing the yellow stability domain. These bands correspond to orbits that become Neptune-crossing only after billions of years of evolution. What is the nature of these weakly unstable orbits?

It has been found [137] that these orbits are, in general, associated either with high-order mean-motion resonances with Neptune (i.e., resonances for which the equivalence $k\mathrm{d}\lambda/\mathrm{d}t = k_{\mathrm{N}}\mathrm{d}\lambda_{\mathrm{N}}/dt$ holds only for large values of the integer coefficients k, k_N) or three-body resonances with Uranus and Neptune (which occur when $k\mathrm{d}\lambda/\mathrm{d}t + k_{\mathrm{N}}\mathrm{d}\lambda_N/\mathrm{d}t + k_U\mathrm{d}\lambda_U/\mathrm{d}t = 0$ occurs for some integers k, k_{N}, and k_{U}).

The dynamics of objects in these resonances is chaotic. The semi-major axis of the objects remains locked at the corresponding resonant value, while the eccentricity of their orbits is slowly modified. In an (a, e)-diagram like Fig. 11, each object's evolution leaves a vertical trace. This phenomenon is called *chaotic diffusion*. Eventually, the growth of the eccentricity can bring the diffusing object above the $q = 35\,\mathrm{AU}$ curve. These resonances are too

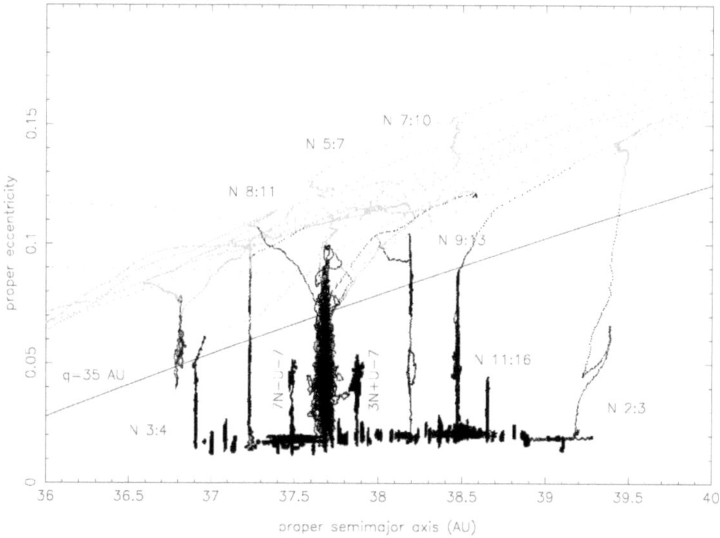

Fig. 11. The evolution of objects initially at $e = 0.015$ and semi-major axes distributed in the 36.5–39.5 AU range. The dots represent the proper semi-major axis and the eccentricity of the objects – computed by averaging their a and e over 10 My time intervals – over the age of the Solar System. They are plotted in gray after the perihelion has decreased below 32 AU for the first time. Labels $Nk : k_{\rm N}$ denote the $k : k_{\rm N}$ two-body resonances with Neptune. Labels $k_{\rm N}{\rm N}+k_{\rm U}{\rm U}+k$ denote the three-body resonances with Uranus and Neptune, corresponding to the equality $k_{\rm N}\dot{\lambda}_{\rm N} + k_{\rm U}\dot{\lambda}_{\rm U} + k\dot{\lambda} = 0$. Reprinted from [137]

weak to offer an effective protection against close encounters with Neptune, unlike the low-order resonances considered above. Thus, once above this critical curve, the encounters with Neptune start to change the semi-major axis of the objects, which leave their original resonance and evolve – from that moment on – in the Scattered disk.

Notice from Fig. 11 that some resonances are so weak that, despite forcing the resonant objects to diffuse chaotically, they cannot reach the $q = 35$ AU curve within the age of the Solar System. Therefore, these objects are "stable" from the astronomical point of view.

Notice also that chaotic diffusion is effective only for selected resonances. The vast majority of the simulated objects are not affected by any macroscopic diffusion. They preserve their initial small eccentricity for the entire age of the Solar System. Thus, the current moderate/large eccentricities and inclinations of most of the Kuiper belt objects cannot be obtained from primordial circular and coplanar orbits by dynamical evolution in the framework of the current orbital configuration of the planetary system. Likewise, the region beyond the 1:2 mean-motion resonance with Neptune is totally stable. Thus, the absence of bodies beyond 48 AU cannot be explained by current dynamical instabilities. Also, the overall mass deficit of the Kuiper belt cannot be due to objects escaping through resonances, because most of the inhabited Kuiper belt is stable for the current planetary architecture. Therefore, all these intriguing properties of the Kuiper belt's structure must, instead, be explained within the framework of the formation and primordial evolution of the Solar System. This will be the topic of Sect. 4.

1.4 Note on the Scattered Disk

We have seen above that the bodies that escape from the Kuiper belt and decrease their perihelion distance below 35 AU, without being protected by a low-order mean-motion resonance, enter the Scattered disk.

Their subsequent evolution has been studied in detail in [107]. It was found that the median dynamical lifetime is ~ 50 My, the typical end-states being transport toward the inner Solar System (and eventual ejection from the Solar System because of an encounter with Jupiter or Saturn; see Sect. 2), a collision with a planet or outward transport toward the Oort cloud (see Sect. 3). This result suggests that the Scattered disk could be a population of transient objects, which is sustained in steady state by a continuous flux of objects escaping from the Kuiper belt. In this case, the Scattered disk would be, relative to the Kuiper belt, what the population of Near Earth Asteroids is, relative to the main asteroid belt.

However, [39] showed that about 1% of the Scattered disk objects can survive on trans-Neptunian orbits for the age of the Solar System. This leads to the possibility that the current Scattered disk is the remnant of a $\sim 100\times$ more massive primordial structure, which presumably formed when the planets removed the left-over planetesimals from their formation regions. In this case, the Scattered disk would not be in steady state, and it would have no direct relationship with the Kuiper belt.

How can we discriminate between these two hypotheses on the origin of the Scattered disk? In the first case, if the Scattered disk is sustained in steady state by the objects leaking out of the Kuiper belt, the number ratio between the Kuiper belt and Scattered disk populations must be large. Indeed:

$$N_{\rm SD} = N_{\rm KB} \times f_{\rm esc} \times L_{\rm SD}$$

where $N_{\rm SD}$ is the number of Scattered disk objects (larger than a given size), $N_{\rm KB}$ is the number of Kuiper belt objects (down to the same size), $f_{\rm esc}$ is the fraction of the Kuiper belt population that escapes into the Scattered disk in the unit time (due to chaotic diffusion or collisional ejection), and $L_{\rm SD}$ is the mean lifetime in the Scattered disk. Both $L_{\rm SD}$ and $f_{\rm esc}$ are small, so that $N_{\rm KB} >> N_{\rm SD}$. In fact, in the case of the main asteroid belt and the NEA population, the number ratio is about 1,000.

In the second case, if the current Scattered disk is the remnant of a much more massive primordial scattered population, there is no causal relationship between N_{KB} and N_{SD}. The current population of the Scattered disk depends only on its primordial population and not on the current Kuiper belt population.

Discovery statistics [168] suggest that the Scattered disk and the Kuiper belt now contain roughly equal populations. This rules out (by orders of magnitude) the possibility that the Scattered disk is sustained in steady state by the Kuiper belt. Only the scenario of [39] remains valid for the origin of the Scattered disk.

2 The Dynamics of Comets

Comets are usually classified in categories according to their orbital period (Fig. 12). Comets with orbital period $P > 200$ years are called *long-period comets* (LPCs); those with shorter period are called *short-period comets* (SPCs). The threshold of 200 years is arbitrary and has been chosen mostly for historical reasons: modern instrumental astronomy is about two centuries old, so that the LPCs that we see now are unlikely to have been observed in the past.

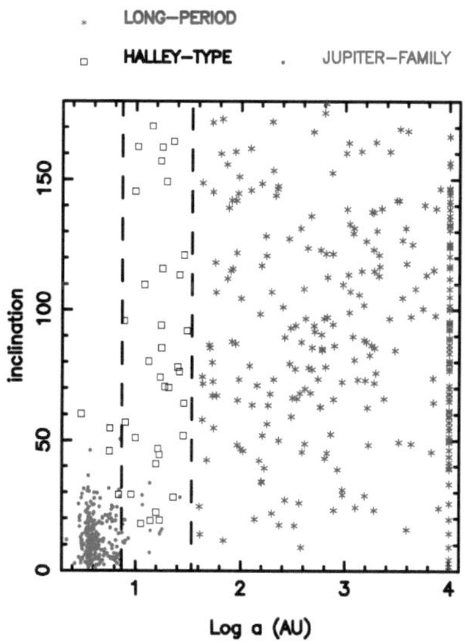

Fig. 12. The distribution of comets according to their orbital semi-major axis and inclination. Here, the orbital elements are defined at the moment of the comet's last aphelion passage. Long period, Halley-type, and Jupiter family comets are plotted as red stars, black squares, and blue dots, respectively. The separation between Halley-types and Jupiter family comets has been made according to the value of their Tisserand parameter, following [105]. The vertical dashed lines correspond to orbital periods $P = 20$ years and $P = 200$ years, respectively. All LPCs with $a > 10{,}000$ AU have been represented on the $\log a = 4$ line

If the orbital distribution of the comets is plotted, like in Fig. 12, using the orbital elements that the comets had when they last passed at *aphelion* – which can be computed through a backward numerical integration – one sees a clustering of long period comets with $a \sim 10^4$ AU. These comets are called *new comets* because they are passing through the region of the giant planets system for the first time. In fact, after a passage through the inner Solar System, it is unlikely that the semi-major axis remains of order 10^4 AU. It either decreases to $\sim 10^3$ AU or the orbit becomes hyperbolic. The reason is that the binding energy of a new comet is $E = -\mathcal{G}M_\odot/2a \sim 10^{-4}$, but typically, during a close perihelion passage, the energy suffers a change of order of the mass of Jupiter relative to the Sun: 10^{-3}. This change is not due to close encounters with the planet (which might not occur). It is because the comet has a barycentric motion when it is far away, an heliocentric motion when it is close, and the distance of the barycentre from the Sun is of the order of the relative mass of Jupiter.

The SPCs are in turn subdivided into *Halley-type* (HTCs) and *Jupiter family* (JFCs). Historically, the partition between the two classes is done according to the orbital period being respectively longer or shorter than 20 years. This threshold has been chosen because there is an evident change in the inclination distribution at the corresponding value of semi-major axis (see Fig. 12). However, comets continuously change semi-major axis as a consequence of their encounters with the planets. In particular, all SPCs had to have a larger semi-major axis in the past, given that they come from the trans-planetary region. Thus, by adopting a partition between HTCs and JFCs based on orbital period, one is confronted with the unpleasant situation of objects changing their classification during their lifetime.

This problem has motivated Levison [105] to re-classify SPCs according to their *Tisserand parameter* relative to Jupiter

$$T_\mathrm{J} = \frac{a_\mathrm{J}}{a} + 2\sqrt{\frac{a}{a_\mathrm{J}}(1-e^2)}\cos i\ . \tag{6}$$

This new classification makes sense, because the Tisserand parameter is quite well preserved during the comet's evolution. In Levison's classification, HTCs and JFCs have T_J, smaller and larger, respectively, than 2. Figure 12 adopts this classification and shows that, for most of the objects, the classifications based on orbital period and on Tisserand parameter are in agreement, but a few objects (those with $P < 20$ years and large inclination or those with $P > 20$ years and low inclination) change their classification depending on the adopted criterion.

Tisserand Parameter

Given the importance of the Tisserand parameter in cometary dynamics, it is useful to derive its expression (which outlines the limitations of its use) and discuss its properties.

The Tisserand parameter is an approximation of the Jacobi constant, which is an invariant of the dynamics of a small body in the framework of the restricted, circular, three-body problem.

The expression of the Jacobi constant is:

$$C_{\rm J} = -(\dot{x}^2 + \dot{y}^2 + \dot{z}^2) + 2\left(\frac{1}{r} + \frac{m_{\rm p}}{\Delta}\right) + 2H_{\rm z} \ , \tag{7}$$

where $\mathcal{G}{\rm M}_{\oplus} = a_{\rm p} = 1$ are assumed, $a_{\rm p}, m_{\rm p}$ are the semi-major axis and mass of the perturbing planet, and $H_{\rm z}$ is the z-component of the small body's angular momentum. The quantity Δ is the distance between the small body and the planet.

We can write the kinetic energy of the small body as a function of its semi-major axis and heliocentric distance:

$$\frac{1}{2}(\dot{x}^2 + \dot{y}^2 + \dot{z}^2) = -\frac{1}{2a} + \frac{1}{r} \ , \tag{8}$$

while the z-component of the angular momentum can be written:

$$H_z = \sqrt{a(1-e^2)}\cos i \ . \tag{9}$$

Substituting (8) and (9) into (7) and neglecting the term $m_{\rm p}/\Delta$ one gets

$$C_{\rm J} \sim T \equiv \frac{1}{a} + 2\sqrt{a(1-e^2)}\cos i \ , \tag{10}$$

where the right-hand side is equivalent to (6), given that a is expressed in units of the planet's semi-major axis.

This derivation of the Tisserand formula shows that the Tisserand parameter is constant as long as the Jacobi constant is preserved, and $m_{\rm p}/\Delta$ is small. This last condition requires that the comet is not in a close encounter with the planet. During a close encounter, the Tisserand parameter has large and abrupt changes, but it returns to the value that it had before the encounter, once the distance to the planet increases back to large values. The conservation of the Jacobi constant, conversely, requires that the conditions of the restricted three-body problem are fulfilled, namely one planet must dominate the comet's evolution, and the effects of the planet's eccentricity must be negligible. This requires that the comet is not in a region where it can have encounters with *two* planets, otherwise the one-planet approximation does not hold. Also, it requires that the comet is not in a secular resonance with the planet, otherwise the effects of the planet's small eccentricity are enhanced.

One can demonstrate that, if a comet intersects the orbit of a planet, the Tisserand parameter T is related to the unperturbed relative velocity U at which it encounters the planet:

$$U = \sqrt{3 - T} \ ,$$

where U is expressed in units of the planet's orbital velocity. The formula is not defined for $T > 3$, which implies that comets with such values of Tisserand parameter cannot intersect the orbit of the planet (obviously for $e_{\rm p}$ 0). Note, however, that comets on non-intersecting the orbits with respect to the planet can have $T < 3$. Only objects with $T < 2\sqrt{2} \sim 2.83$ (the value for a parabolic trajectory with $i = 0$ and $q = a_{\rm p}$) can be ejected onto a hyperbolic orbit in a single encounter with a planet.

2.1 Origin and Evolution of Jupiter Family Comets

The fact that the JFCs have (by definition) a Tisserand parameter with respect to Jupiter that is distinct from that of HTCs and LPCs suggests that the former are not the small semi-major axis end of the distribution of the latter. The average low inclination of the JFCs and the absence of retrograde comets in the JFC population (whichever of the two definitions for JFCs is adopted, see Fig. 12) suggests that the source of the JFCs must be a disk-like structure. In 1980, [44] proposed that the source of the JFCs was the – at the time still putative – Kuiper belt, a hypothesis later supported in [37].

However, today we know that there are two distinct disk-like structures in the trans-Neptunian region: the Kuiper belt and the Scattered disk. Which of the two is the source of JFCs? We have seen in Sect. 1.4 that the Scattered disk is too populous to be sustained in steady state by the objects leaking out of the Kuiper belt. If the Scattered disk is not sustained in steady state, it means that the number of objects that leave the Scattered disk – mostly evolving towards the inner Solar System – is larger than the number of objects entering the Scattered disk from the Kuiper belt. Thus, the Scattered disk must be the dominant source of the JFC population, rather than the Kuiper belt.

The dynamical evolution of objects from the Scattered disk to the JFC region has been studied in detail in [107], with statistics calculated from a large number of numerical simulations. The results illustrated in that paper essentially supersede all the results from the previous literature. Thus, most of what I report below is taken from that source. The origin and dynamics of JFCs has also been exhaustively reviewed in [40].

To evolve from the Scattered disk to the JFC region, a comet has to pass from a Neptune-dominated regime to one controlled by Jupiter (see Fig. 13). Through a transfer process involving multiple planets, the Tisserand parameter is, in principle, not preserved. However, the planetary system is structured in such a way that the transfer chain from Neptune to Jupiter is normally dominated by one single planet at a time (see Fig. 13), and the values of the Tisserand parameter relative to the dominating planets are not very different from each other. For instance, consider a Scattered disk body with Tisserand parameter relative to Neptune $T_{\rm N} = 2.98$. The conservation of the Tisserand parameter implies that the smallest perihelion distance to which Neptune can scatter this object is $q = 17.7\,$AU, just enough to become Uranus-crosser. In this orbit, the body has $T_{\rm U} = 2.96$. If Uranus takes the control of this body,

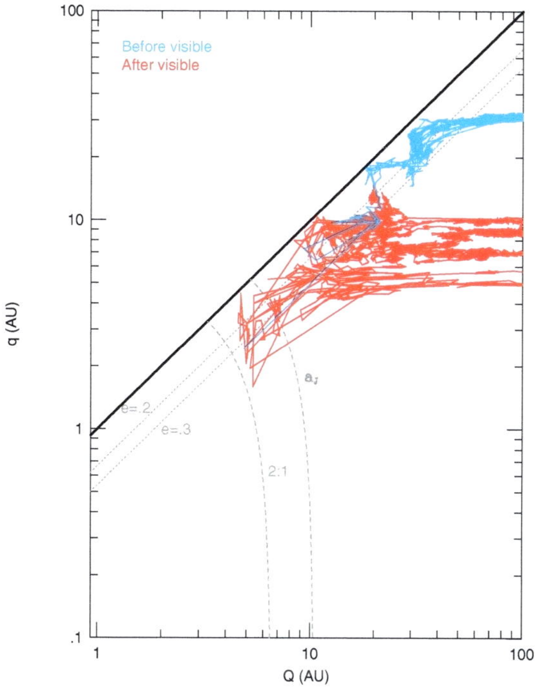

Fig. 13. The evolution of an object from the Scattered disk until its ultimate ejection, projected over the plane representing perihelion vs. aphelion distance. The horizontal structure at $q \sim 30\,\mathrm{AU}$ represents the Scattered disk. When the object evolves along a line q = constant or Q = constant, its dynamics are essentially dominated by one single planet. This happens at least down to 10 AU, and during the final ejection phase. Blue lines denote the evolution before the object becomes a visible JFC, red lines after. The criterion for first visibility is that q has decreased below 2.5 AU for the first time. From [107]

it can scatter it inwards to $q = 9.0\,\mathrm{AU}$, barely a Saturn-crosser. The body has now $T_{\mathrm{S}} = 2.94$ and thus Saturn can lower its q to only 3.8 AU. With such a perihelion, the comet has a Tisserand parameter $T_{\mathrm{J}} = 2.82$. Thus, the body never spends much time in a region where it can encounter two planets, because at each "hand-over," perihelion is converted to aphelion, and the object is taken away from the outer planet (see also [83]). The Tisserand parameter is therefore piece-wise conserved, and the final Tisserand parameter (with respect to Jupiter) is very close to the initial one (with respect to Neptune). Now, the bulk of the observed population in the Scattered disk has $2 < T_{\mathrm{N}} < 3$. Thus, at the end of the transfer chain, the bodies coming from the Scattered disk will have $2 < T_{\mathrm{J}} < 3$, in other words, they will be JFCs.

Because the Tisserand parameter remains close to 3, the inclination cannot grow to large values (because the growth of i would decrease T, see (6)). So, the final inclination distribution is comparable to the inclination distribution in the Scattered disk, i.e. mostly confined within 30°. Figure 14 compares the $(a, i, T_{\rm J})$ distribution of the observed SPCs (top panels) with that obtained in the numerical simulation for the objects coming from the Scattered disk, when their perihelion distance first decreases below 2.5 AU (a criterion for visibility as an active comet). As one can see, the objects with $T_{\rm J} < 2$ (HTCs) are not reproduced, while the observed and simulated distributions of the JFCs agree with each other in a remarkable way.

Nevertheless, a quantitative comparison would show that the inclination distribution of the simulated comets when they first become visible is slightly skewed toward low values relative to the observed distribution. Similarly, the distribution of the distances of the comets' nodes from Jupiter's orbit is also skewed toward small values. However, the dynamical lifetime of comets after they first become visible is of order 10^5 years. As time passes, the conservation of the Tisserand parameter degrades, as a result of the combined effects of Jupiter and Saturn and of secular resonances. Thus, the inclination is puffed up, and the distribution of ω (initially strongly peaked around 0° and 180°) is randomized. As a consequence, the nodal distance distribution is also puffed

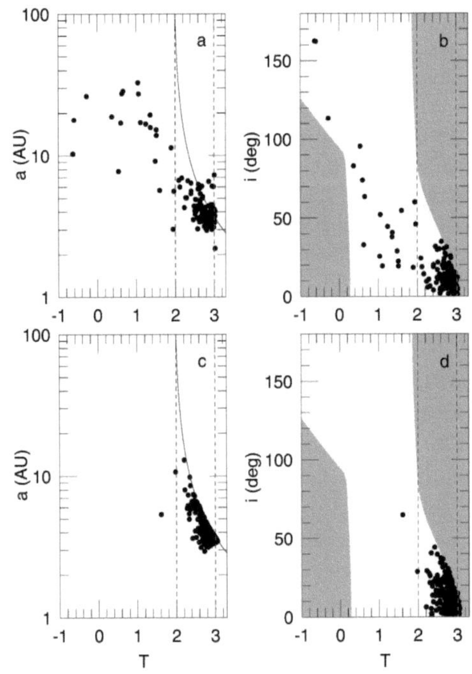

Fig. 14. The distribution of short-period comets projected over the $(T_{\rm J}, a)$ and $(T_{\rm J}, i)$ planes. Top panels: the observed distribution. Bottom panels: the distribution of the objects coming from the Scattered disk, when they are visible ($q < 2.5$ AU) for the first time. From [107]

up[5]. Consequently, the agreement between the observed and simulated distributions first improves with the age of the comets and then eventually degrades. Thus, [39] considered the distribution of all simulated objects, from the time they first become visible up to time τ. Using a Kolmogorov–Smirnov test to measure quantitatively the statistical agreement between simulated and observed distributions, [39] concluded that the best match is achieved – both for the inclination and for the nodal distance distributions – for $\tau \sim 12{,}000$ years. The interpretation of this result is that this value of τ corresponds to the typical physical lifetime of JFCs, after which the comets lose their activity and are no longer observed. Comparing the physical lifetime with the dynamical lifetime, [39] concluded that, if all faded JFCs are dormant objects with asteroidal appearance, the ratio between the number of dormant vs. active JFCs should be $\sim$4.

The comparison between the q distribution of the simulated and observed JFCs suggests that the population of comets is observationally complete up to $q \sim 2$ AU. There are $\sim$40 known JFCs with total absolute magnitude $H_{10} < 9$[6] and $q < 2$ AU. The simulated q distribution indicates that there should be about 100 comets with $q < 2.5$ AU, with the same total magnitude. If all faded JFCs are dormant, then we should expect an additional 400 bodies of asteroidal appearance on similar orbits. About 100 of them should have $q < 1.3$ AU and belong to the NEO population. The size of these putative bodies is badly constrained, because the conversion from total magnitude to nuclear magnitude (i.e. the absolute magnitude of the nucleus, in absence of cometary activity) is poorly known. Published estimates for the nucleus size for $H_{10} = 9$ comets range from $D = 0.8$ km [7] to $D = 4.5$ km [48], with a mean of about 2 km [48]. I will return to the nature of faded comets in Sect. 2.4.

With this estimate of the total number of JFCs, the rate at which Scattered disk bodies become JFCs and the mean lifetime of JFCs measured in their simulations, [39] computed that there should be 4×10^8 such objects (i.e., big enough to have total magnitude $H_{10} < 9$ when active) in the Scattered disk. The extrapolated size distribution obtained from observations of the Scattered disk [10] is roughly consistent with this estimate.

The Orbit of Comet P/Encke

Despite the overall good agreement between the observed and the simulated distribution of JFCs shown in Fig. 14, there is one important difference that should not be overlooked: the orbit of comet P/Encke is not re-produced in

[5]Some comets eventually evolve toward the $T_J < 2$ region, although they never manage to reproduce the (a, i, T_J) distribution illustrated in the top panels of Fig. 14.

[6]The total absolute magnitude is computed from the apparent magnitude V (of nucleus plus coma), the heliocentric and geocentric distances r and Δ by the formula $H_{10} = V + 5\log\Delta + 10\log r$, instead of the usual formula for dormant bodies $H = V + 5\log\Delta + 5\log r$. The coefficient 10, instead of 5, accounts for the fact that the intensity of the activity of the comet is proportional to r^{-2}.

the simulation of [107]. P/Encke is peculiar. It is the only regurlarly active comet with an orbit totally interior to the orbit of Jupiter and $T_J > 3$. In addition, a few asteroids on orbits decoupled from Jupiter are supposed to be dormant cometari nuclei, because of their sporadic activity (such as 4015 Wilson-Harrington) or association with a meteor stream (such as 2201 Oljato). However, the overall number of comets with orbits totally interior to that of Jupiter should be small. In fact, a search for objects with albedo typical of dormant cometary nuclei among the NEOs with $T_J > 3$ [53] has showed that these objects, if they exist, are rare.

The aphelion distance of P/Encke is currently 4.1 AU, so that it is not scattered by Jupiter's encounters. This implies that encounters with Jupiter cannot have emplaced the comet onto its current orbit. It has been proposed that P/Encke reached its orbit from the $T_J < 3$ region because of close encounters with the terrestrial planets, to the effect of non-gravitational forces,[7] or both [50, 73, 145, 174]. Neither of these aspects have been included in the simulations of [107].

A quantitative model of the orbital distribution of JFCs has been recently proposed [115]. This model is an extension of [107], but accounting also for terrestrial planets encounters. According to this model, at any one time, there should be roughly 12 objects in Encke-like orbits. However, it takes roughly 200 times longer to evolve onto an orbit like Encke's than the typical cometary physical lifetime. Thus, all comets decoupled from Jupiter should be inactive! To solve this apparent conundrum, the authors of [115] propose that comet Encke has been recently reactivated, as its perihelion distance is plunging toward the Sun (indeed its future fate is to collide with our star [174]).

2.2 Origin and Evolution of Long-Period Comets

In a historical paper, Oort [140] pointed out that the presence of numerous *new comets* with $a > 10^4$ AU – which appears as a spike in the distribution for $1/a$ of the LPCs (see Fig. 15) – argues for the existence of a reservoir of objects in that distant region. The fact that the inclination distribution of new comets is essentially isotropic, not only in $\cos i$ (from -1 to 1, i.e. including retrograde orbits), but also in ω and Ω, indicates that this reservoir must have a quasi-spherical symmetry, namely it has the shape of a cloud surrounding the Solar System. This cloud is now generally called the *Oort cloud*. In Oort's view, *all* LPCs come from this cloud. The LPCs with $a < 10^4$ AU are returning comets, which originally belonged to the new comet group when they first entered the inner Solar System, but subsequently had their orbit perturbed and acquired a more negative binding energy (smaller semi-major axis). This view remains essentially valid even today.

At such large distances from the Sun, the evolution of the comets in the Oort cloud is strongly affected by the overall gravitational field resulting from

[7] For a recent review on non-gravitational forces acting on comet dynamics see [186].

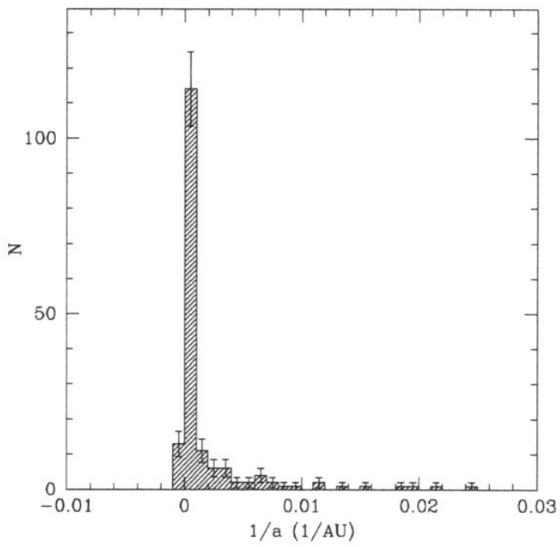

Fig. 15. The differential distribution of LPCs as a function of the inverse semi-major axis. The big spike at $1/a < 10^{-4}$ is due to the new comets and is usually called the *Oort spike.* From [183]

the mass distribution in the galaxy (the so-called *galactic tide*) and by sporadic passing stars and giant molecular clouds (GMCs).

Assuming that the galaxy has a disk-like structure and considering that the Sun is not at the center, the galactic tide has both "disk" and "radial" force components. In a coordinate system centered on the Sun, with x-axis pointing away from the galactic center, y-axis in the direction of the galactic rotation, and z-axis toward the south galactic pole, the radial component of the tide can be expressed with forces along the x and y directions, respectively:

$$F_{\mathrm{x}} = \Omega_0^2 x \ ; \quad F_{\mathrm{y}} = -\Omega_0^2 y \ , \tag{11}$$

where Ω_0 is the frequency of revolution of the Sun around the galaxy. The disk component of the tide can be represented with a force along the z direction:

$$F_{\mathrm{z}} = -4\pi \mathcal{G} \rho_0 z \ , \tag{12}$$

where ρ_0 is the mass density in the solar neighborhood [76]. The disk component dominates over the radial component by a factor 8–10, so that typically only the disk component (12) is considered.

The effect of the disk tide is analogous to the Kozai effect for the dynamics of asteroids with high inclination relative to Jupiter's orbit [101]. In the following, I denote the inclination of the comet relative to the galactic plane by $\tilde{\imath}$ and the argument of perihelion by $\tilde{\omega}$ (not to be confused with the inclination i and the argument of perihelion ω relative to the Solar System plane; the two planes are inclined at 120° relative to one other). The disk tide preserves a and the z-component of the angular momentum $H_z = \sqrt{1-e^2}\cos\tilde{\imath}$ of the comet, while its e and $\tilde{\imath}$ change with the precession of $\tilde{\omega}$. This evolution is periodic; e has a maximum and $\tilde{\imath}$ a minimum when $\tilde{\omega} = 90°, 270°$, while e has a minimum and $\tilde{\imath}$ a maximum when $\tilde{\omega} = 0°, 180°$.[8] The difference between the maximum and the minimum values of e and $\tilde{\imath}$ increases when a increases or H_z decreases. There is no variation of e and $\tilde{\imath}$ if $\tilde{\imath} = 0$.

Thus, Oort cloud comets with high inclination relative to the galactic plane, under the effect of the tide, increase their orbital eccentricity; their perihelion distance decreases and the objects become a planet-crosser. If this evolution is fast enough that q decreases from beyond 10 AU to less than ~ 3 AU within half an orbital period, the comet becomes active during its first dive into the inner Solar System (i.e., without having interacted with Jupiter or Saturn during its previous orbits), namely it appears as a "new comet." The perturbations from the planets remove the planet-crossing comets from the Oort cloud, by either decreasing their semi-major axis or ejecting them from the Solar System on hyperbolic orbits. Thus, the high inclination portion of the Oort cloud is progressively depleted. The role of passing stars and GMCs is to reshuffle the comet distribution in the Oort cloud and to refill the high inclination region where comets are pushed into the planetary region by the disk's tide. Of course, stars and GMCs can also directly deflect the cometary trajectories, injecting the comets into the inner Solar System without the help of the galactic tide. This happens particularly during comet showers caused by close encounters between the Sun and these external perturbers [78, 84]. These directly injected comets do not need to have a large inclination relative to the galactic plane.

The transfer of comets from the Oort cloud to the inner Solar System has been simulated by many authors, in particular by [78,178] and, more recently, [183]. In what follows I will mostly refer to this latter, most modern work.

In [183], the Oort cloud was modeled as a collection of objects with $10{,}000 < a < 50{,}000$ AU, a differential distribution $N(a)\mathrm{d}a \propto a^{-1.5}$ and uniform distribution on each energy hyper-surface, consistent with an earlier model of Oort cloud formation [36]. The evolution of the comets was followed numerically, under the influence of the galactic disk's tide and of the four giant planets, with the latter assumed to be on coplanar circular orbits. Stellar and GMC passages, as well as the radial component of the galactic tide, were neglected. Figure 16a shows the $\cos\tilde{\imath}$ distribution of the simulated comets at their first passage within 3 AU from the Sun (the limit assumed for comet physical activity and visibility). The distribution peaks at $\cos\tilde{\imath} = \pm 0.5$ and is relatively depleted at $\cos\tilde{\imath} = \pm 1$ and 0. This is the signature of the galactic tide. Comets with $\tilde{\imath} \sim 0°$ (or equivalently, $\tilde{\imath} \sim 180°$) have an oscillation in the perihelion distance that is too small to bring them from the trans-planet region into the visibility region. Comets with initial $\tilde{\imath} \sim 90°$ have their inclination decreased to lower values by the time that the perihelion distance is decreased below 3 AU. Similarly, Fig. 17a shows the $\tilde{\omega}$ distribution. The

[8] Here I assume that $\tilde{\imath}$ is defined in the range between $-90°$ and $90°$ (negative $\tilde{\imath}$ corresponding to retrograde orbits relative to the galactic plane), and by "maximum" and "minimum" I mean the maximum and minimum of $|\tilde{\imath}|$

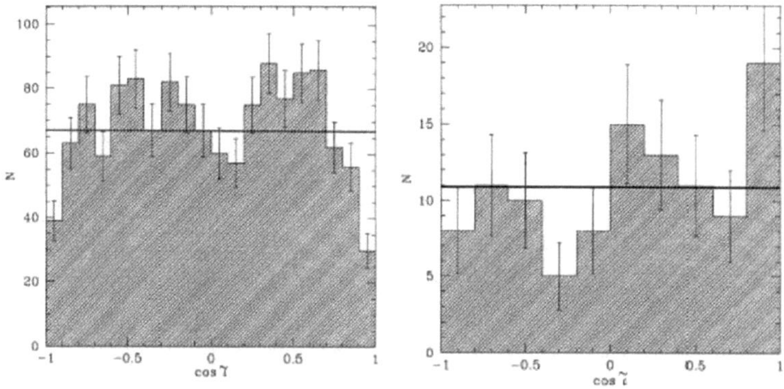

Fig. 16. The inclination distribution relative to the galactic plane for new comets. (a) (left): result of a numerical simulation. (b) (right): the observed distribution. Here $\tilde{\imath}$ is defined in the range between 0° and 180°; values of $\tilde{\imath}$ larger than 90° correspond to retrograde orbits relative to the galactic plane. From [183]

peaks at $\tilde{\omega} \sim 1/4\pi$ and $3/4\pi$ are, again, a signature of the galactic tide. In fact, the precession of $\tilde{\omega}$ is counter-clockwise, and the minimal q is achieved when $\tilde{\omega} = \pi/2, 3/2\pi$. Thus, the perihelion distance decreases below the imposed threshold $q = 3\,\mathrm{AU}$ when $\tilde{\omega}$ is *en route* from 0 to $\pi/2$ or from π to $3/2\pi$. Figures 16b and 17b show the same distributions for the observed new comets. The observed and simulated distributions are quite similar, which confirms the dominant role of the galactic tide. However, the peak and valleys observed in the simulated distributions are not nearly as pronounced as those in the observed dataset. This suggests that the direct injection of comets from the Oort cloud because of passing stars and/or GMCs (neglected in the simulation) has non-negligible importance.

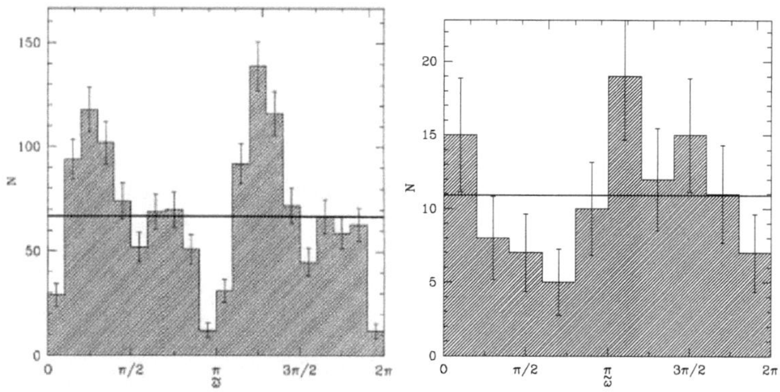

Fig. 17. The same as Fig. 16, but for the distribution of the argument of perihelion relative to the galactic plane. From [183]

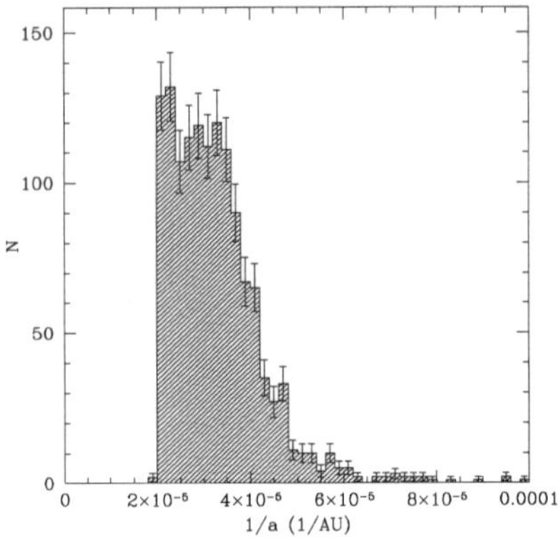

Fig. 18. The distribution of $1/a$ of the comets at their first appearance ($q < 3\,\mathrm{AU}$) from the Oort cloud, according to [183]. The sharp fall-off at $1/a = 2 \times 10^{-5}\,\mathrm{AU}^{-1}$ is due to the choice of the initial conditions ($a < 50,000\,\mathrm{AU}$)

Figure 18 shows the distribution of $1/a$ for the comets at their first apparition, still according to the simulation in [183]. Notice the sharp fall off at $a \lesssim 20{,}000\,\mathrm{AU}$ ($1/a \gtrsim 5 \times 10^{-5}\,\mathrm{AU}^{-1}$) that reproduces the one observed in the $1/a$ distribution of LPCs (see Fig. 15). Thus, essentially all comets at their first apparition have semi-major axes beyond 20,000 AU and therefore would be classified as "new comets" by an observer. This sharp fall off is due to the so-called *Jupiter barrier*. The fact is that new comets must have decreased their q from $>10\,\mathrm{AU}$ to $<3\,\mathrm{AU}$ in less than one orbital period, otherwise they would have encountered Jupiter and Saturn during an earlier revolution, and most likely would have been ejected from the Solar System. This condition is fulfilled only if the semi-major axis is larger than ~20,000 AU. The implication of this result is that LPCs do not probe the Oort cloud inside this semi-major axis threshold, except during rare showers because of a very close encounter between a passing star and the Solar System (which allows a rapid decrease of q even for a <20,000 AU; see [77]). Therefore, our information on the inner Oort cloud does not come from the observations of comets, but solely from models of Oort cloud formation (see Sect. 3).

From the fraction of the Oort cloud population that enters the visibility region per unit time, and the flux of new comets with $H_{10} < 11$ and $q < 3\,\mathrm{AU}$ estimated from observations, [183] concluded that the Oort cloud population with a >20,000 AU and $H_{10} < 11$ is $\sim 10^{12}$. This estimate agrees with [179], and is two times higher than that in [78], which gives a measure of its uncertainty. For the reason explained above, the estimated population in the Oort cloud with smaller semi-major axis is totally dependent on the model of Oort cloud formation.

The evolution of the comets, from their first apparition to their ultimate dynamical elimination, has also been followed in [183]. If the orbital elements of all comets at every passage at $q < 3\,\mathrm{AU}$ are added up (without limitation

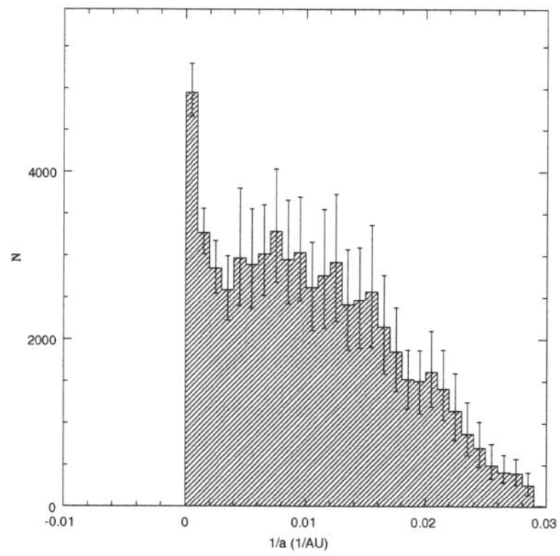

Fig. 19. The distribution of the inverse semi-major axis of all LPCs, independent of the number of perihelion passages within 3 AU, according to the simulation in [183]. This distribution is very different from that observed, illustrated on the same scale in Fig. 15

on the number of perihelion passages that they already suffered), the resulting distribution of $1/a$ (Fig. 19) is very different from the observed distribution (Fig. 15). In particular, the ratio between the number of comets in the Oort spike and the number of returning comets is much smaller than observed. This problem was already pointed out in [140]. As suggested by Oort himself, this mismatch indicates that comets from the Oort cloud have a very limited physical lifetime: after a few perihelion passages, they fade away from visibility, either by becoming inactive or disintegrating. In [183], it was shown that a very good match with the observed distribution of LPCs can be achieved if one assumes that the probability P_m that a comet is still active after m perihelion passages within 3 AU decays as $m^{-0.6}$. This fading law implies that only 10% of the comets survive more than 50 passages and only 1% of them survive more than 2,000 passages. Other equally drastic fading laws, such as $P_m = 1$ for $m \leq 6$ and $p_m = 0.04$ for $m > 6$ [177], can also reproduce the observed distribution of LPCs.

Therefore, the conclusion is that comets from the Oort cloud fade very quickly, in just a few revolutions. This is very different behavior to that of JFCs, which have a physical lifetime of $\sim 10,000$ years (they remain active for about 1,000 revolutions). The fate of faded comets (disruption versus inactivity) for both LPCs and JFCs is discussed in Sect. 2.4.

2.3 Note on Halley-Type Comets

The HTCs have traditionally been considered as the low semi-major axis end of the returning LPC distribution. Indeed, at a first glance, the distribution of HTCs and of returning LPCs (apart from the semi-major axis range that they cover) appear fairly similar.

Under the effect of close encounters with Jupiter and Saturn, some returning comets can have their semi-major axis decreased to less than 34.2 AU. At that point, their orbital period becomes shorter than 200 years, so that, by convention, they are classified as SPCs. They are predominantly HTCs, and not JFCs, because their Tisserand parameter relative to Jupiter is typically smaller than 2. The reason for this is that new comets from the Oort cloud, having $q < 3$, $a \sim \infty$, $e \sim 1$ must have $T_J < 2.15$, and the Tisserand parameter remains roughly conserved during the subsequent evolution down to the SPC region, because of the dominance of Jupiter's perturbations. The transfer of comets from the Oort spike to the HTC region typically requires a large number of revolutions. Thus, the HTCs should belong to the small fraction ($\sim 4\%$) of Oort cloud comets that do not fade away rapidly.

This transfer process from the Oort cloud to the HTC region has been revisited recently in [109], using state-of-the-art numerical simulations. It was found that, although the semi-major axis distribution of the HTCs obtained in the simulations is a reasonable match for the observed distribution, the inclination distributions are profoundly different (Fig. 20). In particular, the median inclination distribution of the observed HTCs is 45°, and 80% have a prograde orbit, whereas the median inclination of the HTCs obtained in the simulation is 120° and only 25% of them have prograde orbit. The reason that the simulated distribution is skewed toward retrograde objects is that such orbits have a longer dynamical lifetime (100,000 years, as opposed to 60,000 years for prograde HTCs).

In [109], to solve the mismatch between the inclination distributions, the authors proposed that part of the HTCs come from the inner Oort cloud ($a <$ 20,000 AU) and that this reservoir has a disk-like structure, with

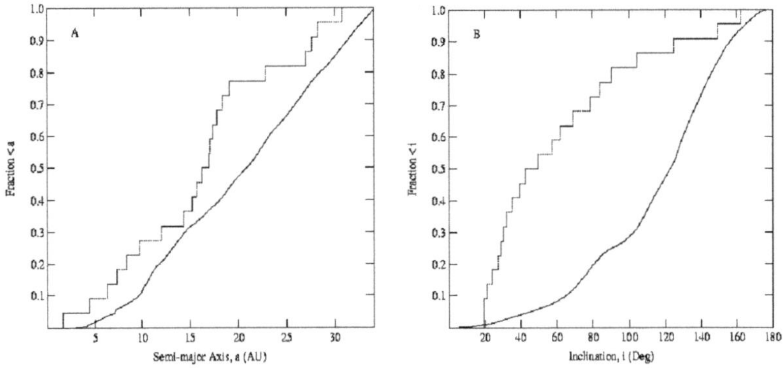

Fig. 20. Comparison between the cumulative orbital element distributions of the observed HTCs (dotted line) and those produced in the integrations of [109] (solid line). (a) Semi-major axis distributions; (b) inclination distributions. Note the significant disagreement in the inclination distributions. Only comets with $q < 1.3$ AU are considered

inclinations within 50° of the ecliptic. However, modern formation models of the Oort Cloud (see Sect. 3 and Fig. 23) show that retrograde orbits in the Oort cloud start to appear beyond 6,000–7,000 AU, and a flattened region can only be found inside this boundary in semi-major axis. However, this region is too tightly bound to the Sun to be an abundant source of comets.

In [116], it has been recently proposed that part of the HTC population comes from the distant end of the Scattered disk. They would be objects that, pushed outward by Neptune, eventually feel the galactic tide and have their perihelion decreased further into the planetary region ($q < 25$ AU). Subsequent encounters with the giant planets then bring these objects into the HTC region. Reminiscent of its Scattered disk origin, this population would be predominantly prograde. The final HTC population would be a combination of this population with that of comets coming from Oort cloud as described above [109]. This would explain why the observed HTC inclination distribution, while ranging from 0 to 180°, is skewed toward prograde values (see Fig. 20).

Probably, the last word on the problem of the inclination distribution of HTCs has not yet been said. It is possible that part of the solution is that HTCs, even if longer-lived than new LPCs, cannot be active for more than $\sim 10,000$ years, as it is the case for JFCs. This would bring the median value of the inclination distribution of the simulated "active" comets down to $\sim 90°$, or even less [46]. Moreover, the median inclination of the currently observed HTCs might be smaller than the real value, because of observational biases and/or small number statistics. In fact, an update of the HTC catalog with respect to that used in [109] shows an increase of the median inclination from 45 to 60°. In addition, the HTC catalog might be contaminated by a few prograde objects coming from the JFC population (see Sect. 2.1). Finally, I notice that in the simulations of [183], 65% of the SPCs were on prograde orbits. Why this result is different from that in [109] (25%) is not clear. The efficiency of transfer of comets from the Oort cloud to the SPC region is very small, so that it is possible that the results of any model based on numerical simulations is dominated by small number statistics. Definitely, the issue of the origin of HTCs needs to be investigated further.

2.4 The Fate of Faded Comets

We have seen in Sects. 2.1 and 2.2 that there is quite strong evidence that comets fade after a limited number of revolutions and that the rate at which they do so is different for JFCs and LPCs. What happens to the faded comets? Do they remain on orbit around the Sun as dormant asteroid-like objects, or do they disintegrate into smaller, undetectable pieces?

To answer this question, it is necessary to look for asteroid-like objects on orbits typical of these comets and compute if their number is consistent with that expected assuming that all faded comets are dormant and accounting for the discovery biases.

Several Near Earth Asteroids (NEA) have been discovered on orbits typical of JFCs, with $2 < T_J < 3$. The NEA model developed in [13], calibrated on Spacewatch discoveries, argues that the asteroid belt is not a sufficient source of these objects. This model implies that, among the NEA population, 60 ± 40 objects with $H < 18$ are dormant JFCs. A similar model [182], developed using the more extended dataset provided by the LINEAR survey, estimates ~ 70 dormant JFCs in the NEA population in the same magnitude range (for comparison, the total number of NEAs with $H < 18$ is estimated to be $\sim 1,200$ [161]). Assuming 4% albedo – typical of cometary nuclei without activity – $H = 18$ corresponds to $D = 1.7$ km. As we have seen in Sect. 2.1, [39] estimate the existence of ~ 100 faded JFCs in the NEO region with diameter of about 2.0 km.

An independent confirmation that many/most NEAs with $T_J < 3$ are dormant comets comes from spectroscopic observations [12, 53], which show that the albedo distributions of the NEAs with, respectively, $T_J > 3$ (the majority of the population) and $T_J < 3$ are totally different. The latter have much darker albedos than the former. In conclusion, there is solid evidence that at least a significant fraction of JFCs become dormant when they fade. It should be remembered, however, that some JFCs have been observed to disintegrate. For instance, comet 3P/Biela split into two parts in 1846 and disappeared 6 years later. Similarly, comet 73P/Schwassmann–Wachmann broke into five big fragments in 1995; eight fragments can now be seen, and they are probably continuing to split into smaller pieces.

The situation is totally different for LPCs and HTCs, as shown in [111]. The steep fading law required to explain the observed number ratio between new and returning comets (see Sect. 2.2) implies that for every active returning comet there should be 20 faded comets. Thus, if all faded comets were dormant, the model in [183] would imply the existence of 4×10^6 objects, with $q < 3$ AU and semi-major axis distribution similar to that of Fig. 19. Again, the absolute magnitude H of these objects, corresponding to comets with $H_{10} < 11$ when in activity, is very uncertain. Assuming that they have $H < 18$ – with cumulative distribution $N(< H) \propto 10^{0.28H}$ as in [180] – [111] estimated that 1 object out of 20,000 should have been discovered by asteroid surveys, namely ~ 200 objects. However, only two "asteroidal" objects have been discovered on LPC orbits. Similarly, if all faded comets were dormant, the model in [109] estimates that there should be 100,000 inactive HTCs with $H < 18$ and $q < 2.5$ AU. Of these, [111] estimated that 1,000 should have been discovered by asteroid surveys. This estimate is again 100 times larger than the number of actual discoveries of "asteroids" on corresponding orbits. Thus, the conclusion seems to be that only $\sim 1\%$ of the comets from the Oort cloud become dormant when they fade. The remaining 99% apparently split in smaller undetectable fragments (like in the case of comet LINEAR C/2001 A2), if not into dust trails.

In summary, the JFCs and LPCs seem to fade away because of different physical processes. This may be surprising, given that both are thought to be similar mixtures of ice and rock. However, evolutionary processes could affect comets' susceptibility to disruption. For example, over long timescales, JFCs could have lost more volatiles than LPCs because they have been stored in the Scattered disk, at closer heliocentric distances and thus higher temperatures than in the Oort cloud. JFCs could be more porous, and thus less susceptible to disruption resulting from volatile pressure buildup, because of a relatively violent collisional environment. Finally, the dynamical pathways that LPCs and JFCs take on their way into the inner Solar System might lead to very different thermal histories for the two populations. To jump over the Jupiter barrier in one orbital period, LPCs have to evolve from very distant orbits (with perihelia outside the planetary region) to orbits that closely approach the Sun. On the other hand, objects from the Scattered disk slowly move through the planetary region, taking $\sim$10 My to evolve onto orbits with $q <$ 2.5 AU [107]. Perhaps LPCs disrupt because of strong thermal gradients or volatile pressure buildup, while JFCs survive because they are warmed more slowly.

3 The Formation of the Oort Cloud

To explain the formation of the Oort cloud, it is intuitive to invoke the mechanism described in the previous section for the origin of LPCs, but "played" in "reverse mode." Imagine an early time when the Oort cloud was still empty and the giant planets' neighborhoods were full of icy planetesimals. The scattering action of the planets dispersed the planetesimals throughout the Solar System. Some were moved onto eccentric orbits with large semi-major axis, but with perihelion distance still in the planetary region. Those that reached a semi-major axis of $\sim$10,000 AU started to feel a galactic tide strong enough to modify their orbit on a timescale of one orbital period. During the scattering process, these planetesimals remained relatively close to the ecliptic plane, so that their inclination relative to the galactic plane $\tilde{\imath}$ was $\sim 120°$. Because of their large e and $\tilde{\imath}$, the effect of the tide on the evolution of $e, \tilde{\imath}$ was large. The planetesimals with $\tilde{\omega}$ between 90° and 180° (or, symmetrically, between 270° and 360°) had their eccentricity decreased. This lifted their perihelion distances beyond the planets' reach, so that they could not be scattered any more: they became Oort cloud objects. The precession of $\tilde{\Omega}$ and the occasional passage of rogue stars randomized the planetesimals' distribution, resulting in the Oort cloud structure that is inferred from observations of LPCs.

This scenario, originally proposed in [104], was first simulated in [42, 43] using a Monte Carlo method to represent the effects of repeated, uncorrelated encounters of the planetesimals with the giant planets and passing stars (the role of the galactic tide was not yet taken into account). The first simulation of Oort cloud formation using direct numerical simulations and accounting for the galactic tide was done in [36]. To save computing time, however, the simulations were started with comets already on low inclination, high eccentricity orbits: initially having $a = 2{,}000\,\mathrm{AU}$ and q uniformly distributed between 5 and 35 AU.

The formation of the Oort cloud has recently been revisited in [33] (see also [32]), using more modern numerical simulation techniques. The authors started with more realistic initial conditions, assuming planetesimals initially distributed in the 4–40 AU zone with small eccentricities and inclinations. The giant planets were assumed to be on their current orbits, and the migration of planets in response to the dispersion of the planetesimals (see Sect. 4) was neglected. The evolution of the planetesimals was followed for 4 Gy, under the gravitational influence of the four giant planets, the galactic tide (both radial and disk components – see (11), (12)), and passing stars. Both the tide and the statistics of passing stars were calibrated using the current galactic environment of the Sun. A stellar density of $0.041 M_{\odot}/\mathrm{pc}^3$ was assumed, with stellar masses distributed in the range 0.11–18.24 $M_{\odot}$ and relative velocities between 1.7 and $158\,\mathrm{km\,s^{-1}}$ (with a median value of $46\,\mathrm{km\,s^{-1}}$). In total, the simulation described in [33] recorded ∼50,000 stellar encounters within 1 pc of the Sun in 4 Gy. In the following discussion of Oort cloud formation, I mostly refer to the results of this work.

Figure 21 shows an example of the evolution of a comet from the neighborhood of Neptune to the Oort cloud. Through a sequence of encounters, the object is first scattered by Neptune to larger semi-major axis, while keeping the perihelion distance slightly beyond 30 AU (typical of Scattered disk bodies). After about 700 My, the random walk brings the body's semi-major axis to $\sim 10{,}000\,\mathrm{AU}$. At this time the galactic tide starts to be effective, and the perihelion distance is rapidly lifted above 45 AU. Neptune's perturbations cease to be important and further changes in semi-major axis are due to the effects of distant stellar encounters. When the body starts to feel the galactic tide, its inclination relative to the galactic plane is 120°. As the perihelion distance is lifted (the eccentricity decreases), the inclination decreases towards 90°.[9] A stellar passage causes a sudden jump of $\tilde{\imath}$ to 65° just before $t = 1$ Gy. This allows the effect of the tide to become more pronounced, bringing the perihelion distance of the object beyond 1,000 AU and the inclination $\tilde{\imath}$ up to 80°. This configuration is reached at $t = 1.7$ Gy, when $\tilde{\omega}$ is 180°. From this time onward, the galactic tide reverses its action, decreasing q and $\tilde{\imath}$. In principle, the action of the galactic tide is periodic, so that the object's perihelion should be decreased back to planetary distances. However, the jumps in $a, q, \tilde{\imath}$ caused by stellar encounters break this reversibility. The oscillation of q becomes more shallow and the object never returns to the planetary region within the age of the Solar System. Notice finally that during this evolution,

[9] Notice that, for the dynamical evolution forced by the galactic disk tide, the decrease of $\tilde{\imath}$ from 120 to 90° is equivalent to an increase from 60 to 90°, in agreement with what has been said in the previous section on the anti-correlation of the evolutions of eccentricity and inclination.

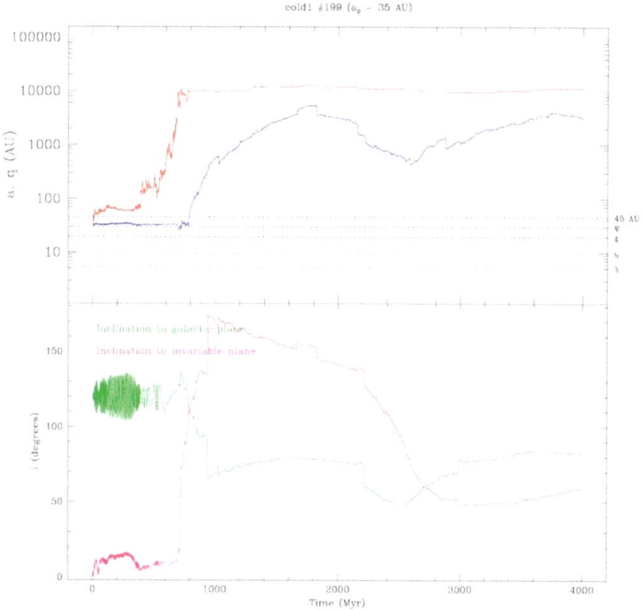

Fig. 21. An example of evolution of a comet from the vicinity of Neptune into the Oort cloud, from [33]. The top panel shows the evolution of the object's semi-major axis (red) and perihelion distance (blue). The bottom panel shows the inclinations relative to the galactic plane (green) and Solar System invariable plane (the plane orthogonal to the total angular momentum of the planetary system; in magenta)

the inclination relative to the invariable plane is strongly changed. It is turned to retrograde, and then back to prograde values, as the longitude of galactic node $\tilde{\Omega}$ precesses.

Not all particles follow this evolution, though. Those that interact closely with Jupiter and Saturn are mostly ejected from the Solar System. Those that have distant encounters with Saturn are transported more rapidly and further out in semi-major axis compared with the evolution shown in Fig. 21. The strength of the galactic tide increases with a; thus, for the comets that are scattered to $a \sim 20,000\,\mathrm{y}$ or beyond, the oscillation period of q and $\tilde{i}$ is shorter than for the particle in Fig. 21.

Figures 22 and 23 give a global illustration of the Oort cloud formation process, showing snapshots of the (a, q) and (a, i) distributions of all planetesimals at 0 (initial conditions), 1, 10, 100 My and 1, 4 Gy. The planetesimals in these plots are color-coded according to their initial position: Jupiter region objects are magenta; Saturn region objects are blue; Uranus region objects are green; Neptune region objects are red, and trans-Neptunian objects are black. Figure 22 shows that, after only 1 My, a Scattered disk is formed by Jupiter and Saturn, out of particles initially in the Jupiter–Uranus region. This Scattered disk differs from the current Scattered disk because most of

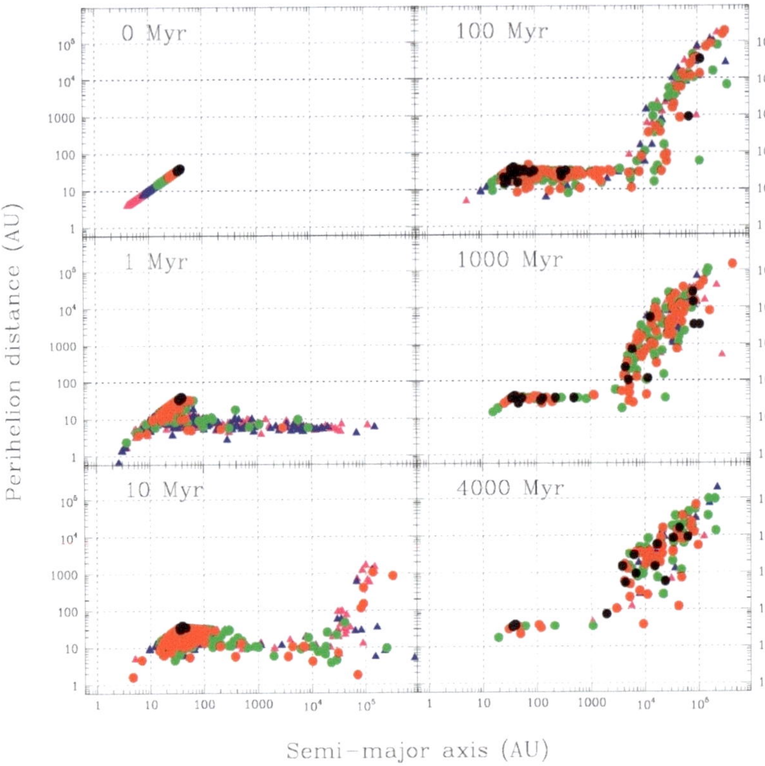

Fig. 22. Scatter plot of osculating barycentric pericenter distance vs. osculating barycentric semi-major axis, at various times in the Oort cloud formation simulations of [33]. The points are color-coded to reflect the region in which the simulated comets formed. Each panel is labeled by the simulation time that it corresponds to

its objects have $q < 10\,\mathrm{AU}$. Particles originally in Neptune's region or beyond have not yet been scattered. At 10 My, a signature of the galactic tide starts to be visible. The Oort cloud begins to form. Particles with $a > 30,000$, mostly from the Jupiter–Saturn region, have their perihelia lifted beyond the orbits of the planets. Neptune's particles start to populate the Scattered disk. From 100 My to 1 Gy, particles continue to enter the Oort cloud from the Scattered disk. The population of the Oort cloud peaks at 840 My, at which time 7.55% of the initial particles occupy the cloud. Objects from the Uranus–Neptune region gradually replace those from the Jupiter–Saturn zone. The latter have been lost during stellar encounters, as they predominantly occupied the very outer part of the Oort cloud ($a > 30{,}000\,\mathrm{AU}$). Because of the longer time over which the galactic tide has acted and to stellar encounters, the population of bodies with perihelion distances above 100 AU can have semi-major axes as low as 3,000 AU. The Oort cloud with $a < 20{,}000\,\mathrm{AU}$ is usually called the inner Oort cloud, or Hills cloud from [80]. The last panel in Fig. 22, representing the

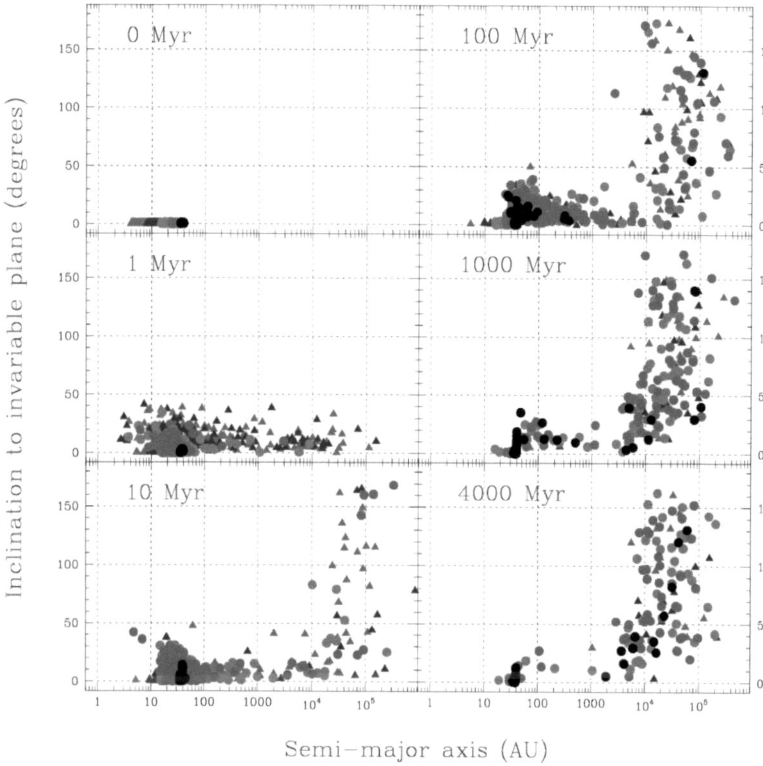

Fig. 23. The same as Fig. 22 but for osculating barycentric inclination relative to the Solar System mid-plane vs. osculating barycentric semi-major axis. From [33]

distribution at 4 Gy, should correspond to the current structure of the Oort cloud. The distribution remains nearly the same as that at 1 Gy, but the Oort cloud population has declined slightly in number.

Figure 23 shows the evolution of the inclinations of the particles. After 1 My, the planets have scattered the comets into moderately inclined orbits. After 10 My, the particles with $a > 30{,}000$ AU have been perturbed by the galactic tide and passing stars into a nearly isotropic distribution of inclinations. As time passes, tides affect the inclinations of particles closer to the Sun, so that at 4,000 My inclinations are clearly isotropic for $a > 20{,}000$ AU.

The final Oort cloud contains roughly equal populations in the inner and outer parts, with radial distribution $N(r)\mathrm{d}r \propto 1/r^3$. About 5–9% of the planetesimals initially in the Uranus–Neptune–transneptunian region remain in the Oort cloud at the end of the simulation. Conversely, only 2% of the planetesimals originally in the Jupiter–Saturn region do so. The scattering action of these planets is too strong to deposit a large fraction of planetesimals in the Oort cloud. The reason is the same as that invoked to explain the Jupiter barrier for the new-LPC distribution (see Sect. 2.2). In energy space, the Oort cloud is 10^{-4} wide, whereas the random walk in energy of particles scattered by Jupiter and Saturn has steps of width $\sim 10^{-3}$ (in other words, proportional to the masses of these planets relative to that of the Sun). Thus, most of the particles scattered by these planets go directly from a Scattered disk orbit (Energy $< -10^{-3}$) to unbound orbit (Energy > 0), without passing through the Oort cloud ($-10^{-4} <$ Energy $< 10^{-3}$).

Figure 24 shows the evolution of the mass in the Oort cloud as a function of time. The formation and the erosion of the Oort cloud are not separate processes. Throughout the history of the Solar System, new planetesimals have entered the Oort cloud from the Scattered disk, while other comets left the cloud, as a result of the galactic tide pushing their perihelia back to the planetary region, and through the perturbations of passing stars putting them onto hyperbolic orbits. Therefore, the flux of LPCs started as soon as the first planetesimals reached ~10,000 AU (10 My), and the supply of new objects to the Oort cloud is still ongoing today [51]. However, as mentioned above, the mass in the cloud peaks at about 800 My. Before this date, the formation process dominated over the erosion process. Then – because the mass of the Scattered disk dropped – the erosion process became dominant, and the total mass in the cloud decayed to ~5.5% of the mass originally in the planetesimal disk. The outer Oort cloud formed faster than the inner cloud – because of the contribution of planetesimals from Jupiter–Saturn region – but then eroded faster because its objects are less tightly bound to the Sun.

3.1 Problems with the Classical Scenario

The classical scenario of Oort cloud formation discussed above meets two problems when confronted with the quantitative constraints provided by the current Solar System.

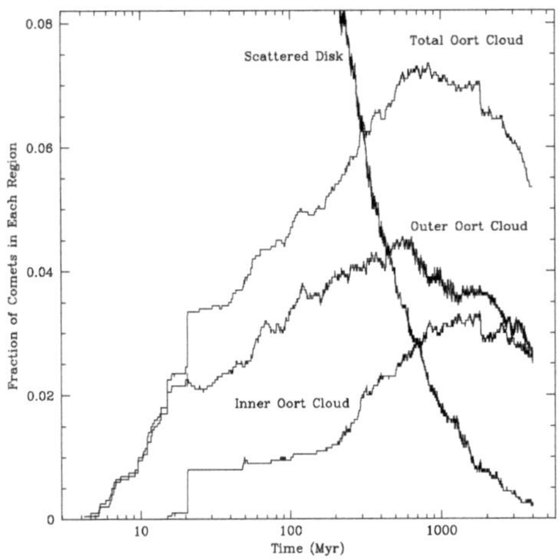

Fig. 24. Fraction of the initial planetesimal population that is in the Oort cloud, in its inner and outer parts and in the Scattered disk, as a function of time. From [33]

As we have seen in Sect. 2.2, the outer Oort cloud should currently contain 10^{12} comets with $H_{10} < 11$. The estimates of the nuclear size of a $H_{10} = 11$ comet range from 1 km [8] to 2.3 km [179]. Assuming, as in [111], that a $H_{10} = 11$ comet has $D \sim 1.7$ km, and assuming also a cumulative size distribution proportional to D^{-2} and a density of $0.6\,\mathrm{g\,cm^{-3}}$ (as for P/Halley), one obtains a total mass of 3×10^{28} g, namely $3\,\mathrm{M}_{\oplus}$. Because the overall efficiency of formation of the outer Oort cloud is small (2.5%), this implies that the original planetesimal disk in the Jupiter–Neptune region was $\sim 100\,\mathrm{M}_{\oplus}$. This seems rather high compared with the total mass of solids associated with the minimum mass solar nebula [69]. Also, numerical simulations show that a planetesimal disk more massive than 30–50 $\mathrm{M}_{\oplus}$ would have driven Neptune beyond 30 AU and that, in such a disk, Jupiter and Saturn would have passed across their mutual 2:5 mean-motion resonance (see Sects. 4 and 5). The uncertainty in the conversion between H_{10} magnitude and size, however, allows enough room for us to make consistent estimates. For instance, if the nuclear size of $H_{10} = 11$ comets is 1.3 km (instead of the assumed 1.7), the required mass of the planetesimal disk falls to a more reasonable value of $50\,\mathrm{M}_{\oplus}$.

A second, more severe problem concerns the number ratio between the comet populations in the Oort cloud and in the Scattered disk. We have seen in Sect. 2.1 that the Scattered disk, to be a sufficient source of JFCs, has to contain 4×10^{8} comets with $H_{10} < 9$. The number of comets with $H_{10} < 11$ depends on the exponent of the H_{10} distribution of comets, which is still highly debated. Using the largest value available in the literature (0.7 [48]), the Scattered disk should have 10^{10} $H_{10} < 11$ comets. Using the exponent for the nuclear magnitude distribution in [111, 180] (0.28) and assuming a linear scaling between nuclear magnitude and H_{10}, the number of $H_{10} < 11$ comets in the Scattered disk reduces to 1.5×10^{9}. Because the number of comets in the outer Oort cloud with $H_{10} < 11$ is 10^{12}, the comet number ratio *inferred from observations* between the outer Oort cloud and the Scattered disk is in the range 100–1,000. However, in the simulations in [33], the final ratio is ~ 10 (see Fig. 24).

The way out of this problem is much more difficult than for the total mass problem. The discrepancy does not depend on assumed relationships between total magnitude and size, nor on density. It cannot be alleviated with any reasonable assumption of the exponent of the H_{10} distribution. Also, different assumptions of the initial planetesimal distribution in the disk would not help. The point is that most of the Oort cloud is made of planetesimals from the Uranus–Neptune–trans-Neptunian zone, which have to pass through the Scattered disk to reach the cloud. Thus, there is a causal relationship between the final numbers of comets in the Scattered disk and Oort cloud. To change this relationship, it would be necessary that a much larger number of planetesimals could reach the Oort cloud without passing through the Scattered disk. This requires that Jupiter and Saturn were more effective in the real Oort cloud-building process than in the simulations of [33]. A possible scenario in which this can occur is discussed below.

3.2 Oort Cloud Formation in a Dense Galactic Environment

It is now known that most stars form in clusters. In [47], it was pointed out that a denser galactic environment would have exerted a stronger tide on the scattered planetesimals. In addition, stellar encounters would have been more effective, because of the slower relative velocities and smaller approach distances typical of a cluster environment. As a consequence, the threshold semi-major axis value beyond which planetesimals could be decoupled from the planets would have been ∼1,000 AU, instead of the current value of ∼10,000 AU. In other words, the Oort cloud would have extended closer to the Sun, covering the region with binding energy down to $\sim -10^{-3}$ in normalized units. Because this width is of the same order as the energy change suffered by planetesimals crossing the orbits of Jupiter and Saturn, the role of these gas giants in building the Oort cloud would be greatly enhanced.

Simulations of Oort cloud formation in a dense environment have been done in [49]. Three kinds of environments were considered: (i) a loose cluster with 10 stars pc^{-3}; (ii) a dense cluster with 25 stars pc^{-3}; and (iii) a superdense cluster with 100 stars pc^{-3}. In all cases, all stars were assumed to have a solar mass (compare with the current stellar density of $0.041\,\mathrm{M}_\odot\,\mathrm{pc}^{-3}$ [33]). The average relative velocity among the stars was assumed to be $1\,\mathrm{km\,s}^{-1}$, typical of star clusters [11] (instead of the current $\sim 40\,\mathrm{km\,s}^{-1}$). In addition, a placental molecular cloud containing 10^5 molecules of Hydrogen per cm^3 was assumed (the current molecular density is $\sim 3\,\mathrm{g\,cm}^{-3}$). The initial conditions of the planetesimals were similar to those in [36]. Comets were placed on initial orbits with $100 < a < 250$ AU and q ranging from 4 to 30 AU.

Figure 25 shows the result of these simulations. As expected, the denser the cluster, the more tightly the resulting Oort cloud is bound to the Sun. Notice, however, that the outer part of the cloud (beyond 10^4 AU) becomes totally empty, because all comets beyond this limit are stripped off by the

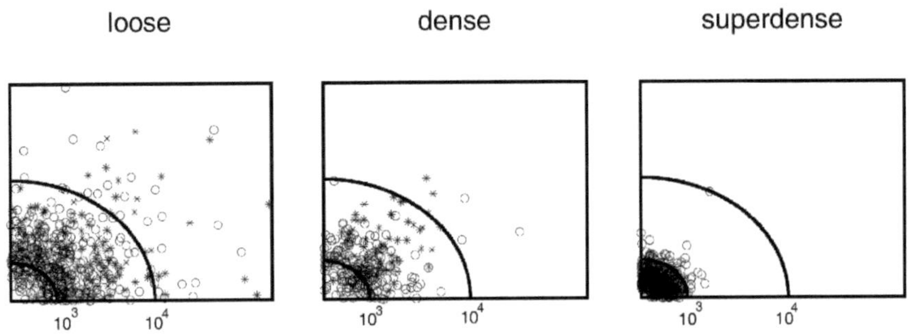

Fig. 25. A sketch showing how comets trapped in the Oort cloud would appear distributed in the circumsolar space, for three kinds of star clusters surrounding the Sun. The radii of the circles are expressed in AU. Stars denote comets coming from Jupiter–Saturn zone, while open circles denote bodies from the Uranus–Neptune zone. From [49]

passing stars. Thus, a mechanism would be required to transfer the comets from the massive inner Oort cloud to the outer cloud, to explain the current flux of LPCs (which come from the outer cloud only). Less effective stellar encounters, occurring during the dispersal of the cluster and in the current galactic environment, might be responsible for this process.

In terms of efficiency of Oort cloud formation, [49] found that about 30% of the initial planetesimals were trapped in the cloud, a factor of 6 higher than in [33]. However, this new efficiency is of the same order of that found in [36], which used initial conditions similar to those in [49], but no star cluster. Thus, it is unclear if the difference in efficiency between [49] and [33] is due to the different choice of initial conditions (in which case the efficiency in [33] is more accurate because the initial conditions are more realistic) or to the presence of the cluster. Moreover, a totally unexpected result was that the final contribution of Jupiter and Saturn to the formation of the Oort cloud (i.e., the fraction of the planetesimal population with initial $q < 10\,\mathrm{AU}$ that ended in the cloud) was minimal. This happened because the planetesimals scattered by Jupiter and Saturn typically ended up in the outer part of the cloud and were subsequently stripped away by the numerous stellar encounters.

More recently, [15] revisited the problem and simulated the evolution of particles initially on circular and coplanar orbits in the Jupiter–Saturn region in presence of a local clusters of various densities. The authors found that, for clusters with densities (gas plus stars) of 5×10^3–$10^4\,\mathrm{M}_\odot\,\mathrm{pc}^{-3}$ in the vicinity of the Sun, about 10–15% of the simulated particles are trapped in the inner Oort cloud (extending from a few 100 AU to ∼10,000 AU) at the end of the simulation.

The study of the formation of the Oort cloud in a dense environment is not finished, however. It is still necessary to quantify which mechanism could transfer the comets from the massive inner Oort cloud – produced in the dense environment – to the outer Oort cloud – where comets must reside at the current time to produce LPCs, and the efficiency of this process. Moreover, it would be more realistic to re-do the simulations in [15], taking into account the effect of gas drag, given that the gas-disk was present for most of the time that the Sun spent in the cluster. Gas drag could protect comets from ejection (Levison, private communication), thus increasing further the fraction of planetesimals from the Jupiter–Saturn zone that are trapped in the cloud.

Sedna: An Inner Oort Cloud Object?

One piece of evidence for a moderate stellar cluster surrounding the early Sun is provided by Sedna. The distribution of the Extended Scattered disk bodies shows a clear tendency. In the 50–60 AU region 2004 XR_{190} has $q \sim 50\,\mathrm{AU}$, but this region is affected by many resonances that can raise the perihelion distance (see Sect. 4.2). Further out, the perihelion distance is larger for bodies with larger semi-major axis: up to ∼200 AU the Extended Scattered disk bodies have $q < 41.5\,\mathrm{AU}$; 2000 CR_{105} ($a = 222\,\mathrm{AU}$) has $q = 44.3\,\mathrm{AU}$ and

Sedna ($a = 495\,\mathrm{AU}$) has $q = 76\,\mathrm{AU}$. Although only a few such bodies are known – and one should be careful about small number statistics – the lack of objects with perihelion distances comparable to those of 2000 CR_{105} and Sedna but smaller semi-major axes seems significant. In fact, observational biases (given an object's perihelion distance and absolute magnitude, and a survey's limiting magnitude of detection) sharply favor the discovery of objects with smaller semi-major axes. So, it would be unlikely that the first two discovered bodies with $q > 44\,\mathrm{AU}$ have $a > 200\,\mathrm{AU}$ if the real semi-major axis distribution in the Extended Scattered disk were skewed toward smaller a.

Assuming that the Extended Scattered disk bodies belonged to the Scattered disk until a perturbation lifted their perihelion distance beyond Neptune's reach, the fact that q increases with a is a clear signature that the perturbation had to grow in magnitude with increasing heliocentric distance. Passing stars produce this very signature [49, 129, 146]. In particular, it was shown in [129] that an encounter with a solar mass star at 800 AU with an unperturbed relative velocity of $1\,\mathrm{km\,s^{-1}}$ (see Fig. 26) would have produced

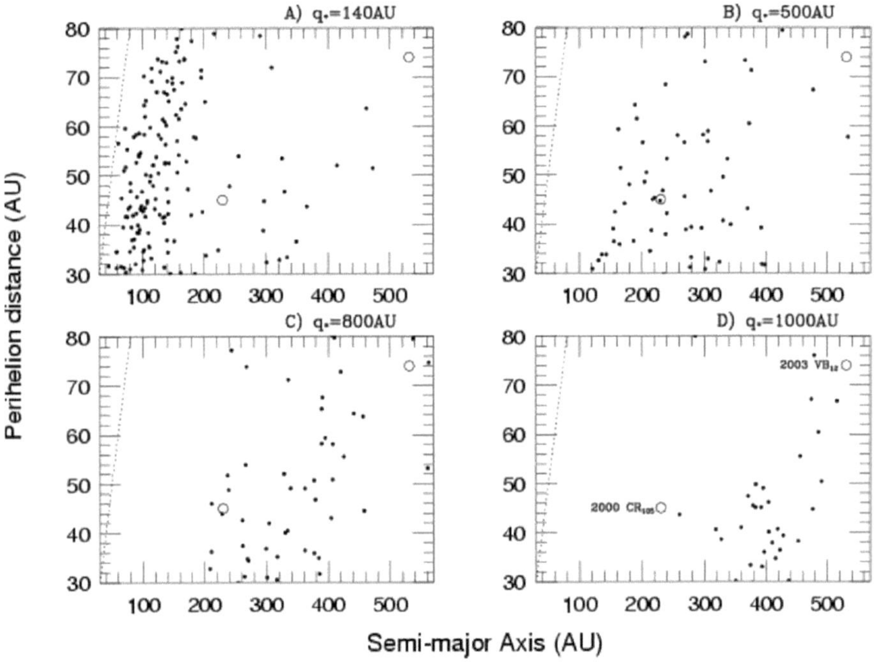

Fig. 26. The Extended Scattered disk that results from passing stars. In all cases, the passing star had $1\,\mathrm{M}_\odot$ and was on a hyperbolic orbit with relative velocity of $1\,\mathrm{km\,s^{-1}}$. Only the perihelion distance of the stellar orbit is varied from panel to panel. The particles were intially in the Scattered disk created in the simulation of [33], at $t = 10^5$ y. The two open circles show the orbits of Sedna and 2000 CR_{105}. From [129]

a distribution of Extended Scattered disk objects that overlaps the orbits of Sedna and 2000 CR_{105} and does not extend to smaller semi-major axes. Closer stellar encounters would still produce a distribution overlapping the orbits of Sedna and 2000 CR_{105}, but such distributions would extend to smaller semi-major axes, inconsistent with the lack of detections of large-q bodies at small a. More distant encounters would not reproduce the orbits of Sedna and/or 2000 CR_{105} (see Fig. 26). The best "fit distance" of the stellar encounter depends on the stellar mass. A star with $\mathrm{M} = (1/4)\,\mathrm{M}_\odot$ should have passed at $\sim$400 AU to produce a distribution similar to that in Fig. 26C.

Stellar encounters at such short distances from the Sun are statistically reasonable only if the Sun was embedded in a cluster, which supports the necessity of developing models of Oort cloud formation in the framework of a dense galactic environment. In fact, in the simulations of [15], the required encounters are obtained, statistically, for clusters with a density of about $10^4\,\mathrm{M}_\odot\,\mathrm{pc}^{-3}$ in the vicinity of the Sun. If this view is correct, then the outer part of the Extended Scattered disk smoothly joins the inner Oort cloud. In particular, Sedna could be considered the first discovered object in the inner Oort cloud!

4 The Primordial Sculpting of the Kuiper Belt

In Sect. 1, I showed that many properties of the Kuiper belt cannot be explained in the framework of the current Solar System:

i) the existence of the resonant populations,
ii) the excitation of the eccentricities in the classical belt,
iii) the co-existence of a cold and a hot population with different physical properties,
iv) the presence of an outer edge at the location of the 1:2 mean-motion resonance with Neptune,
v) the mass deficit of the Kuiper belt,
vi) the existence of the Extended Scattered disk population (with the exception of 2000 CR_{105} and Sedna, whose orbits can be explained by the formation of the Oort cloud in a dense galactic environment, as discussed above).

These puzzling aspects reveal that the trans-Neptunian population was sculpted when the Solar System was different, because of mechanisms that are no longer at work. Like detectives at the scene of a crime, trying to reconstruct what happened from the available clues, astronomers have tried to reconstruct how the Solar System formed and evolved from the traces left in the structure of the Kuiper belt. A large number of mechanisms have been proposed to explain some of the properties of the Kuiper belt listed above. To save space, here I debate only those which, in my opinion – in light of our current observational knowledge of the Kuiper belt – played a role in in the primordial sculpting of the trans-Neptunian population. I will try to put the various scenarios together, to build a consistent view of the primordial sculpting of the Kuiper belt. For a more exhaustive review, see [131].

4.1 The Origin of the Resonant Populations

It was shown in [45] that, while scattering away the primordial planetesimals from their neighboring regions, the giant planets had to migrate in semi-major axis as a consequence of angular momentum conservation. Given the configuration of the giant planets in our Solar System, this migration should have had a general trend. As discussed when we considered the formation of the Oort cloud, the ice giants have difficulty ejecting planetesimals onto hyperbolic orbits. Apart from the few percent of planetesimals that they can permanently store in the Oort cloud or in the Scattered disk, the remaining planetesimals (the large majority) are eventually scattered inward, toward Saturn and Jupiter. Thus, the ice giants, by reaction, have to move outward. Jupiter, on the other hand, eventually ejects from the Solar System almost all of the planetesimals that it encounters: thus, it has to move inward. The fate of Saturn is more difficult to predict, a priori. However, modern numerical simulations show that this planet also moves outward, although only by a few tenths of an AU for reasonable disk's masses [62, 70].

In [118,119], it was realized that, following Neptune's migration, the mean-motion resonances with Neptune also migrated outward, sweeping through the primordial Kuiper belt until they reached their present positions. From adiabatic theory [79], some of the Kuiper belt objects over which a mean-motion resonance swept were captured into resonance; they subsequently followed the resonance through its migration, with ever increasing eccentricities. This model accounts for the existence of the large number of Kuiper belt objects in the 2:3 mean-motion resonance with Neptune (and also in other resonances) and explains their large eccentricities (see Fig. 27). Neptune had to migrate ~7 AU to reproduce quantitatively the observed range of eccentricities of the resonant bodies. In [119], it was also showed that the bodies captured in the 2:3 resonance can acquire large inclinations, comparable to those of Pluto and other objects. The mechanisms that excite the inclination during the capture process have been investigated in detail in [59], who concluded that, although large inclinations can be achieved, the resulting proportion of high inclination vs. low inclination bodies, as well as their distribution in the eccentricity vs. inclination plane, does not reproduce the observations well. According to [60] (see Sect. 4.2), most high inclination Plutinos were captured from the Scattered disk population during Neptune's migration, rather than from an originally cold Kuiper belt (as in [119]).

The mechanism of adiabatic capture into resonance requires that Neptune's migration happened very smoothly. If Neptune had encountered a significant number of large bodies (Lunar mass or more), its jerky migration would have jeopardized the capture into resonances. For instance, direct simulations

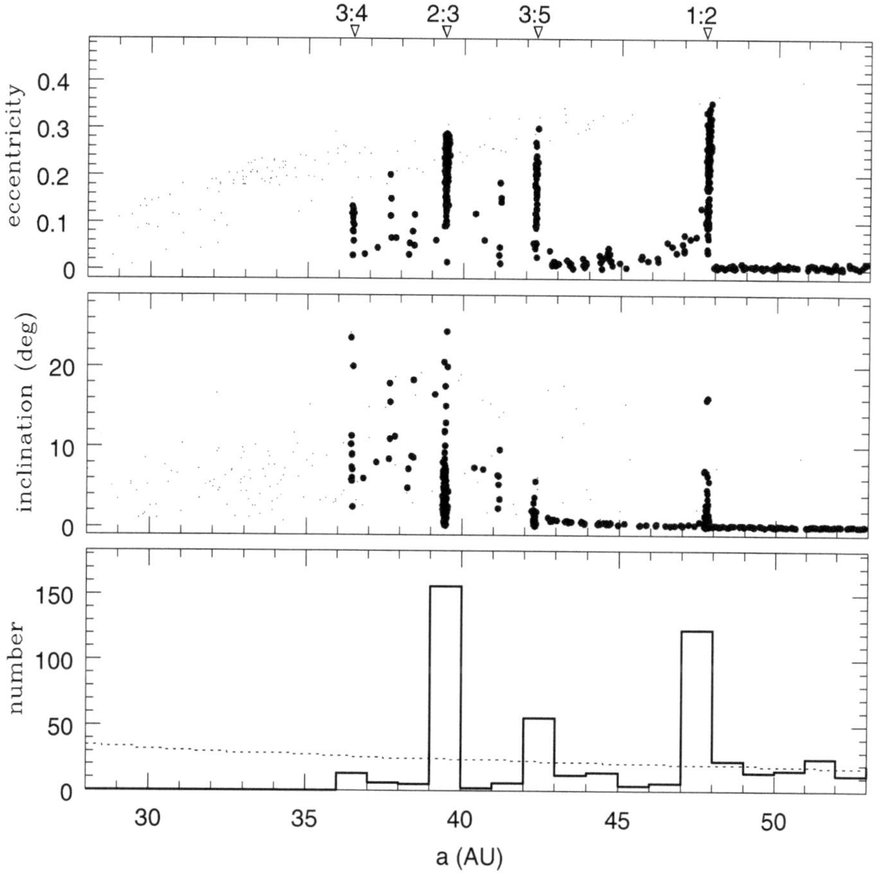

Fig. 27. The final distribution of Kuiper belt bodies according to the sweeping resonances scenario (courtesy of R. Malhotra). This simulation was done by numerically integrating, over a 200 My time-span, the evolution of 800 test particles on initially quasi-circular and coplanar orbits. The planets are forced to migrate by a quantity Δa (equal to -0.2 AU for Jupiter, 0.8 AU for Saturn, 3 AU for Uranus, and 7 AU for Neptune) and approach their current orbits exponentially as $a(t) = a_\infty - \Delta a \exp(-t/4\,My)$, where a_∞ is the current semi-major axis. Large solid dots represent "surviving" particles (i.e., those that have not suffered any planetary close encounters during the integration time); small dots represent the "removed" particles at the time of their close encounter with a planet (e.g., bodies that entered the Scattered disk and whose evolution was not followed further). In the lowest panel, the solid line is the histogram of semi-major axes of the "surviving" particles; the dotted line is the initial distribution. The locations of the main mean-motion resonances are indicated above the top panel

of Neptune's migration in [70] – which modeled the disk with Lunar to Martian-mass planetesimals – did not result in any permanent captures. Adiabatic captures into resonance can be seen in numerical simulations only if the disk is modeled using many more planetesimals with smaller masses [60,62,71]. The constraint set by the capture process on the maximum size of the planetesimals that made up the bulk of the mass in the disk has been recently estimated in [134].

4.2 The Origin of the Hot Population

An appealing mechanism for the origin of the hot population, also in the framework of the planet migration scenario, has been proposed in [60]. Like in [60, 70] simulated Neptune's migration by the interaction with a massive planetesimal disk, extending from beyond Neptune's initial position. But, taking advantage of improved computer technology, 10,000 particles were used to simulate the disk population, with individual masses roughly equal to twice Pluto's mass. For comparison, [70] used only 1,000 particles, with Lunar to Martian masses. Moreover, Neptune was started at ∼15 AU, instead of 23 AU (as in [70]).

In the simulations of [60], during its migration Neptune scattered the planetesimals and formed a massive Scattered disk. Some of the scattered bodies decoupled from the planet, decreasing their eccentricities through interactions with some secular or mean-motion resonance. If Neptune were not migrating, the decoupled phases would have been transient – as often observed in the integrations of [39]. In fact, the dynamics are reversible, so that the eccentricity would have eventually increased back to Neptune-crossing values. However, Neptune's migration broke the reversibility, and some of the decoupled bodies managed to escape from the resonances and remained permanently trapped in the Kuiper belt. As shown in Fig. 28, the current Kuiper belt would therefore be the result of the superposition in (a, e)-space of these bodies with the local population, originally formed beyond 30 AU, which stays dynamically cold because the objects there were only moderately excited (by the resonance sweeping mechanism, as in Fig. 27).

The migration mechanism is sufficiently slow (several 10^7 years) that the scattered particles have enough time to acquire very large inclinations, consistent with the observed hot population. The resulting inclination distribution of the bodies in the classical belt is bimodal, because it results from the superposition of two different populations, each having its own inclination distribution. If the number of objects in the cold population is properly scaled[10], the resulting distribution can quantitatively reproduce the de-biased inclination distribution computed in [16] from the observations.

[10] The cold population is not depleted in [60], while only a fraction of a percent of the Scattered disk remains trapped in the hot population. So the former would outnumber the latter by orders of magnitude unless some other mechanism trimmed it down; see Sect. 4.4.

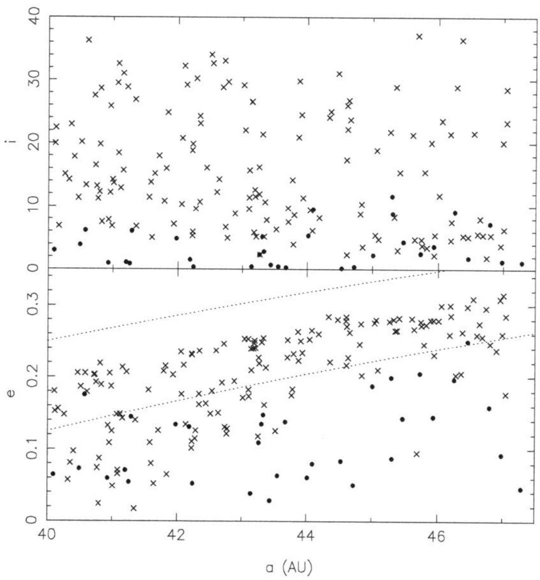

Fig. 28. The orbital distribution in the classical belt according to the simulations in [60]. The dots denote the population that formed locally, which is only moderately dynamically excited. The crosses denote the bodies that were originally inside 30 AU. Therefore, the resulting Kuiper belt population is the superposition of a dynamically cold population and a dynamically hot population, which gives a bimodal inclination distribution comparable to that observed. The dotted curves in the eccentricity vs. semi-major axis plot correspond to $q = 30$ AU and $q = 35$ AU. Courtesy of R. Gomes

Assuming that the bodies' color varied in the primordial disk with heliocentric distance, the scenario proposed in [60] qualitatively explains why the scattered objects and hot classical belt objects – which mostly come from regions inside ~30 AU – appear to have similar color distributions, while the cold classical objects – the only ones that actually formed in the trans-Neptunian region – have a different distribution. Similarly, assuming that the maximum size of the objects was a decreasing function of the heliocentric distance at which they formed, the scenario also explains why the biggest Kuiper belt objects are all in the hot population.

The mechanism found in [60] would also have important implications for two other trans-Neptunian sub-populations: the Plutinos and the Extended Scattered disk. In the simulations, some scattered objects also reached stable Plutino orbits, with orbital properties remarkably similar to those of the observed objects. Because the final (e, i) distribution of the Plutinos captured by resonance sweeping from the cold population is not consistent with observations [59], this suggests that the Plutinos have been predominantly captured from the Scattered disk. The fact that the Plutinos have a color distribution similar to that of the hot population (see [171]), without a predominant red component typical of the cold population, also supports this scenario.

An Extended Scattered disk is also formed in [60] (see also [61,63]), beyond 50 AU. Objects on orbits similar to that of 2004 XR_{190} are produced in [63]. However, orbits similar to that of Sedna are not achieved in these simulations. Some orbits like that of 2000 CR_{105} are obtained in [61], but the resulting population with q ~45 AU is skewed toward small semi-major axes, which – as discussed before – is probably inconsistent with observations. It is probable that the large perihelion distance population with $a > 100$ AU simulated in [61] really exists, but it is very small in number, so that none of these objects has yet been discovered. In this case, 2000 CR_{105} (and Sedna of course) would be representative of a more conspicuous population with $a > 200$ AU, decoupled from the planets by a stellar encounter during the formation of the Oort cloud [15, 129]. Conversely, the observed Extended Scattered disk bodies with a ~50–100 AU most likely achieved their current orbits as shown in [60, 61, 63].

4.3 The Origin of the Outer Edge of the Kuiper Belt

The existence of an outer edge of the Kuiper belt is very intriguing. Several mechanisms for its origin have been proposed, none of which have resulted in a general consensus between the experts in the field. These mechanisms can be grouped in three classes.

Class I: Destroying the Distant Planetesimal Disk

It has been shown, with numerical simulations in [19], that a Martian mass body residing for 1 Gy on an orbit with $a \sim 60$ AU and $e \sim 0.15$–0.2 could have scattered most of the Kuiper belt bodies originally in the 50–70 AU range into Neptune-crossing orbits, leaving this region strongly depleted and dynamically excited. As shown in Fig. 6, the apparent edge at 50 AU might simply be the inner edge of such a gap in the distribution of Kuiper belt bodies. A problem with this scenario is that there are no evident dynamical mechanisms that would ensure the later removal of the massive body from the system. In other words, the massive body should still be present, somewhere in the ~50-70 AU region. A Mars-size body with 4% albedo at 70 AU would have apparent magnitude brighter than 20. In addition its inclination should be small, both in the scenario where it was originally a Scattered disk object whose eccentricity (and inclination) were damped by dynamical friction [19], and in that where the body reached its required heliocentric distance by migrating through the primordially massive Kuiper belt [62]. Thus, in view of its brightness and small inclination, it is unlikely that the putative Mars-size body could escape detection in the numerous wide field ecliptic surveys that have been performed up to now, and in particular in that led by Trujillo and Brown [171].

A second possibility is that the planetesimal disk was truncated by a close stellar encounter. The eccentricities and inclinations of the planetesimals resulting from a stellar encounter depend critically on a/D, where a is their semi-major axis of the planetesimal and D is the heliocentric distance of the stellar encounter [85, 99]. A stellar encounter at ∼200 AU would make most of the bodies beyond 50 AU so eccentric that they intersect the orbit of Neptune, which would eventually produce the observed edge [123]. An interesting constraint on the time at which such an encounter occurred is set by the existence of the Oort cloud. It was shown in [113] that the encounter had to occur much earlier than ∼10 My after the formation of Uranus and Neptune; otherwise most of the existing Oort cloud would have been ejected to interstellar space. Moreover, many of the planetesimals in the Scattered disk at that time would have had their perihelion distance lifted beyond Neptune, decoupling them from the planet. As a consequence, the Extended Scattered disk population, with $a > 50$ AU and $40 < q < 50$ AU, would have had a mass comparable or larger than that of the resulting Oort cloud, hardly compatible with the few detections of Extended Scattered disk objects achieved up to now. As discussed in Sect. 3.2, a close encounter with a star during the first million years of planetary formation is a possible event if the Sun formed in a stellar cluster. However, at such an early time, the Kuiper belt objects were presumably not yet fully formed [92, 153] (unless they accreted very rapidly by gravitational instability). In this case, the edge of the belt would be at a heliocentric distance corresponding to a post-encounter eccentricity excitation of ∼0.05, a threshold value below which collisional damping is efficient and accretion can recover and beyond which the objects rapidly grind down to dust [95].

An edge-forming stellar encounter could not be responsible for the origin of the peculiar orbit of Sedna, unlike the scenario proposed in [96]. In fact, such a close encounter would also produce a relative overabundance of bodies with perihelion distance similar to that of Sedna but with semi-major axes in the 50–200 AU range [129]. These bodies have never been discovered, although they would be favored by observational biases.

Class II: Forming a Bound Planetesimal Disk from an Extended Gas-dust Disk

In [176], it was suggested that the outer edge of the Kuiper belt is the result of two facts: i) accretion takes longer with increasing heliocentric distance and ii) small planetesimals drift inward because of gas drag. This leads to a steepening of the radial surface density gradient of solids. The edge effect is augmented because, at whatever distance large bodies can form, they capture the approximately metre-sized bodies spiraling inward from farther out. The net result of the process, as shown by numerical modeling in [176], is the production of an effective edge, where both the surface density of solid matter and the mean size of planetesimals decrease sharply with distance.

A variant of this scenario has been proposed in [187]. In their model, planetesimals could form by gravitational instability in the regions where the local solid/gas ratio was 2–10 times that corresponding to cosmic abundances. According to the authors, this large ratio could be achieved because of a radial variations of orbital drift speeds of millimeter-sized particles induced by gas drag. However, this mechanism would have worked only within some threshold distance from the Sun, so that the resulting planetesimal disk would have had a natural edge.

A third possibility is that planetesimals formed only within a limited heliocentric distance, because of the effect of turbulence. If turbulence in proto-planetary disks is driven by magneto-rotational instability (MRI), one can expect that it was particularly strong in the vicinity of the Sun and at large distances (where solar and stellar radiation could more easily ionize the gas), while it was weaker in the central, optically thick region of the nebula, known as the "dead zone" [158]. The accretion of planetesimals should have been inhibited by strong turbulence, because the latter enhanced the relative velocities of the grains. Consequently, the planetesimals could have formed only in the dead zone, with well-defined outer (and inner) edge(s).

Class III: Truncating the Original Gas Disk

The detailed observational investigation of star formation regions has revealed the existence of many *proplyds* (anomalously small proto-planetary disks). It is believed that these disks were originally much larger, but in their distant regions, the gas was photo-evaporated by highly energetic radiation emitted by the massive stars of the cluster [1]. Thus, it has been proposed that the outer edge of the Kuiper belt reflects the size of the original Solar System proplyd [81].

A Remark on the Location of the Kuiper Belt Edge

In all the scenarios discussed above, the location of the edge can be adjusted by tuning the relevant parameters of the corresponding model. In all cases, however, Neptune plays no direct role in the edge formation. In this context, it is particularly important to remark (as seen in Fig. 3) that the edge of the Kuiper belt appears to coincide precisely with the location of the 1:2 mean-motion resonance with Neptune. This strongly suggests that, whatever mechanism formed the edge, the planet was able to adjust the final location of the outer boundary through gravitational interactions. I will return to this in Sect. 4.5.

4.4 The Mass Deficit of the Cold Population

The scenario proposed in [60] (see Sect. 4.2) confines the problem of the mass depletion of the Kuiper belt to just the cold population. In fact, in [60] only $\sim$0.2% of the bodies initially in the disk swept by Neptune remained in the Kuiper belt on stable high-i orbits at the end of Neptune's migration. This naturally explains the current low mass of the hot population. However, the population originally in the 40–50 AU range – which would constitute the cold population in the scenario of [60] – should have been only moderately excited and not dynamically depleted, so that it should have preserved most of its primordial mass.

Two general mechanisms have been proposed for the mass depletion: the dynamical ejection of most of the bodies from the Kuiper belt to the Neptune-crossing region, and the collisional comminution of most of the mass of the Kuiper belt into dust.

The dynamical depletion mechanism was proposed in [127] and later revisited in [141]. In this scenario, a planetary embryo, with mass comparable to that of Mars or the Earth, was scattered by Neptune onto a high-eccentricity orbit that crossed the Kuiper belt for $\sim 10^8$ years. The repeated passage of the embryo through the Kuiper belt excited the eccentricities of the Kuiper belt bodies, the vast majority of which became Neptune crossers and were subsequently dynamically eliminated by the planets' scattering action. The integrations in [141], however, treated the Kuiper belt bodies as test particles, and therefore, their encounters with Neptune did not alter the position of the planet. Thus, similar simulations have been re-run in [62], in the framework of a more self-consistent model accounting for planetary migration in response to planetesimal scattering. As expected, the dynamical depletion of the Kuiper belt greatly enhanced Neptune's migration. The reason for this is that, thanks to the dynamical excitation of the distant disk provided by the embryo, Neptune interacted not only with the portion of the disk in its local neighborhood, but with the entire mass of the disk at the same time. As shown in Fig. 29, even a low mass disk of 30 $M_\oplus$ between 10 and 50 AU (just 7.5 $M_\oplus$ in the Kuiper belt) could drive Neptune well beyond 30 AU. Halting Neptune's migration at ~30 AU requires a disk mass of ~15 $M_\oplus$ or less (depending on Neptune's initial location). Such a mass and density profile would imply only 3.75 $M_\oplus$ of material between 40 and 50 AU as the Kuiper belt formed, which is less than the mass required (10–30 $M_\oplus$) by models of the accretion of Kuiper belt bodies [94, 154].

A priori, for the migration of Neptune, there is no evident difference between the case where the Kuiper belt is excited to Neptune-crossing orbits by a planetary embryo or by some other mechanism, such as the primordial secular resonance sweeping proposed in [135]. Therefore, we conclude that Neptune never "saw" the missing mass of the Kuiper belt. The remaining possibility for a dynamical depletion of the Kuiper belt is that the Kuiper belt objects were kicked directly to hyperbolic or Jupiter-crossing orbits and consequently were eliminated without interacting with Neptune. Only the passage of a star through the Kuiper belt seems to be capable of such an extreme excitation [99].

The collisional grinding scenario was proposed in [29, 30, 155] and then pursued in [93, 95, 97]. In essence, a massive Kuiper belt with large eccentricities and inclinations would experience a very intense collisional activity. Consequently, most of the mass originally in bodies smaller than 50–100 km in

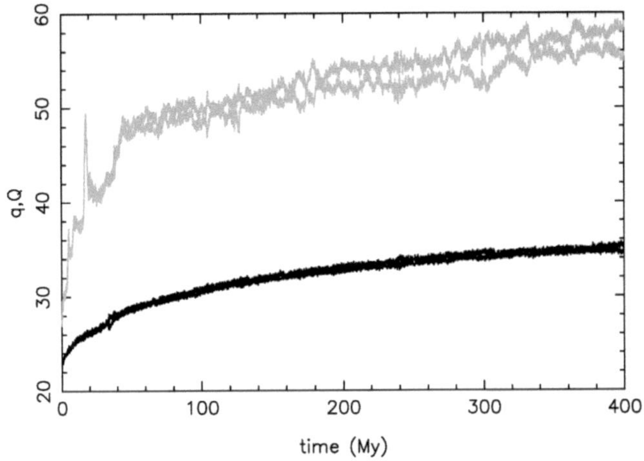

Fig. 29. A self-consistent simulation of the scenario proposed in [141] for the excitation and dynamical depletion of the Kuiper belt (from [62]). Neptune is originally placed at ~23 AU and an Earth-mass embryo at ~27 AU. Both planets are embedded in a 30 $M_\oplus$ disk, extending from 10 to 50 AU with an r^{-1} surface density profile (7.5 $M_\oplus$ between 40 and 50 AU). The black curve shows the evolution of Neptune's semi-major axis (its eccentricity remains negligible), while the gray curves refer to the perihelion and aphelion distances of the embryo. Notice that the embryo is never scattered by Neptune, unlike in [141]. It migrates through the disk faster than Neptune, up to the disk's outer edge. Neptune interacts with the entire mass of the disk, thanks to the dynamical excitation of the disk because of the presence of the embryo. Therefore, it migrates much further than it would if the embryo were not present, reaching a final position well beyond 30 AU (40 AU after 1 Gy)

size could be comminuted into dust and then evacuated by radiation pressure and Poynting–Robertson drag, causing a substantial mass depletion.

To work, the collisional erosion scenario requires that two essential conditions are fulfilled. First, it requires a peculiar primordial size distribution, such that all of the missing mass was contained in small, easy-to-break objects, while the number of large objects was essentially identical to that in the current population. Some models support the existence of such a size distribution at the end of the accretion phase [92, 94]. However, the collisional formation of the Pluto–Charon binary [20], the capture of Triton onto a satellite orbit around Neptune [2], and the discovery of 2003 UB_{313} in the Extended Scattered disk [17] suggest that the number of big bodies was much larger in the past, with as many as 1,000 Pluto-sized objects [151]. In principle, it is possible that all of these large bodies were in the planetesimal disk inside 30 AU, swept by Neptune's migration, while the primordial Kuiper belt contained only the number of large bodies inferred from the current discovery statistics, but this would require that the size distribution in the planetesimal disk had a very sensitive dependence on heliocentric distance.

The second essential condition for substantial collisional grinding is that the massive primordial Kuiper belt had a large eccentricity and inclination excitation, comparable to the current one ($e \sim 0.25$ and/or $i \sim 7°$). However, as reported at the beginning of this section, in light of [60], the mass depletion problem concerns only the cold Kuiper belt, and the dynamical excitation of the cold population is significantly smaller than that required by the collisional grinding models.

Moreover, it must be said that even assuming that the two conditions above are fulfilled, the collisional grinding models still have problems in reducing the total mass of the belt down to the current values of a few percent of an Earth mass. As the mass decreases, the collisional grinding process progressively slows down and eventually effectively stops when the total mass is still about $1\,M_\oplus$. The most advanced of the collisional models [97] can reduce the total mass to few $0.01\,M_\oplus$ only if a very low specific disruption energy Q_* is assumed; if more reasonable values of Q_* (similar to those obtained in hydro-code experiments [9]) are adopted, the final mass achieved by collisional grinding is at least one-tenth of the initial mass, namely about $1\,M_\oplus$ or more.

It is very difficult to reach a firm conclusion on the possibility of collisional grinding of the Kuiper belt from the collisional models alone, because of the sensitivity of these models on the assumed parameters. Perhaps the best strategy is to assume that the collisional grinding was effective, explore its general consequences and compare them with the available constraints. This work is mostly in progress, but I can briefly outline its preliminary results.

First, most of the binaries in the cold population would not have survived the collisional grinding phase [143]. In fact, given that the observed Kuiper belt binaries have large separations, it can be easily computed that the impact of a projectile just 1% the mass of the satellite at $1\,\mathrm{km\,s^{-1}}$ would give the satellite an impulse velocity sufficient to escape to an unbound orbit. If the collisional activity was strong enough to cause an effective reduction of the overall mass of the Kuiper belt, these kind of collisions had to be extremely common, so that we would not expect a significant fraction of widely separated binary objects in the current population.

Second, if the conditions favorable for collisional grinding in the Kuiper belt are assumed for the entire planetesimal disk (5–50 AU), the Oort cloud would not have formed: the planetesimals would have been destroyed before being ejected as in [156] (Charnoz private communication).

Third, as the Kuiper belt mass decreased during the grinding process, the precession frequencies of Neptune and the planetesimals had to change. Consequently, secular resonances had to move, potentially sweeping the belt. Assuming that, when Neptune reached 30 AU, the disk was already depleted inside 35 AU but was still massive in the 35–50 AU region, [62] showed that the ν_8 secular resonance would have started sweeping through the disk as soon as the mass decreased below $10\,M_\oplus$. The ν_8 resonance sweeping would have excited the eccentricity of the bodies to Neptune-crossing values and – given the large mass that the Kuiper belt would have still had when this phenomenon started – Neptune would have continued its radial migration well beyond its current location.

4.5 Pushing out the Kuiper Belt

Given the difficulties of the collisional grinding scenario for the cold Kuiper belt, a dynamical way to solve the mass depletion problem has been proposed in [112]. In this scenario, the primordial edge of the massive proto-planetary disk was somewhere around 30–35 AU and the *entire* Kuiper belt population – not only the hot component as in [60] – formed within this limit and was transported to its current location during Neptune's migration. The transport process for the cold population had to be different from the one found in [60] for the hot population (but still work in parallel with it), because the inclinations of the hot population were excited, while those of the cold population were not.

In the framework of the classical migration scenario [119] [62], the mechanism proposed in [112] was the following: the cold population bodies were initially trapped in the 1:2 resonance with Neptune; then, as they were transported outward by the resonance, they were progressively released because of the non-smoothness of the planetary migration. In the standard adiabatic migration scenario [119], there would be a resulting correlation between the eccentricity and the semi-major axis of the released bodies. However, this correlation was broken by a secular resonance embedded in the 1:2 mean-motion resonance. This secular resonance was generated because the precession rate of Neptune's orbit was modified by the torque exerted by the massive proto-planetary disk that drove the migration.

Simulations of this process can match the observed (a, e) distribution of the cold population fairly well (see Fig. 30), while the initially small inclinations are only very moderately perturbed. In this scenario, the small mass of the current cold population is simply because only a small fraction of the massive disk population was initially trapped in the 1:2 resonance and then released on stable non-resonant orbits. The preservation of the binary objects would not be a problem because these objects were moved out of the massive disk in which they formed by a gentle dynamical process. The final position of Neptune would simply reflect the primitive truncation of the proto-planetary disk, as in [62]. Most important, this model explains why the current edge of the Kuiper belt is at the 1:2 mean-motion resonance with Neptune, although none of the mechanisms proposed for the truncation of the planetesimal disk involves Neptune in a direct way (see Sect. 4.3). The location of the edge was modified by the migration of Neptune by the migration of its 1:2 resonance.

On the flip side, the model in [112] re-opens the problem of the origin of the different physical properties of the cold and hot populations, because both would have originated within 35 AU, although in somewhat different parts of the disk.

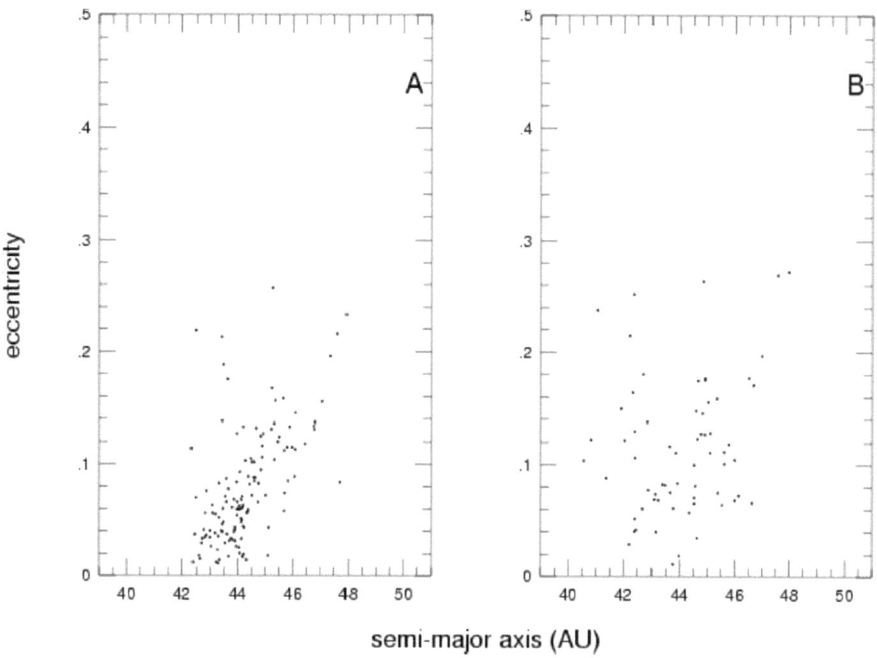

Fig. 30. Left: the observed semi-major axis vs. eccentricity distribution of the cold population. Only bodies with multi-opposition orbits and $i < 4°$ are taken into account. Right: the resulting orbital distribution in the scenario proposed in [112]

I stress, however, that the strength of [112] is in the idea that pushing out the cold Kuiper belt could solve both the problems related to mass deficit and edge location. The specific mechanism for pushing out the cold belt depends on the particular model of giant planet evolution that is adopted. The classical planet migration scenario used in [112] might not reflect the real evolution of the system (see Sect. 5). In this case, alternative push-out mechanisms should be investigated. Whatever the preferred mechanism, it will have to give a predominant role to the 1:2 mean-motion resonance with Neptune to explain the current location of the Kuiper belt edge.

5 Origin of the Late Heavy Bombardment of the Terrestrial Planets

The models proposed in the previous sections for the formation of the Oort cloud and the sculpting of the Kuiper belt seem to offer a quite complete view of the formation and evolution of the Solar System. But they are not entirely satisfactory, because they ignore an important fact in the history of the Solar System: the late heavy bombardment (LHB) of the terrestrial planets.

Below, I review the observational constraints on the LHB, then describe the models proposed in the past to explain a spike in the bombardment rate. Finally, I will focus on an emerging view of what happened ∼650 My after the formation of the planets. In Sect. 6, I will discuss how our understanding of Oort cloud and Kuiper belt formation needs to be modified in light of the LHB evidence and will point to open problems and prospects for future research.

Evidence for a Late Cataclysmic Bombardment

The crust of the Moon crystallized around 4.44 Gy ago, and the morphology of its highlands records a dense concentration of impact craters, excavated before the emplacement – around 3.8 Gy ago – of the first volcanic flows in the mare plains [184]. Thus, a period of intense bombardment – the LHB – occurred in the first 600–700 My of the Moon's history. However, the magnitude and the chronology of the collisions between 4.5 and 4 Gy remain a topic of controversy.

Two explanations have been proposed. According to [74, 184], the frequency of impacts declined slowly and progressively after the end of the accretion period, up to 3.9 Gy ago. In this view, the LBH is not an exceptional event. Rather, it is a 600 My tail of the collisional process that built the terrestrial planets.

Another view advocates a rapid decline in the frequency of impacts after the formation of the Moon, down to a value comparable to the current one. This was followed by a cataclysmic period between ∼4.0 and ∼3.8 Gy ago, marked by an extraordinarily high rate of collisions [27, 147–149, 164].

Today, the majority of authors favor the cataclysmic scenario of the LHB. This theory is supported by a series of arguments:

i) 600 million years of continual impacts should have left an obvious trace on the Moon. So far, no such trace has been found. The isotopic dating of the samples returned by the various Apollo and Luna missions revealed no impact melt-rock older than 3.92 Gy [147, 148]. The lunar meteorites confirm this age limit. The meteorites provide a particularly strong argument because they likely originated from random locations on the Moon [27], unlike the lunar samples collected directly on its surface. A complete resetting of all ages all over the Moon is possible [67] but highly unlikely, considering the difficulties of completely resetting isotopic ages at the scale of a full planet [31]. The U-PB and Rb-Sr isochrones of lunar highland samples indicate a single metamorphic event at 3.9 Gy ago, and between 3.85 and 4 Gy, ago respectively [164]. There is no evidence for these isotopic systems being reset by intense collisions between 4.4 and 3.9 Gy.

ii) The old upper crustal lithologies of the Moon do not show the expected enrichment in siderophile elements (in particular the Platinum Group Elements) implied by an extended period of intense collisions [148].

iii) If the elevated mass accretion documented in the period around 3.9 Gy is considered to be the tail end of an extended period of collisions, the whole Moon should have accreted at about 4.1 Gy ago instead of 4.5 Gy [100, 149].

iv) The 15 largest impact structures on the Moon, the so-called basins, with diameters between 300 and 1200 km, have been dated to have formed between 4.0 and 3.9 Gy ago. If the bombardment had declined monotonically since 4.5 Gy ago, it appears strange that the largest impacts all occurred at the end of the period.
v) On Earth, the oxygen isotopic signature of the oldest known zircons (age: 4.4 Gy) indicates formation temperatures compatible with the existence of liquid water [173]. This argument seems contradictory with an extended period of intense collisions, which would have raised the Earth's temperature to exceed the water evaporation threshold.
vi) These same zircons retain secondary overgrowths developed after primary core crystallization during their 4.4 Gy long crystal residence times. The rim overgrowths can record discrete thermal events subsequent to zircon formation and provide a unique window in crustal processes before the beginning of the terrestrial rock record. In [167], all these rim overgrowths have been dated to be $\sim$3.9 Gy old. No (preserved) older rim overgrowths, associated to more primoridal events, have been found. This suggests that the thermal events were associated to impacts and that these impacts were concentrated in time about 3.9 Gy ago.

Therefore, it can be concluded that there is strong evidence for a cataclysmic Late Heavy Bombardment event around 3.9 Gy ago. This cataclysm did not just affect the Moon, but has now been clearly established throughout the inner Solar System [102]. The exact duration of the cataclysm is difficult to estimate, however. On the basis of the cratering record of the Moon, it lasted between 20 and 200 My, depending on the mass flux estimate used in the calculation.

Early Models of LHB Origin

The occurrence of a cataclysmic LHB challenges our naive view of a Solar System gradually evolving from chaos to order. Several ideas have been proposed to examine what could have abruptly changed the evolution of the system, causing a spike in the bombardment rate.

The possibility of a stochastic break-up of an asteroid close to a resonance in the main belt has been investigated in [188]. The flux of projectiles inferred from the crater density would require the break-up of an object larger than Ceres. This event is very implausible and would have left a huge asteroid family in the main belt, of which we see no trace.

If a stochastic break-up is ruled out, then the remaining possibility is that a reservoir of small bodies, which remained stable up to the time of the LHB, suddenly became unstable, with most of its objects achieving planet crossing orbits.

A comet shower from the Oort cloud, possibly triggered by a stellar encounter, is a first possibility. However, a new LPC has a probability to collide with the Earth of about 10^{-9}. Because the mass hitting the Earth during the LHB is estimated to be $\sim 10^{-5}\,M_\oplus$ [75], this would require an Oort cloud initially containing 10^4 Earth masses, which – as discussed in Sect. 3 – is impossible.

In [21], it was proposed that a fifth terrestrial planet, with a mass comparable to that of Mars, became unstable after ∼600 My of evolution, and crossed the asteroid belt before being dynamically removed. Invaded by this new perturber, the asteroid belt became unstable and most of its objects acquired planet crossing eccentricities. The simulations presented in [21] show that a late instability of a 5-planet terrestrial system is indeed possible, but it requires that the rogue planet was initially at about 1.9 AU, with an inclination of ∼15°. Whether this initial configuration is consistent with terrestrial planet formation models was not discussed. Similarly, the resulting orbital distribution in the asteroid belt, after the removal of the rogue planet, was not investigated. Moreover, in most simulations, the rogue planet was removed by a collision with Mars, and the red planet does not show any sign of such a gigantic strike.

In [114], it was proposed that the LHB was associated with the "late appearance" of Uranus and Neptune in the planetesimal disk. That paper showed that the planetesimals scattered away from the neighborhoods of the ice giants would have been sufficient to cause a bombardment on the Moon with a magnitude comparable to that of the LHB. Moreover, the dynamical removal of these planetesimals would have caused a radial migration of Jupiter and Saturn, which in turn would have forced the ν_6 secular resonance to sweep across the main asteroid belt [58]. Their eccentricities being excited by the resonance passage, most asteroids would have acquired planet-crossing orbits. Consequently, they would have contributed to – or even dominated – the terrestrial planets cratering process. The problem in this work was that the "late appearance" of Uranus and Neptune was postulated, rather than explained. The authors argued that these planets might have formed very slowly, although this seems implausible given that they accreted hydrogen atmospheres of 1–2 Earth masses from the proto-solar nebula [66], which should have dissipated within ∼10 My [72]. Later, in [110], it was proposed that Uranus and Neptune formed in between Jupiter and Saturn. The system remained stable for 600 My, until an instability was produced by the gradual evolution of the planetary orbits. Consequently, Uranus and Neptune were scattered outward by Jupiter and Saturn. After this, interactions with the disk eventually damped their eccentricities and parked them on stable orbits. As a by-product of this process, the planetesimal disk was destroyed as in [114]. The simulations in [110] showed that a late instability of a Jupiter–Uranus–Neptune–Saturn system is indeed possible. However, the instability time depends critically on the initial conditions, and it is unclear if those adopted in the successful simulations could be consistent with giant planet formation models. More importantly, the scattering of Uranus and Neptune by Jupiter and Saturn would have destabilized the regular satellite systems of all the planets. Finally, the massive planetesimal disk required to stabilize the orbits of Uranus and Neptune would have forced the latter to migrate well beyond its current position. Thus, as admitted by the authors themselves, this scenario has to be considered as a "fairy tale."

The Great Comet-Asteroid Alliance: An Emerging View of the Origin of the LHB

Starting from two key considerations:

i) giant planet migration through the planetesimal disk induces a bombardment of the terrestrial planets of sufficient magnitude to explain the LHB (from [114]),
ii) at the end of the migration phase, the Solar System is essentially identical to the current one (namely there are no more reservoirs of planetesimals to destabilize),

it was realized in [130] that solving the problem of the LHB origin required a plausible mechanism to be found that would trigger planet migration at a late time.

Pursuing this goal, in [64] the authors remarked that, in all previous simulations, planet migration started immediately because planetesimals were placed close enough to the planets to be violently unstable. While this type of initial condition was reasonable for the goals of those works, it is unlikely to have been the case in reality. In fact, planetesimal-driven migration is probably not important for planetary dynamics as long as the gaseous massive nebula exists (the nebula accounts for about 100 times more mass than the planetesimals). The initial conditions in simulations of the planetesimal-driven migration should therefore represent the system that existed at the time the nebula dissipated. Thus, the planetesimal disk should contain only those particles that had dynamical lifetimes longer than the lifetime of the solar nebula (a few million years), because the planetesimals initially on orbits with shorter dynamical lifetimes should have been eliminated earlier, during the nebula era. If this constraint on the initial conditions is fulfilled, then the resulting migration is necessarily slow, because it depends on the rate at which disk particles evolve onto planet-crossing orbits, which is long by definition. If the planetary system, in the absence of planetesimals, is stable, this slow migration can continue for a long time, slightly accelerating or damping depending on the disk's surface density [62]. Conversely, if the planet system is – or becomes – unstable, then the planets tend to increase their orbital separation. The outermost planet penetrates into the disk, and this starts a fast migration, similar to that obtained in previous simulations, where the planets are embedded in the disk from the very beginning. Thus, the problem of triggering the LHB is reduced to the problem of understanding how the giant planets, during their slow migration, could pass from a stable configuration to an unstable one.

A solution to this problem has been proposed in [172]. This work postulated that, at the time of the dissipation of the gas disk, the four giant planets were in a compact configuration, with quasi-circular, quasi-coplanar orbits with radii ranging from 5.5 to 13–17 AU. Saturn and Jupiter were close enough to have a ratio of orbital periods less than 2. This choice of the initial conditions for the two giant planets is supported by simulations of their evolution during the gas-disk phase [122] [132]. The assumption of initial small eccentricities and inclinations is consistent with planet formation models. The small eccentricities ensure the stability of such a compact planet configuration. In the scenario of [172], during their migration in divergent directions, Jupiter and Saturn eventually crossed their mutual 1:2 mean-motion resonance. This resonance crossing excited their eccentricities to values comparable to those currently observed (for eccentricity excitation because of resonance crossing, see also [24]). The acquired eccentricities of Jupiter and Saturn destabilized the planetary system as a whole. The planetary orbits became chaotic and started to approach each other. Thus, a short phase of encounters followed the resonance-crossing event. Consequently, both ice giants were scattered outward, deep into the disk. As discussed above, this abruptly increased the migration rates of the planets. During this fast migration phase, the eccentricities and inclinations of the planets decreased as a result of the dynamical friction exerted by the planetesimals and the planetary system was finally stabilized.

With a planetesimal disk of about 35 $M_\oplus$, the simulations in [172] reproduced the current architecture of the orbits of the giant planets remarkably well, in terms of semi-major axes, eccentricities, and inclinations. In particular, this happened in the simulations where at least one of the ice giants encountered Saturn (see Fig. 31). Conversely, in the simulations where encounters with Saturn never occurred, Uranus typically ended its evolution on an orbit too close to the Sun, and the final eccentricities and inclinations of all the planets involved were too small.

With this result, [64] could put all the elements together in a coherent scenario for the LHB origin. Assuming an initial planetary system like that described in [172], the planetesimal disk fulfilled the lifetime constraint discussed above only if its inner edge was located about 1 AU beyond the position of the last planet. With this kind of disk, the 1:2 resonance crossing event that destabilized the planetary system occurred at a time ranging from 192 to 875 My (see Fig. 32). Modifying the planetary orbits also led to changes in the resonance-crossing time, pushing it up to 1.1 Gy after the beginning of the simulation. This range of instability times well brackets the estimated date of the LHB from lunar data.

The top panel of Fig. 33 shows the giant planets' evolution in a representative simulation of [64]. Initially, the giant planets migrated slowly because of the leakage of particles from the disk. This phase lasted 875 My, at which point Jupiter and Saturn crossed their 1:2 resonance. At the resonance crossing event, as in [172], the orbits of the ice giants became unstable, and they were scattered into the disk by Saturn. They disrupted the disk and scattered objects all over the Solar System, including the inner regions. Eventually,

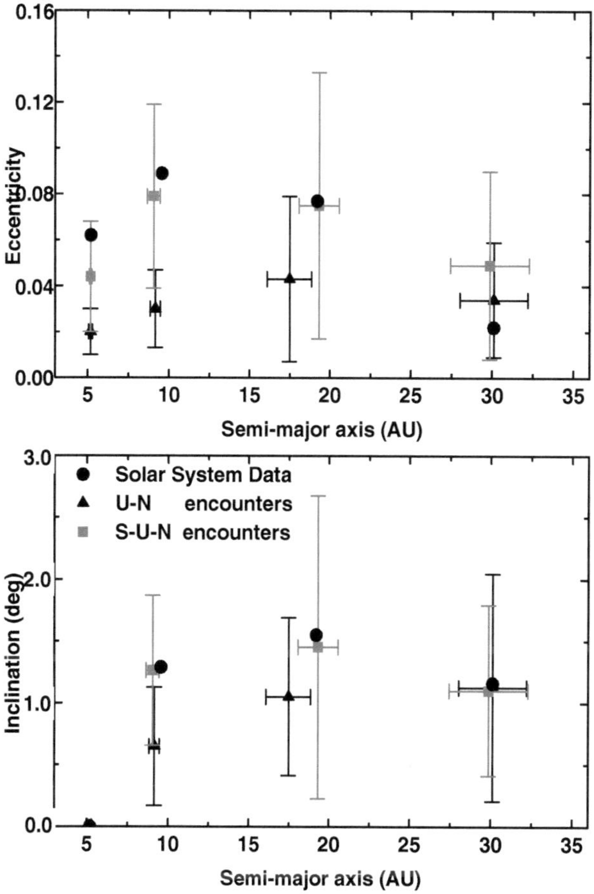

Fig. 31. Comparison of the synthetic final planetary systems obtained in [172] with the real outer Solar System. Top: Proper eccentricity vs. semi-major axis. Bottom: Proper inclination vs. semi-major axis. Here, proper eccentricities and inclinations are defined as the maximum values acquired over a 2 My time-span and were computed from numerical integrations. The inclinations are measured relative to Jupiter's orbital plane. The values for the real planets are presented as filled black dots. The gray squares mark the mean of the proper values for the runs with no planetary encounters involving Saturn, while the black triangles mark the same quantities for the runs where at least one ice giant encountered the ringed planet (about 15 runs in each case). The error bars represent one standard deviation of the measurements. From [172]

they stabilized on orbits very similar to the current ones, at ~20 and ~30 AU respectively. The solid curve in the bottom panel shows the amount of material that struck the Moon as a function of time. As predicted in [114], the amount of material hitting the Moon after resonance crossing is consistent

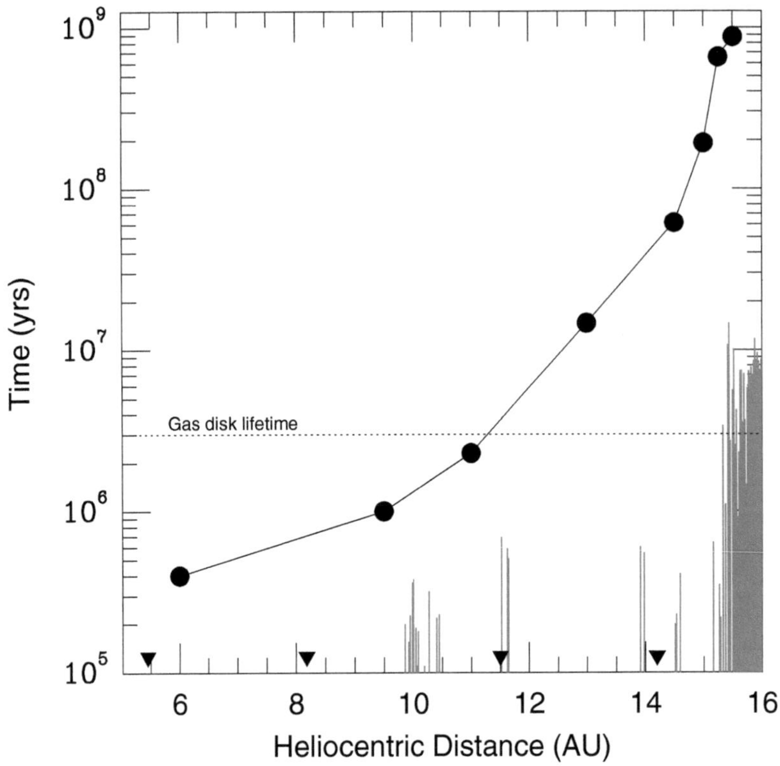

Fig. 32. Disk location and LHB timing. The histogram reports the average dynamical lifetime of massless test particles placed in a planetary system with Jupiter, Saturn, and the ice giants on nearly circular, coplanar orbits at 5.45, 8.18, 11.5, and 14.2 AU, respectively (marked as black triangles on the plot). The dynamical lifetime was computed by placing 10 particles with $e = i = 0$ and random mean anomaly in each semi-major axis bin. Each vertical bar in the plot represents the average lifetime for those 10 particles, after having removed stable Trojan cases. The "lifetime" is defined as the time required for a particle to encounter a planet within one Hill radius. A comparison between the histogram and the putative lifetime of the gaseous nebula [72] suggests that, when the nebula dissipated, the inner edge of the planetesimal disk had to be about 1–1.5 AU beyond the outermost ice giant. The time at which Jupiter and Saturn crossed their 1:2 mean-motion resonance, as a function of the location of the planetesimal disk's inner edge, is shown with filled dots. From [64]

with the mass (6×10^{21} g) estimated from the number and size distribution of lunar basins that formed around the time of the LHB epoch [75].

As discussed in [114], however, the planetesimals from the distant disk – which can be identified as "comets" – were not the only ones to hit the terrestrial planets. The radial migration of Jupiter and Saturn forced the secular resonances ν_6 and ν_{16} to sweep across the asteroid belt [58], exciting the

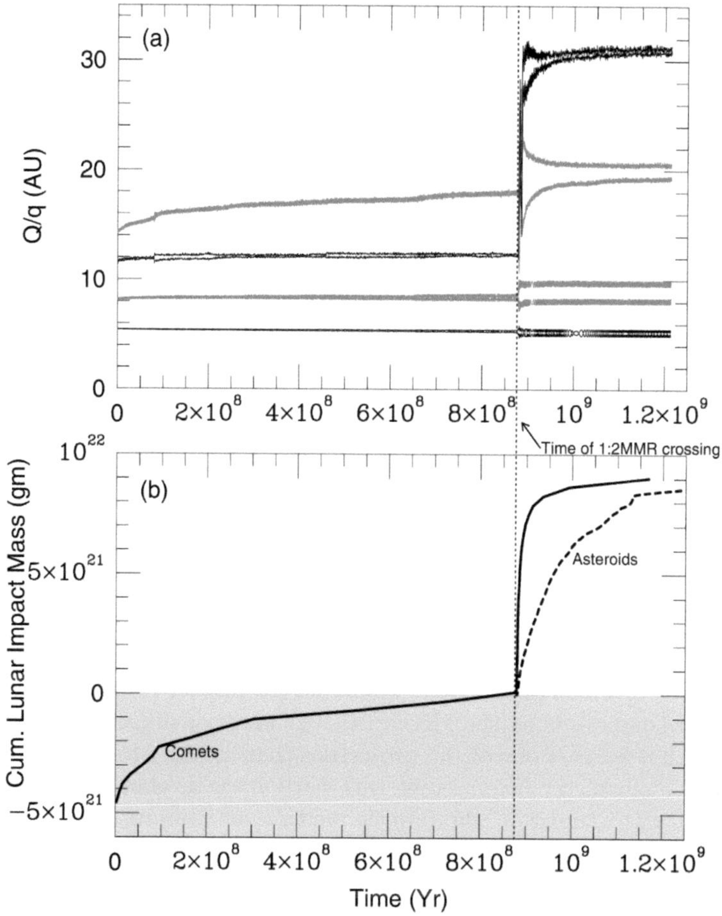

Fig. 33. Planetary migration and the associated mass flux toward the inner Solar System from a representative simulation of [64]. Top: the evolution of the four giant planets. Each planet is represented by a pair of curves – the top and bottom curves are the aphelion and perihelion distances, respectively. Jupiter and Saturn cross their 1:2 mean-motion resonance at 880 My. Bottom: the cumulative mass of comets (solid curve) and asteroids (dashed curve) accreted by the Moon. The comet curve is offset, so that the value is zero at the time of 1:2 resonance crossing. The estimate of the total asteroidal contribution is very uncertain but should be roughly of the same order of magnitude as the cometary contribution. However, it should occur over a longer time-span. From [64]

eccentricities and the inclinations of asteroids. The fraction of the main belt population that acquired planet-crossing eccentricities depends quite crucially on the orbital distribution that the belt had before the LHB, which is not well known. The asteroid belt could not be a massive, dynamically cold disk at the time of the LHB. If it were, essentially all of the asteroids would have been ejected onto planet-crossing orbits, the bombardment of the Moon would have been orders of magnitude more intense than that recorded by the LHB [114], and the few asteroids surviving in the belt after the secular resonance sweeping would have an orbital distribution inconsistent with that currently observed [58]. Presumably, the asteroid belt underwent a first phase of dynamical depletion and excitation at the time of terrestrial planet formation [142, 181] and then a second dynamical depletion at the time of the LHB. If, at the end of the first phase, the orbital distribution in the belt was comparable to the current one, then the secular resonance sweeping at the time of the LHB would have left ~10% of the objects in the asteroid belt [64]. Assuming this figure, the pre-LHB main belt contained roughly $5 \times 10^{-3}\,\mathrm{M}_{\oplus}$ (10 times its current mass) and the total mass of the asteroids hitting the Moon was comparable to that of the comets (see Fig. 33). However, slight changes in the pre-LHB asteroid distribution, and the migration rate of Jupiter and Saturn (also highly variable from simulation to simulation, depending on the chaotic evolution of Neptune), can change this result for the asteroidal contribution to the Lunar cratering rate by a factor of several. In conclusion, the model in [64] cannot state whether asteroids or comets dominated the impact flux on the terrestrial planets. What it can say, however, is that the asteroidal contribution came later and more slowly than the cometary contribution (see Fig. 33), possibly erasing much of the signature of the cometary bombardment.

The issue of which population dominated the impact rate can be solved by looking for constraints on the Moon. In [102], analysis of Lunar impact melts indicated that at least one of the projectiles that hit the Moon, and probably more, had a chemistry inconsistent with carbonaceous chondrites or comets. In [162], it was found that the impact melt at the landing site of Apollo 17 was caused by a projectile of LL-chondritic composition. These results imply that the bombardment was dominated by asteroids typical of the inner belt.

In [159], the comparison of size distributions of the craters formed at the time of the LHB on Mercury, Mars, and the Moon allowed the calculation of the ratios of the impact velocities on these planets, leading to the conclusion that most projectiles had a semi-major axis between 1 and 2 AU. Comets never acquire such a small semi-major axis during their evolution, so this argument again favors a dominant contribution from the inner main belt. More recently, [160] found that the crater size distribution on the lunar highlands is consistent with the size distribution of objects currently observed in the main belt.

Taken altogether, these results point with little doubt to asteroids being the dominating (or, possibly, latest-arriving) projectile population for the terrestrial planets at the time of the LHB. However, they do not imply that the asteroids *triggered* the LHB. On the contrary, the result in [160] implies that the LHB was triggered by a distant disk of comets (as in [64]), for the reasons explained below.

The remarkable match between the size distributions of craters and the main belt asteroids, pointed out in [160], implies that – at the LHB time – asteroids were ejected from the main belt onto planet-crossing orbits in proportions independent of their size .[11] Only the sweeping of secular resonances can give a size-independent ejection throughout the main belt. At the time of the LHB, the gas disk was already totally dissipated. Thus, secular resonance sweeping could only be caused by the radial displacement of Jupiter and Saturn. Now, even assuming that the entire LHB on the terrestrial planets was caused by asteroids, from the mass hitting the Moon at that time [75] and the collision probability typical of NEAs with the Moon, one can easily compute that the total asteroid mass on planet-crossing orbits was about $0.01\,M_{\oplus}$. This mass was too small to cause a significant migration of the giant planets. In conclusion, a more massive disk – which could only be trans-Neptunian – had to trigger and drive planet migration. Comets mandated the bombardment, and asteroids executed it.

A Note on the Trojans and the Satellites of the Giant Planets

To validate or reject a model, it is important to look at the largest possible number of constraints. Two populations immediately come to mind when considering the LHB scenario proposed in [64]: the Trojans and the satellites of the giant planets. Is their existence consistent with this scenario?

Jupiter has a conspicuous population of Trojan objects. These bodies, usually referred to as "asteroids," follow essentially the same orbit as Jupiter, but lead or trail that planet by an angular distance of $\sim 60°$, librating around the Lagrange triangular equilibrium points. The first Trojan of Neptune was recently discovered [23]; and detection statistics imply that the Neptune Trojan population could be comparable in number to that of Jupiter, and possibly even ten times larger [25].

The simulations in [64, 172] led to the capture of several particles on long-lived Neptunian Trojan orbits (2 per run, on average, with a lifetime larger than 80 My). Their eccentricities, during their evolution as Trojans, reached values smaller than 0.1. These particles were eventually removed from the Trojan region, but this is probably an artifact of the graininess of Neptune's migration in the simulation, because of the quite large individual mass of the planetesimals [71].

Jovian Trojans are a more subtle issue that is described in detail in [133]. There is a serious argument in the literature against the idea that Jupiter and Saturn crossed their 1:2 mean-motion resonance: if the crossing had happened, any pre-existing Jovian Trojans would have become violently unstable, and Jupiter's co-orbital region would have emptied [58, 124]. However, the dynamical evolution of a gravitating system of objects is time reversible. Thus, if the original objects can escape the Trojan region when it becomes unstable, other bodies can enter the same region and be temporarily trapped. Consequently,

[11] unlike the current Near Earth Asteroids (NEAs) which, escaping from the belt because of size-dependent non-gravitational forces, have a size distribution significantly steeper than that of the main belt population.

a transient Trojan population can be created if there is an external source of objects. In the framework of the scenario in [64], the source consists of the very bodies that are forcing the planets to migrate, which must be a large population given how far the planets must migrate. When Jupiter and Saturn move far enough from the 1:2 resonance that the co-orbital region becomes stable, the population that happens to be there at that time remains trapped. It then becomes the population of permanent Jovian Trojans still observable today.

This possibility has been tested with numerical simulations in [133]. Of the particles that were Jupiter or Saturn crossers during the critical period of Trojan instability, a fraction between 2.4×10^{-6} and 1.8×10^{-5} remained permanently trapped as Jovian Trojans. More importantly, at the end of the simulations, the distribution of the trapped Trojans in the space of the three fundamental quantities for Trojan dynamics – the *proper* eccentricity, inclination, and libration amplitude [125] – was remarkably similar to the current distribution of the observed Trojans, as illustrated in Fig. 34. In particular, this is the only model proposed so far that explains the inclination distribution of the Jovian Trojans. The origin of this distribution was considered to be the hardest problem in the framework of the classical scenario, according to which the Trojans formed locally and were captured at the time of Jupiter's growth [121].

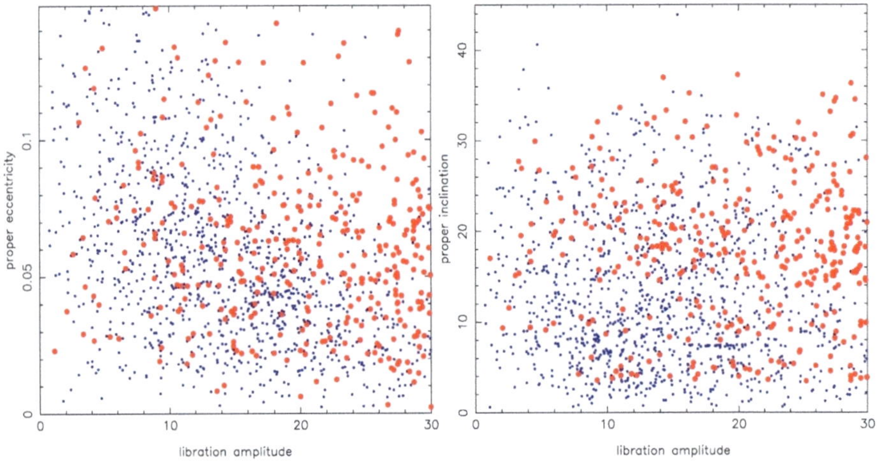

Fig. 34. Comparison of the orbital distribution of Trojans between the simulations in [133] and observations. The simulation results are shown as red circles and the observations as blue dots in the planes of proper eccentricity vs. libration amplitude (left) and proper inclination vs. libration amplitude (right). The distribution of the simulated Trojans is somewhat skewed toward large libration amplitudes, relative to the observed population. However, this is not a serious problem because a fraction of the planetesimals with the largest amplitudes would leave the Trojan region during the subsequent 4 Gy of evolution [106], leading to a better match. The similarity between the two inclination distributions provides strong support for the LHB model in [64]

The capture probabilities reported above allowed [133] to conclude that the total mass of the captured Trojan population was between $\sim 4 \times 10^{-6}$ and $\sim 3 \times 10^{-5}\,M_{\oplus}$. Previous estimates from detection statistics [88] concluded that the current mass of the Trojan population is $\sim 10^{-4}\,M_{\oplus}$. However, taking into account modern, more refined knowledge of the Trojans absolute magnitude distribution (discussed in [133]), mean albedo [52] and density [120], the estimate of the current mass of the Trojan population is reduced to $7 \times 10^{-6}\,M_{\oplus}$, which is consistent with the simulations in [133]. The bulk density of $0.8^{+0.2}_{-0.1}\,\mathrm{g\,cm^{-3}}$, measured for the binary Trojan 617 Patroclus [120] is an independent confirmation of the model of chaotic capture of Trojans from the original trans-Neptunian disk. In fact, this density is significantly smaller than any density measured so far in the asteroid belt, including for the most primitive objects, while it is essentially identical to the bulk densities inferred for the trans-Neptunian objects Varuna [89] and 1997 CQ_{29} [138].

In conclusion, the properties of Jovian Trojans are not simply consistent with the LHB model of [64]: they constitute a strong indication – if not a smoking gun – in support of the 1:2 mean-motion resonance crossing of Jupiter and Saturn, which is at the core of the model in [64].

I now briefly come to the satellites of the giant planets. As discussed above, the non-survivability of the regular satellite systems is one of the killing arguments against the exotic scenario proposed in [110]. Because Saturn, Uranus, and Neptune also have encounters with each other in the model of [64,172], it is important to look at the satellites' fates in this new framework. This issue has been addressed in [172]. The authors recorded all encounters deeper than one Hill-radius occurring in eight simulations. Then, they integrated the evolution of the regular satellite systems of Saturn, Uranus, and Neptune during a re-enactment of these encounters. They found that, in half of the simulations, all of the satellite systems survived the entire suite of encounters with final eccentricities and inclinations smaller than 0.05. The difference in comparison to the case of [110] is that, in the latter model, both ice giants had to have close and strong encounters with Jupiter or Saturn, whereas in the simulations of [64, 172], encounters with Jupiter never occur, and encounters with Saturn are typically distant, with moderate effects. Thus, the survivability of the regular satellites is not a problem for the LHB model. However, the more distant, irregular satellites would not survive the planetary encounters. Thus, if the LHB model is correct, they must have been captured at the time of the LHB (see Sect. 6).

6 Building a Coherent View of Solar System History: Perspectives for Future Work

From the emerging view of the events that led to the origin of the LHB, it appears that the evolution of the Solar System was characterized by three main phases:

(i) *the planetary accretion phase.* The giant planets formed within a few million years, in a compact orbital configuration embedded in a gas disk. The terrestrial planets presumably formed on a timescale of several 10^7 y [3,150,185]. Planetesimals formed out to a threshold distance of ∼30–35 AU. The asteroid belt underwent a first dynamical depletion and excitation during this phase [142], while planetesimals in the vicinity of the giant planets were removed, leaving a massive planetesimal disk that was only present beyond the orbit of the outermost giant planet [64].
(ii) *a long quiescent phase.* lasting around 600 My, during which the distant planetesimal disk was gradually eroded at its inner edge by planetary perturbations, leading to a slow migration of the giant planet orbits [64].
(iii) *the current phase.* which has lasted since 3.8 Gy ago, during which the Solar System has maintained essentially the same structure [65].

The LHB marks the cataclysmic transition between phase (ii) and phase (iii).

From this template of the history of the Solar System, I will attempt to put the various scenarios discussed in the previous sections for the origin of the cometary reservoirs into a new context and to suggest new directions for future research.

The Oort cloud should have formed in two stages. The first stage occurred as soon as (or even during the time that) the giant planets formed. This occurred very early, when the system was still rich in gas, and presumably the Solar System was still embedded in a stellar cluster. Appropriate simulations should thus account for a dense galactic environment, close and frequent stellar encounters (as in [15, 49]), but accounting also for gas drag. The decoupling of Sedna and 2000 CR_{105} from the scattering action of the planets should have occurred in this phase. The second stage occurred at the time of the LHB, when the original outer planetesimal disk was destroyed and a massive Scattered disk was formed. The classical simulations discussed in Sect. 3 are pertinent for this second phase. The inferred ratio between the number of comets currently in the Oort cloud and in the Scattered disk (see Sect. 3.1) argues that the first stage was more effective than the second.

The Kuiper belt took shape at the time of the LHB. As the outer planetesimal disk was destroyed by the eccentric and migrating ice giants, a fraction of a percent of the planetesimals managed to be pushed outward and were implanted in a region of orbital space that became stable when the planets finally settled onto their current orbits. Thus, the principle of the push-out scenario for the Kuiper belt should remain valid, although the simulations discussed in Sects. 4.2 and 4.5 are not really pertinent. In fact, these simulations assumed a smooth, long-range migration of Neptune, which is not what the LHB simulations in [64] show. Simulations in progress seem to indicate that the mechanism proposed in [60] for the origin of the hot population still applies (Gomes, private communication). For the cold population, the mechanism proposed in [112] has to be replaced by a new one. It turns out that, during the short phase when Neptune's orbit is eccentric, the Kuiper belt is totally unstable up to the 1:2 mean-motion resonance with the planet. It can therefore be visited by planetesimals coming from inside the outer edge of the disk. This builds a sort of steady-state population in the Kuiper belt region, which remains permanently trapped when Neptune's eccentricity is damped by dynamical friction, and the Kuiper belt becomes stable again [117]. This process would therefore be analogous to that leading to the capture of Jovian Trojans. If the damping of Neptune's eccentricity occurred sufficiently fast, as in the LHB model of [64], the planetesimals that remained trapped in the Kuiper belt by this mechanism would not have had enough time to develop large inclinations, and therefore, the population trapped by this process would match the cold Kuiper belt. In this scenario, the current size distribution of the Kuiper belt should be a fossil record of that acquired during the ~600 My time-span that the objects spent in the massive planetesimal disk, before being pushed out [139].

The irregular satellites of (at least) Saturn, Uranus, and Neptune, if they existed before the LHB, would have been lost during the phase of encounters among the planets. Thus, those currently observed had to be captured later, from the flux of planetesimals coming from the distant disk. At this late stage, the capture process could not be related to gas drag, nor to a fast growth of the planetary masses (the so-called pull-down scenario); it is instead probably related to three-body interactions (i.e., interactions between planetesimals within the Hill's sphere of a planet), although the exact mechanism has not been demonstrated yet (see however [2] for a description of such a mechanism for the capture of Triton). This view is consistent with that proposed in [90], from the comparative analysis of the size distributions of the irregular satellite populations of the four giant planets. Moreover, in this scenario, the irregular satellites should have the same composition as Kuiper belt objects, given that both populations were extracted from the same primordial planetesimal disk. The recent data collected on the satellite Phoebe by the Cassini mission suggest a composition lying in this direction [26, 91].

The new LHB scenario also has important implications for aspects of Solar System formation and primordial evolution not discussed in this chapter. The formation of the terrestrial planets should be revisited, accounting for giant planets on more compact, circular orbits, as required in [64]. Similarly, the evolution of the asteroid belt should also be re-assessed. As mentioned before, the belt should have suffered two phases of dynamical depletion and excitation: The first one during the formation of the terrestrial planets and the second one during the LHB. Therefore, during the 600 My period between the end of terrestrial planet accretion and the LHB, the asteroid belt should have remained about 10–20 times more massive than the current belt, in a dynamically excited state. The collisional evolution during this period should have been very important, and the current size distribution in the main belt should be a fossil of the one that was developed during this phase. A study similar to [14] should be done, but taking into account this two-stage evolution of the belt.

Finally, the LHB scenario constrains the orbital architecture of the giant planets at the end of the gas disk phase. Future simulations of the formation of these planets and of their interactions with the nebula will have to meet these constraints. In particular, the compact configuration of the planetary orbits and the presence of a massive disk of planetesimals outside the orbit of the outermost planet constrain the maximum range of radial migration that the giant planets could suffer during the gas phase. For instance, if Jupiter had formed, say, at 30 AU and migrated down to 5 AU during the gas-disk lifetime, the outer planetesimal disk required to trigger the LHB would have been destroyed. Most probably, the cores of all giant planets formed within 10–15 AU from the Sun [166] and, for some reason not yet totally clear, never migrated substantially.

References

1. Adams, F. C., Hollenbach, D., Laughlin, G., Gorti, U. 2004. Photoevaporation of Circumstellar Disks Due to External Far-Ultraviolet Radiation in Stellar Aggregates. Astroph. J. **611**, 360–379.
2. Agnor, C. B., Hamilton, D. P. 2006. Neptune's capture of its moon Triton in a binary-planet gravitational encounter. Nature **441**, 192–194.
3. Allegre, C. J., Manhes, G., Gopel, C. 1995. The Age of the Earth. Geochimica et Cosmochimica Acta. **59**, 1445–1456.
4. Allen, R. L., Bernstein, G. M., Malhotra, R. 2001. The Edge of the Solar System. Astroph. J. **549**, L241–L244.
5. Allen, R. L., Bernstein, G. M., Malhotra, R. 2002. Observational Limits on a Distant Cold Kuiper Belt. Astron. J. **124**, 2949–2954.
6. Astakhov, S. A., Lee, E. A., Farrelly, D. 2005. Formation of Kuiper-Belt Binaries Through Multiple Chaotic Scattering Encounters with Low-mass Intruders. Monthly Notices of the Royal Astronomical Society **360**, 401–415.
7. Bailey, M. E., Clube, S. V. M., Hahn, G., Napier, W. M., Valsecchi, G. B. 1994. Hazards Due to Giant Comets: Climate and Short-term Catastrophism. In *Hazards Due to Comets and Asteroids* (T. Gehrels, M. S. Matthews eds.), Univ Arizona Press, Tucson, Arizona, 479–533.
8. Bailey, M. E., Stagg, C. R. 1988. Cratering Constraints on the Inner Oort Cloud - Steady-State Models. Monthly Notices of the Royal Astronomical Society **235**, 1–32.
9. Benz, W., Asphaug, E. 1999. Catastrophic Disruptions Revisited. Icarus **142**, 5–20.
10. Bernstein, G. M., Trilling, D. E., Allen, R. L., Brown, M. E., Holman, M., Malhotra, R. 2004. The Size Distribution of Trans-Neptunian Bodies. Astron. J. **128**, 1364–1390.
11. Binney, J., Tremaine, S. 1987. *Galactic Dynamics.* Princeton University Press, Princeton, NJ, 747 p.
12. Binzel, R. P., Rivkin, A. S., Stuart, J. S., Harris, A. W., Bus, S. J., Burbine, T. H. 2004. Observed Spectral Properties of Near-Earth Objects: Results for Population Distribution, Source Regions, and Space Weathering Processes. Icarus **170**, 259–294.

13. Bottke, W. F., Morbidelli, A., Jedicke, R., Petit, J.-M., Levison, H. F., Michel, P., Metcalfe, T. S. 2002. Debiased Orbital and Absolute Magnitude Distribution of the Near-Earth Objects. Icarus **156**, 399–433.
14. Bottke, W. F., Durda, D. D., Nesvorný, D., Jedicke, R., Morbidelli, A., Vokrouhlický, D., Levison, H. 2005. The fossilized size distribution of the main asteroid belt. Icarus **175**, 111–140.
15. Brasser, R., Duncan, M., Levison, H. F. 2006. Embedded Star Clusters and the formation of the Oort cloud. Icarus, In press.
16. Brown M. 2001. The Inclination Distribution of the Kuiper Belt. Astron. J. **121**, 2804–2814.
17. Brown, M. E., Trujillo, C. A., Rabinowitz, D. L. 2005. Discovery of a Planetary-sized Object in the Scattered Kuiper Belt. Astroph. J. **635**, L97–L100.
18. Brown, M. E., Schaller, E. L., Roe, H. G., Rabinowitz, D. L., Trujillo, C. A. 2006. Direct Measurement of the Size of 2003 UB313 from the Hubble Space Telescope. Astroph. J. **643**, L61–L63.
19. Brunini A., Melita M. 2002. The Existence of a Planet Beyond 50 AU and the Orbital Distribution of the Classical Edgeworth Kuiper Belt Objects. Icarus **160**, 32–43.
20. Canup, R. M. 2005. A Giant Impact Origin of Pluto-Charon. Science **307**, 546–550.
21. Chambers, J. E., Lissauer, J. J. 2002. A New Dynamical Model for the Lunar Late Heavy Bombardment. Lunar and Planetary Institute Conference Abstracts 33, 1093.
22. Chiang, E. I., Brown, M. E. 1999. Keck Pencil-beam Survey for Faint Kuiper Belt Objects. Astron. J. **118**, 1411–1422.
23. Chiang E. I., Jordan A. B., Millis R. L., Buie M. W., Wasserman L. H., Elliot J. L., Kern S. D., Trilling D. E., Meech K. J., and Wagner R. M. 2003. Resonance Occupation in the Kuiper Belt: Case Examples of the 5:2 and Trojan Resonances. Astron. J. **126**, 430–443.
24. Chiang, E. I. 2003. Excitation of Orbital Eccentricities by Repeated Resonance Crossings: Requirements. Astroph. J. **584**, 465–471.
25. Chiang, E. I., Lithwick, Y. 2005. Neptune Trojans as a Test Bed for Planet Formation. Astroph. J. **628**, 520–532.
26. Clark, R. N., and 25 colleagues 2005. Compositional maps of Saturn's moon Phoebe from imaging spectroscopy. Nature **435**, 66–69.
27. Cohen, B. A., Swindle, T. D., Kring, D. A. 2000. Support for the Lunar Cataclysm Hypothesis from Lunar Meteorite Impact Melt Ages. Science **290**, 1754–1756.
28. Danby J. M. A. 1962. *Fundamentals of Celestial Mechanics.* Willmann-Bell Inc. Richmond, Virginia.
29. Davis D. R., Farinella P. 1998. Collisional Erosion of a Massive Edgeworth-Kuiper Belt: Constraints on the Initial Population. In *Lunar Planet.* Science Conference. **29**, 1437–1438.
30. Davis D. R., Farinella P. 1997. Collisional Evolution of Edgeworth-Kuiper Belt Objects. Icarus **125**, 50–60.
31. Deutsch, A., Schrarer, U. 1994. Dating Terrestrial Impact Craters. Meteoritics **29**, 301–322.
32. Dones, L., Weissman, P. R., Levison, H. F., Duncan, M. J. 2004. Oort Cloud Formation and Dynamics. Comets II 153–174.

33. Dones, L., Levison, H. F., Duncan, M. J., Weissman, P. R. 2005. Simulations of the Formation of the Oort Cloud I. the reference model. Icarus, in press.
34. Doressoundiram A., Barucci M. A., Romon J., Veillet C. 2001. Multicolor Photometry of Trans-neptunian Objects. Icarus **154**, 277–286.
35. Doressoundiram, A., Peixinho, N., Doucet, C., Mousis, O., Barucci, M. A., Petit, J. M., Veillet, C. 2005. The Meudon Multicolor Survey (2MS) of Centaurs and Trans-Neptunian Objects: Extended Dataset and Status on the Correlations Reported. Icarus **174**, 90–104.
36. Duncan, M., Quinn, T., Tremaine, S. 1987. The Formation and Extent of the Solar System Comet Cloud. Astron. J. **94**, 1330–1338.
37. Duncan, M., Quinn, T., Tremaine, S. 1988. The Origin of Short-period Comets. Astroph. J. **328**, L69–L73.
38. Duncan, M. J., Levison, H. F., Budd, S. M. 1995. The Long-Term Stability of Orbits in the Kuiper Belt, Astron. J. **110**, 3073–3083.
39. Duncan, M. J., Levison, H. F. 1997. Scattered Comet Disk and the Origin of Jupiter family Comets. Science **276**, 1670–1672.
40. Duncan M., Levison H. F., Dones L. 2004. Dynamical Evolution of Ecliptic Comets. In *Comet II* (Festou et al. eds.), University Arizona Press, Tucson, Azizona, 193–204.
41. Elliot, J. L., and 10 colleagues 2005. The Deep Ecliptic Survey: A Search for Kuiper Belt Objects and Centaurs. II. Dynamical Classification, the Kuiper Belt Plane, and the Core Population. Astron. J. **129**, 1117–1162.
42. Fernandez, J. A. 1978. Mass Removed by the Outer Planets in the Early Solar System. Icarus **34**, 173–181.
43. Fernandez, J. A. 1980. Evolution of Comet Orbits under the Perturbing Influence of the Giant Planets and Nearby Stars. Icarus **42**, 406–421.
44. Fernandez, J. A. 1980. On the Existence of a Comet Belt Beyond Neptune. Monthly Notices of the Royal Astronomical Society **192**, 481–491.
45. Fernandez, J. A., Ip, W.-H. 1984. Some Dynamical Aspects of the Accretion of Uranus and Neptune – The Exchange of Orbital Angular Momentum with Planetesimals. Icarus **58**, 109–120.
46. Fernandez, J. A., Gallardo, T. 1994. The Transfer of Comets from Parabolic Orbits to Short-period orbits: Numerical Studies. Astronomy and Astrophysics **281**, 911–922.
47. Fernandez, J. A. 1997. The Formation of the Oort Cloud and the Primitive Galactic Environment. Icarus **129**, 106–119.
48. Fernández, J. A., Tancredi, G., Rickman, H., Licandro, J. 1999. The Population, Magnitudes, and Sizes of Jupiter Family Comets. Astronomy and Astrophysics **352**, 327–340.
49. Fernández, J. A., Brunini, A. 2000. The Buildup of a Tightly Bound Comet Cloud Around an Early Sun Immersed in a Dense Galactic Environment: Numerical Experiments. Icarus **145**, 580–590.
50. Fernández, J. A., Gallardo, T., Brunini, A. 2002. Are There Many Inactive Jupiter-Family Comets Among the Near-Earth Asteroid Population?. Icarus **159**, 358–368.
51. Fernández, J. A., Gallardo, T., Brunini, A. 2004. The Scattered Disk Population as a Source of Oort Cloud Comets: Evaluation of its Current and Past Role in Populating the Oort Cloud. Icarus **172**, 372–381.
52. Fernández, Y. R., Sheppard, S. S., Jewitt, D. C. 2003. The Albedo Distribution of Jovian Trojan Asteroids. Astron. J. **126**, 1563–1574.

53. Fernández, Y. R., Jewitt, D. C., Sheppard, S. S. 2005. Albedos of Asteroids in Comet-Like Orbits. Astron. J. **130**, 308–318.
54. Funato, Y., Makino, J., Hut, P., Kokubo, E., Kinoshita, D. 2004. The Formation of Kuiper-Belt Binaries Through Exchange Reactions. Nature **427**, 518–520.
55. Gladman, B., Kavelaars, J. J., Nicholson, P. D., Loredo, T. J., Burns, J. A. 1998. Pencil-Beam Surveys for Faint Trans-Neptunian Objects. Astron. J. **116**, 2042–2054.
56. Gladman B., Holman M., Grav T., Kaavelars J. J., Nicholson P., Aksnes K., Petit J. M. 2002. Evidence for an Extended Scattered Disk. Icarus **157**, 269–279.
57. Goldreich P., Lithwick Y., Sari R. 2002. Formation of Kuiper-belt Binaries by Dynamical Friction and Three-body Encounters. Nature **420**, 643–646.
58. Gomes, R. S. 1997. Dynamical Effects of Planetary Migration on the Primordial Asteroid Belt. Astron. J. **114**, 396–401.
59. Gomes R. S. 2000. Planetary Migration and Plutino Orbital Inclinations Astron. J. **120**, 2695–2707.
60. Gomes, R. S. 2003. The origin of the Kuiper Belt high-inclination population. Icarus **161**, 404–418.
61. Gomes, R. 2003. The Common Origin of the High Inclination TNO's. Earth Moon and Planets **92**, 29–42.
62. Gomes, R. S., Morbidelli, A., Levison, H. F. 2004. Planetary migration in a planetesimal disk: why did Neptune stop at 30 AU? Icarus **170**, 492–507.
63. Gomes, R. S., Gallardo, T., Fernández, J. A., Brunini, A. 2005. On the Origin of the High-Perihelion Scattered Disk: The Role of the Kozai Mechanism and Mean Motion Resonances. Celestial Mechanics and Dynamical Astronomy **91**, 109–129.
64. Gomes, R., Levison, H. F., Tsiganis, K., Morbidelli, A. 2005. Origin of the Cataclysmic Late Heavy Bombardment Period of the Terrestrial Planets. Nature **435**, 466–469.
65. Grieve, R. A., Shoemaker, E. M. 1994. The Record of Past Impacts on Earth. In *Hazards Due to Comets and Asteroids* (T. Gehrels, M. S. Matthews, eds.), University of Arizona press, Tucson, Arizona, 417–462.
66. Guillot, T. 1999. Interior of Giant Planets Inside and Outside the Solar System. Science **286**, 72–77.
67. Grinspoon D. 1989. Large Impact Events and Atmospheric Evolution on the Terrestrial Planets. Ph.D. thesis, University of Arizona, Tucson, Arizona.
68. Hainaut, O. 2002 http://www.sc.eso.org/ ohainaut/MBOSS/
69. Hayashi C. 1981. Structure of the Solar Nebula, Growth and Decay of Magnetic Fields and Effects of Magnetic and Turbulent Viscosities on the Nebula. Prog. Theor. Phys. Suppl. **70**, 35–53.
70. Hahn, J. M., Malhotra, R. 1999. Orbital Evolution of Planets Embedded in a Planetesimal Disk. Astron. J. **117**, 3041–3053.
71. Hahn, J. M., Malhotra, R. 2005. Neptune's Migration into a Stirred-Up Kuiper Belt: A Detailed Comparison of Simulations to Observations. Astron. J. **130**, 2392–2414.
72. Haisch K. E., Lada E. A., Lada C. J. 2001. Disk Frequencies and Lifetimes in Young Clusters. Astroph. J. **553**, L153–156
73. Harris, N. W., Bailey, M. E. 1998. Dynamical Evolution of Cometary Asteroids. Monthly Notices of the Royal Astronomical Society **297**, 1227–1236.

74. Hartmann, W. K. 1975. Lunar Cataclysm: A Misconception?: Icarus **24**, 181–187.
75. Hartmann, W. K., Ryder, G., Dones, L., Grinspoon, D. 2000. The Time-Dependent Intense Bombardment of the Primordial Earth/Moon System. In Origin of the Earth and Moon (R. M. Canup, K. Righter, 69 collaborating authors, eds.) University of Arizona Press, Tucson, 493–512.
76. Heisler, J., Tremaine, S. 1986. The Influence of the Galactic Tidal Field on the Oort Comet Cloud. Icarus **65**, 13–26.
77. Heisler, J., Tremaine, S., Alcock, C. 1987. The Frequency and Intensity of Comet Showers from the Oort Cloud. Icarus **70**, 269–288.
78. Heisler, J. 1990. Monte Carlo Simulations of the Oort Comet Cloud. Icarus **88**, 104–121.
79. Henrard, J. 1982. Capture into Resonance - An Extension of the Use of Adiabatic Invariants. Cel. Mech. **27**, 3–22.
80. Hills, J. G. 1981. Comet Showers and the Steady-state Infall of Comets from the Oort Cloud. Astron. J. **86**, 1730–1740.
81. Hollenbach, D., Adams, F. C. 2004. Dispersal of Disks Around Young Stars: Constraints on Kuiper Belt Formation. ASP Conf. Ser. 324: Debris Disks and the Formation of Planets 324, 168.
82. Holman, M. J., Wisdom, J. 1993. Dynamical Stability in the Outer Solar System and the Delivery of Short Period Comets. Astron. J. **105**, 1987–1999.
83. Horner, J., Evans, N. W., Bailey, M. E., Asher, D. J. 2003. The Populations of Comet-like Bodies in the Solar System. Monthly Notices of the Royal Astronomical Society **343**, 1057–1066.
84. Hut, P., Alvarez, W., Elder, W. P., Kauffman, E. G., Hansen, T., Keller, G., Shoemaker, E. M., Weissman, P. R. 1987. Comet Showers as a Cause of Mass Extinction. Nature **329**, 118–126.
85. Ida S., Larwood J., Burkert A. 2000. Evidence for Early Stellar Encounters in the Orbital Distribution of Edgeworth-Kuiper Belt Objects. Astroph. J. **528**, 351–356.
86. Jewitt, D. C., Luu, J. X. 1993. Discovery of the Candidate Kuiper Belt Object 1992 QB1, Nature **362**, 730–732.
87. Jewitt, D., Luu, J., Trujillo, C. 1998. Large Kuiper Belt Objects: The Mauna Kea 8K CCD Survey. Astron. J. **115**, 2125–2135.
88. Jewitt, D. C., Trujillo, C. A., Luu, J. X. 2000. Population and Size Distribution of Small Jovian Trojan Asteroids. Astron. J. **120**, 1140–1147.
89. Jewitt, D. C., Sheppard, S. S. 2002. Physical Properties of Trans-Neptunian Object (2000) Varuna. Astron. J. **123**, 2110–2120.
90. Jewitt, D., Sheppard, S. 2005. Irregular Satellites in the Context of Planet Formation. Space Sci. Rev. **116**, 441–455.
91. Johnson, T. V., Lunine, J. I. 2005. Saturn's Moon Phoebe as a Captured body from the Outer Solar System. Nature **435**, 69–71.
92. Kenyon, S. J., Luu, J. X. 1998. Accretion in the Early Kuiper Belt: I. Coagulation and Velocity Evolution. Astron. J. **115**, 2136–2160.
93. Kenyon, S. J., Luu, J. X. 1999a. Accretion in the Early Kuiper Belt: II. Fragmentation. Astron. J. **118**, 1101–1119.
94. Kenyon, S. J., Luu, J. X. 1999b. Accretion in the Early Outer Solar System. Astrophys. J **526**, 465–470.
95. Kenyon S. J, Bromley, B. C. 2002. Collisional Cascades in Planetesimal Disks. I. Stellar Flybys. Astron. J. **123**, 1757–1775.

96. Kenyon, S. J., Bromley, B. C. 2004. Stellar Encounters as the Origin of Distant Solar System Objects in Highly Eccentric Orbits. Nature **432**, 598–602.
97. Kenyon, S. J., Bromley, B. C. 2004. The Size Distribution of Kuiper Belt Objects. Astron. J. **128**, 1916–1926.
98. Knežević Z., Milani A., Farinella P., Froeschlé Ch., Froeschlé C. 1991. Secular resonances from 2 to 50 AU. Icarus **93**, 316.
99. Kobayashi H., Ida S. 2001. The Effects of a Stellar Encounter on a Planetesimal Disk. Icarus **153**, 416–429.
100. Koeberl, C. 2004. The Late Heavy Bombardment in the Inner Solar System. Earth, Moon and Planets **92**, 79–87.
101. Kozai, Y. 1962. Secular Perturbations of Asteroids with High Inclination and Eccentricity. Astron. J. **67**, 579.
102. Kring, D. A., Cohen, B. A. 2002. Cataclysmic Bombardment Throughout the Inner Solar System 3.9-4.0 Ga. J. Geoph. Res. (Planets) **107**, 4–1.
103. Kuchner M. J., Brown M. E., Holman M. 2002. Long-Term Dynamics and the Orbital Inclinations of the Classical Kuiper Belt Objects. Astron. J. **124**, 1221–1230.
104. Kuiper, G. P. 1951. On the Origin of the Solar System. In *Astrophysics*, (J.A. Hynek, ed.), McGraw-Hill, New York, 357 pp.
105. Levison, H. F. 1996. Comet Taxonomy. ASP Conf. Ser. 107: Completing the Inventory of the Solar System **107**, 173–191.
106. Levison, H., Shoemaker, E. M., Shoemaker, C. S. 1997. The Dispersal of the Trojan Asteroid Swarm. Nature **385**, 42–44.
107. Levison, H. F., Duncan, M. J. 1997. From the Kuiper Belt to Jupiter-Family Comets: The Spatial Distribution of Ecliptic Comets, Icarus **127**, 13–32.
108. Levison H. F., Stern S. A. 2001. On the Size Dependence of the Inclination Distribution of the Main Kuiper Belt. Astronomical J. **121**, 1730–1735.
109. Levison, H. F., Dones, L., Duncan, M. J. 2001. The Origin of Halley-Type Comets: Probing the Inner Oort Cloud. Astron. J. **121**, 2253–2267.
110. Levison, H. F., Dones, L., Chapman, C. R., Stern, S. A., Duncan, M. J., Zahnle, K. 2001. Could the Lunar "Late Heavy Bombardment" Have Been Triggered by the Formation of Uranus and Neptune? Icarus **151**, 286–306.
111. Levison, H. F., Morbidelli, A., Dones, L., Jedicke, R., Wiegert, P. A., Bottke, W. F. 2002. The Mass Disruption of Oort Cloud Comets. Science **296**, 2212–2215.
112. Levison, H. F., Morbidelli, A. 2003. The formation of the Kuiper belt by the outward transport of bodies during Neptune's migration. Nature **426**, 419–421.
113. Levison, H. F., Morbidelli, A., Dones, L. 2004. Sculpting the Kuiper Belt by a Stellar Encounter: Constraints from the Oort Cloud and Scattered Disk. Astron. J. **128**, 2553–2563.
114. Levison, H. F., Thommes, E., Duncan, M. J., Dones, L. 2004. A Fairy Tale about the Formation of Uranus and Neptune and the Lunar Late Heavy Bombardment. ASP Conf. Ser. 324. Debris Disks and the Formation of Planets 324, 152.
115. Levison, H. F., Terrell, D., Wiegert, P. A., Dones, L., Duncan, M. J. 2006. On the Origin of the Unusual Orbit of Comet 2P/Encke. Icarus **182**, 161–168.
116. Levison, H. F., Duncan, M. J., Dones, L., Gladman, B. J. 2006. The Scattered Disk as a Source of Halley-Type Comets. Icarus, in press.

117. Levison, H. F., Morbidelli, A., Gomes, R., Backman, D. 2006. Planet Migration in Planetesimal Disks. In *Protostars Planets V*, University of Arizona Press, Tucson, Arizona, in press.
118. Malhotra, R. 1993. The Origin of Pluto's Peculiar Orbit. Nature **365**, 819.
119. Malhotra, R. 1995. The Origin of Pluto's Orbit: Implications for the Solar System Beyond Neptune. Astron. J. **110**, 420.
120. Marchis, F., 17 colleagues. 2006. A low density of $0.8\,\mathrm{g\,cm^{-3}}$ for the Trojan binary asteroid 617 Patroclus. Nature **439**, 565–567.
121. Marzari, F., Scholl, H., Murray, C., Lagerkvist, C. Origin and Evolution of Trojan Asteroids 2002. In *Asteroids III*, (W. F. Bottke, A. Cellino, P. Paolicchi, R. P. Binzel eds.), University of Arizona Press, 725–738
122. Masset, F., Snellgrove, M. 2001. Reversing Type II Migration: Resonance Trapping of a Lighter Giant Protoplanet. Monthly Notices of the Royal Astronomical Society **320**, L55–L59.
123. Melita, M., Larwood, J., Collander-Brown, S., Fitzsimmons, A., Williams, I. P., Brunini, A. 2002. The Edge of the Edgeworth-Kuiper Belt: Stellar Encounter, Trans-Plutonian Planet or Outer Limit of the Primordial Solar Nebula? In *Asteroid, Comet, Meteors*, ESA Spec. Publ. series, 305–308.
124. Michtchenko, T. A., Beaugé, C., Roig, F. 2001. Planetary Migration and the Effects of Mean Motion Resonances on Jupiter's Trojan Asteroids. Astron. J. **122**, 3485–3491.
125. Milani, A. 1993. The Trojan Asteroid Belt: Proper Elements, Stability, Chaos and Families. Celestial Mechanics and Dynamical Astronomy **57**, 59–94.
126. Morbidelli, A., Thomas, F., Moons, M. 1995a. The Resonant Structure of the Kuiper Belt and the Dynamics of the First Five Trans-Neptunian Objects. Icarus **118**, 322.
127. Morbidelli, A., Valsecchi, G. B. 1997. Neptune Scattered Planetesimals could have Sculpted the Primordial Edgeworth–Kuiper Belt. Icarus **128**, 464–468.
128. Morbidelli, A. 2002. Modern Celestial Mechanics: aspects of Solar System dynamics. In *Advances in Astronomy and Astrophysics*, Taylor & Francis, London.
129. Morbidelli, A., Levison, H. F. 2004. Scenarios for the Origin of the Orbits of the Trans-Neptunian Objects 2000 CR_{105} and 2003 VB_{12} (Sedna). Astron. J. **128**, 2564–2576.
130. Morbidelli, A. 2004. How Neptune Pushed the Boundaries of Our Solar System. Science **306**, 1302–1304.
131. Morbidelli A., Brown M. 2004. The Kuiper Belt and the Primordial Evolution of the Solar System. In *Comet II*, (Festou et al. eds.), University Arizona Press, Tucson, Arizona, 175–192.
132. Morbidelli, A., Crida, A., Masset, F. 2005. Preventing Type II Migration of Jupiter and Saturn. AAS/Division for Planetary Sciences Meeting Abstracts 37.
133. Morbidelli, A., Levison, H. F., Tsiganis, K., Gomes, R. 2005. Chaotic Capture of Jupiter's Trojan Asteroids in the Early Solar System. Nature **435**, 462–465.
134. Murray-Clay, R. A., Chiang, E. I. 2005. Stochastic Migration in Planetesimal Disks. preprint.
135. Nagasawa, M., Ida, S. 2000. Sweeping Secular Resonances in the Kuiper Belt Caused by Depletion of the Solar Nebula. Astron. J. **120**, 3311–3322.
136. Nesvorný, D., Roig, F. 2000. Mean Motion Resonances in the Trans-Neptunian Region: Part I: The 2:3 Resonance with Neptune. Icarus **148**, 282–300.

137. Nesvorný, D., Roig, F. 2001. Mean Motion Resonances in the Trans-Neptunian Region: Part II: the 1:2, 3:4 and Weaker Resonances. Icarus **150**, 104–123.
138. Noll, K. S., Stephens, D. C., Grundy, W. M., Osip, D. J., Griffin, I. 2004. The Orbit and Albedo of Trans-Neptunian Binary (58534) 1997 CQ_{29}. Astron. J. **128**, 2547–2552.
139. O'Brien, D. P., Morbidelli, A., Bottke, W. F. 2005. Collisional Evolution of the Primordial Trans-Neptunian Disk: Implications for Planetary Migration and the Current Size Distribution of TNOs. AAS/Division for Planetary Sciences Meeting Abstracts 37 .
140. Oort, J. H. 1950. The Structure of the Cloud of Comets Surrounding the Solar System and a Hypothesis Concerning its Origin. Bulletin of the Astronomical Institute of the Netherlands **11**, 91–110.
141. Petit J. M., Morbidelli A., Valsecchi, G. B. 1999. Large Scattered Planetesimals and the Excitation of the Small Body Belts. Icarus **141**, 367–387.
142. Petit, J.-M., Morbidelli, A., Chambers, J. 2001. The Primordial Excitation and Clearing of the Asteroid Belt. Icarus **153**, 338–347.
143. Petit, J.-M., Mousis, O. 2004. KBO Binaries: How Numerous were they? Icarus **168**, 409–419.
144. Petit, J.-M., Holman, M. J., Gladman, B. J., Kavelaars, J. J., Scholl, H., Loredo, T. J. 2006. The Kuiper Belt Luminosity Function from m_{R}= 22 to 25. Monthly Notices of the Royal Astronomical Society **365**, 429–438.
145. Pittich, E. M., D'Abramo, G., Valsecchi, G. B. 2004. From Jupiter-family to Encke-like orbits. The Rôle of Non-gravitational Forces and Resonances. Astronomy and Astrophysics **422**, 369–375.
146. Rickman, H., Froeschlé, C., Froeschlé, C., Valsecchi, G. B. 2004. Stellar perturbations on the scattered disk. Astronomy and Astrophysics **428**, 673–681.
147. Ryder, G. 1990. Lunar Samples, Lunar Accretion and the Early Bombardment of the Moon: Eos Transactions AGU **71**, 313–323.
148. Ryder, G., Koeberl, C., Mojzsis, S. J. 2000. Heavy Bombardment on the Earth at 3.85: The Search for Petrographical and Geochemical Evidence. In (R.M., Canup, K., Righter, eds.), *Origin of the Earth and Moon.* University of Arizona Press, Tucson, Arizona, 475–492.
149. Ryder, G. 2002. Mass Flux in the Ancient Earth-Moon system and the Benign Implications for the origin of life on Earth. J. Geophy. Res.-Planets **107**, 6–14.
150. Sasaki, T., Abe, Y. 2004. Partial Resetting on Hf-W System by Giant Impacts. Lunar and Planetary Institute Conference Abstracts 35, 1505.
151. Stern, S. A. 1991, On the Number of Planets in the Outer Solar System - Evidence of a Substantial Population of 1000-km Bodies. Icarus **90**, 271–281.
152. Stern, S. A. 1995. Collisional Time Scales in the Kuiper Disk and their Implications. Astron. J. **110**, 856–868.
153. Stern, S. A. 1996. On the Collisional Environment, Accretion Time Scales, and Architecture of the Massive, Primordial Kuiper Belt. Astron. J. **112**, 1203–1210.
154. Stern, S. A., Colwell, J. E. 1997a. Accretion in the Edgeworth-Kuiper Belt: Forming 100–1000 KM Radius Bodies at 30 AU and Beyond. Astron. J. **114**, 841–849.
155. Stern, S. A., Colwell, J. E. 1997b. Collisional Erosion in the Primordial Edgeworth-Kuiper Belt and the Generation of the 30–50 AU Kuiper Gap. Astroph. J. **490**, 879–885.

156. Stern, S. A., Weissman, P. R. 2001. Rapid Collisional Evolution of Comets During the Formation of the Oort Cloud. Nature **409**, 589–591.
157. Stern, S.A. 2002. Evidence for a Collisonal Mechanism Affecting Kuiper Belt Object Colors. Astron. J. **124**, 2300–2304.
158. Stone, J. M., Gammie, C. F., Balbus, S. A., Hawley, J. F. 1998. In *Protostars and Planets IV* (V. Mannings, A. P. Boss, S. S. Russell, eds.) University of Arizona Press, Tucson, 589.
159. Strom, R. G., Neukum, G. 1988. *The Cratering Record on Mercury and the Origin of Impacting Objects.* Mercury, University of Arizona Press, 336–373.
160. Strom, R. G., Malhotra, R., Ito, T., Yoshida, F., Kring, D. A. 2005. The Origin of Planetary Impactors in the Inner Solar System. Science **309**, 1847–1850.
161. Stuart, J. S. 2001. A Near-Earth Asteroid Population Estimate from the LINEAR Survey. Science **294**, 1691–1693.
162. Tagle, R. 2005. LL-Ordinary Chondrite Impact on the Moon: Results from the 3.9 Ga Impact Melt at the Landing Site of Apollo 17. 36th Annual Lunar and Planetary Science Conference 36, 2008.
163. Tegler, S. C., Romanishin, W. 2000. Extremely red Kuiper-belt objects in near-circular orbits beyond 40 AU. Nature **407**, 979–981.
164. Tera, F., Papanastassiou, D. A., Wasserburg, G. J. 1974. Isotopic Evidence for a Terminal Lunar Cataclysm. Earth and Planetary Science Letters **22**, 1–21.
165. Thebault, P., Doeressoundiram, A. 2003. A Numerical Test of the Collisional Resurfacing Scenario. Could Collisional Activity Explain the Spatial Distribution of Color-index Within the Kuiper Belt? Icarus **162**, 27–37.
166. Thommes, E. W., Duncan, M. J., Levison, H. F. 2003. Oligarchic Growth of Giant Planets. Icarus **161**, 431–455.
167. Trail, D., Mojzsis, S. J., Harrison, T. M., Levison, H. F. 2006. Do Hadean Zircons Retain a Record of the Late Heavy Bombardment on Earth?. 37th Annual Lunar and Planetary Science Conference 37, 2139.
168. Trujillo, C. A., Jewitt, D. C., Luu, J. X. 2001. Properties of the Trans-Neptunian Belt: Statistics from the Canada-France-Hawaii Telescope Survey. Astron. J. **122**, 457–473.
169. Trujillo, C. A., Brown, M. E. 2001. The Radial Distribution of the Kuiper Belt. Astroph. J **554**, 95–98.
170. Trujillo, C. A., Brown, M. E. 2002. A Correlation Between Inclination and Color in the Classical Kuiper Belt. Astroph. J **566**, 125–128.
171. Trujillo, C. A. 2003. The Caltech Wide Area Sky Survey: Beyond (50000) Quaoar. In *Proceedings of the First Decadal Review of the Edgeworth-Kuiper Belt Meeting in Antofagasta*, Chile, Earth Moon and Planets **92**, 99–112.
172. Tsiganis, K., Gomes, R., Morbidelli, A., Levison, H. F. 2005. Origin of the Orbital Architecture of the Giant Planets of the Solar System. Nature **435**, 459–461.
173. Valley, J. W., Peck, W. H., King, E. M., Wilde, S. A. 2002. A Cool Early Earth. Geology **30**, 351–354.
174. Valsecchi, G. B., Morbidelli, A., Gonczi, R., Farinella, P., Froeschle, C., Froeschle, C. 1995. The Dynamics of Objects in Orbits Resembling that of P/Encke. Icarus **118**, 169.
175. Weidenschilling S. 2002. On the Origin of Binary Transneptunian Objects. Icarus **160**, 212–215.

176. Weidenschilling, S. 2003. Formation of Planetesimals/Cometesimals in the Solar nebula. In *Comet II*, (Festou et al. eds.), University Arizona Press, Tucson, Arizona, 97–104.
177. Weissman, P. R. 1978. Physical and Dynamical Evolution of Long-Period Comets. Ph.D. thesis, University of California, Los Angeles.
178. Weissman, P. R. 1990. The Oort cloud. Nature **344**, 825–830.
179. Weissman, P. R. 1996. The Oort Cloud. ASP Conf. Ser. 107: Completing the Inventory of the Solar System **107**, 265–288.
180. Weissman, P. R., Lowry, S. C. 2003. The Size Distribution of Jupiter-Family Cometary Nuclei. Lunar and Planetary Institute Conference Abstracts 34.
181. Wetherill, G. W. 1992. An Alternative Model for the Formation of the Asteroids. Icarus **100**, 307–325.
182. Whitman, K., Morbidelli, A., Jedicke, R. 2006. The Size-Frequency Distribution of Dormant Jupiter Family Comets. Icarus, in press.
183. Wiegert, P., Tremaine, S. 1999. The Evolution of Long-Period Comets. Icarus **137**, 84–121.
184. Wilhems, D. E. 1987. Geologic History of the Moon. US Geological Survey, Professional Paper 1348. 302 p. Reston, Virginia.
185. Yin, Q., Jacobsen, S. B., Yamashita, K., Blichert-Toft, J., Télouk, P., Albarède, F. 2002. A Short Timescale for Terrestrial Planet Formation from Hf-W Chronometry of Meteorites. Nature **418**, 949–952.
186. Yeomans, D. K., Chodas, P. W., Sitarski, G., Szutowicz, S., Królikowska, M. 2004. Cometary orbit determination and nongravitational forces. In *Comets II* (M. Festou et al. eds.), University of Arizona press, Tucson. 137–151.
187. Youdin, A. N., Shu, F. H. 2002. Planetesimal Formation by Gravitational Instability. Astroph. J. **580**, 494–505.
188. Zappala, V., Cellino, A., Gladman, B. J., Manley, S., Migliorini, F. 1998. NOTE: Asteroid Showers on Earth after Family Breakup Events. Icarus **134**, 176–179.

Comets

H. Rauer

1 Introduction

Bright comets form spectacular phenomena in the night sky (see Fig. 1), and they have always been subject of attention and fascination. Today, it is generally believed that comets are the least modified bodies in our solar system, although they are certainly not unmodified. A cometary nucleus consists of a mixture of volatile ices (H_2O, CO, CO_2, ...) and silicate dust particles. Their icy nature indicates that comets have been preserved at cold temperatures since the early stages of our solar system. Determining the chemical composition and physical structure of cometary nuclei to better understand these early phases in solar system history is therefore a primary goal of cometary science. Several key questions need to be answered:

- How have comets been formed?
- What is the composition of cometary nuclei?
- Are all comets the same?
- Has their composition been modified since their formation?

Comets come into the inner solar system from at least two reservoirs. The Kuiper–Edgeworth belt [71, 137], or also called trans-Neptunian belt, beyond about 40 astronomical units (AU) is a ring-like reservoir of bodies concentrated near the ecliptic plane. It is believed that the Kuiper belt is the source region for most short-period comets, especially for comets belonging to the Jupiter family. These short-period comets have orbital periods around 5 years and an aphelion near Jupiter's orbit. Transition objects on orbits between Kuiper belt objects and comets in the inner solar system are called centaurs. Long-period comets, with orbital periods >200 years, come into the solar system from the Oort-cloud [171]. This shell of objects surrounding the solar system at distances of several 10^4 AU has been postulated based on the orbital parameters of comets coming into the solar system for the first time on elongated, very long-period orbits. The complex structure and dynamical

H. Rauer, *Comets*. In: K. Altwegg et al., Trans-Neptunian Objects and Comets, Saas-Fee Advanced Courses, pp. 165–254 (2008)
DOI 10.1007/978-3-540-71958-8_3

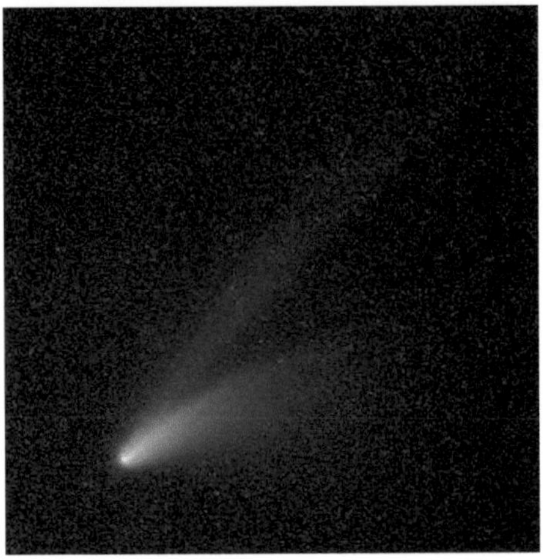

Fig. 1. Image of comet Hale–Bopp during its perihelion in 1997. The straight ion tail, blue in the light of CO^+ ions, and the slightly curved and diffuse dust tail are clearly visible. The dust tail can be seen because small dust particles efficiently scatter the incoming solar radiation

evolution of these reservoir regions are described in detail in the chapter of Morbidelli [160] in this book.

Cometary nuclei are small bodies with radii below 15 km, most of them even below 5 km. Our knowledge of the physical properties of cometary nuclei is outlined and compared to the population of Kuiper belt objects in the chapter by Dave Jewitt. Here, we concentrate on the dynamical and chemical processes of the cometary gas and dust component.

When a comet approaches the Sun along its orbit (Fig. 2), heating by absorption of sunlight leads to sublimation of its icy components (see Sect. 2). A neutral gas coma forms around the nucleus, extending typically a few 10^5 km nucleocentric distance. The molecules sublimating from the nucleus ices are called "parent species." Chemical destruction of the parent species in the coma leads to the formation of daughter products: neutral radicals, atoms, and ions (see Sect. 5).

Ionized molecules interact with the solar magnetic field and form the ion, or plasma, tail extending a few 10^7 km in lengths (Fig. 1; Sect. 3.4). In addition to the ion tail, a neutral tail can be observed in active comets consisting of atoms accelerated by solar radiation pressure. This tail is well visible in case of sodium atoms. Another example for a neutral tail is the neutral hydrogen cloud surrounding comets. The molecules, atoms, and ions in the comae and tails are visible because they emit radiation at wavelengths from the UV up to the radio range (Sect. 4) that can be observed with ground- and space-based telescopes.

The silicate dust particles embedded in the nucleus ices are lifted from the surface by the sublimating volatiles. They form the dust coma in the

Fig. 2. Activity around the Sun [53]

nucleus vicinity and finally the cometary dust tail under the influence of solar gravity and radiation pressure (Sect. 3.3). The dust coma and tail (Fig. 1) are visible because small dust particles scatter the solar light very efficiently (Sect. 7).

A long history of ground-based observations of cometary comae, dust, and ion tails exists. However, observations from Earth or Earth-orbit are unable to resolve the small nucleus embedded in the gas and dust coma. In situ investigations by space missions have therefore been made since the mid-1980s (Table 1), with the highlight of five spacecrafts visiting comet Halley in 1986, providing the first images of a cometary nucleus and a wealth of data on the coma surrounding it. Up to today, Halley is the comet for which we have the most detailed knowledge. However, even future space missions, with the exception of landers such as ESA's Rosetta mission, do not investigate the nucleus directly, but analyze its coma on fly-bys or orbiting trajectories. Thus, a very good understanding of the dynamical and chemical processes in

Table 1. Overview of past and future space missions to comets

Year	Space mission	Comet	Comment
1985	ICE	Giacobini-Zinner	plasma tail
1986	Suisei, Sakigake	Halley	solar wind
1986	Vega 1, 2	Halley	
1986	Giotto	Halley	
2001	Deep Space 1	Borrelly	
2004	Stardust	Wild 2	sample return
2005	Deep Impact	Tempel 1	impactor
2014	Rosetta	Churyumov-Gerasimenko	orbiter+lander

the coma and of the nucleus surface is required for the interpretation of these measurements.

Several reviews on various aspects of cometary physics have been published in the past. A very good recent compendium of reviews is the book *Comet II* [80]. Lecture books on comets are rare, only two have been published so far [79,200]. The aim of this article is to aid students and scientists newly entering the field of cometary physics to obtain an overview on the basic ideas of cometary science and guide them to sources of deeper information in the field of their specialized interest.

2 Sublimation Processes

The cometary nucleus itself can be investigated directly only by landing spacecraft, such as the lander Philae on ESA's Rosetta mission in 2014. Until then, we have to infer the nucleus composition and structure from the analysis of the molecules and dust particles in the coma. However, models of the sublimation process predict that the relative parent molecule abundance ratios measured in the coma may differ from the composition of ices in the nucleus. Furthermore, the nucleus itself may be inhomogeneous, and the upper surface layers, that can be accessed by a lander, may not have the pristine composition and structure. We therefore need to understand the sublimation processes to be able to interpret the gas abundance ratios measured in the coma by in situ spacecraft, by ground-based telescopes and the Rosetta lander. However, measurements in the coma can also help us to constrain the sublimation models.

The outgassing of cometary nuclei depends on the dust/ice ratio, the composition of volatile ices and on the internal physical structure of the nucleus (e.g., the presence of amorphous and/or crystalline ices, pore sizes, heat conduction). In this section, a brief overview on the currently proposed concepts of the sublimation processes of cometary nuclei is given. In addition, measurements of gas production rates are discussed and compared to the model predictions. For a detailed discussion of the chemical coma processes, we refer to the following sections. How the gas production rates are derived from coma observations is described in Sect. 6.

2.1 General Overview

The sublimation of nucleus ices is governed by the balance between the effective incident solar radiation energy on the surface, the reflected radiation, the thermally re-radiated energy, and the energy used for volatile sublimation and internal heat conduction. The sublimated gas produces a drag force on the dust particles at the surface, which then expand into the coma with the gas (Fig. 3, see also Sect. 3). In the coma, the parent gas molecules are subject to various chemical destruction processes (see Sect. 5).

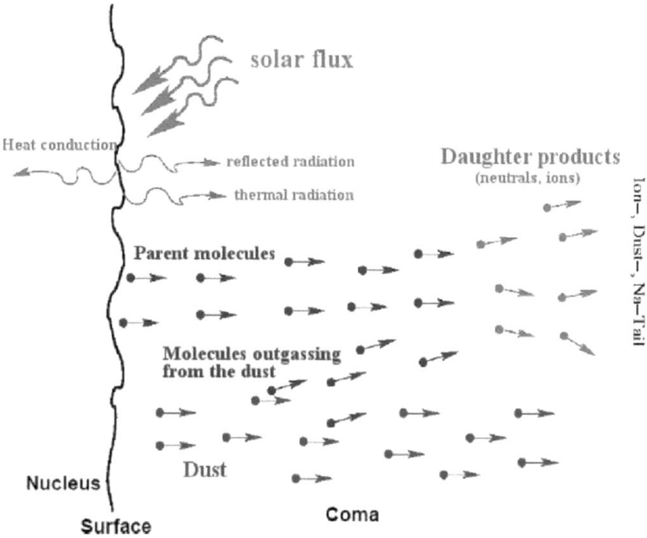

Fig. 3. Sketch to illustrate the surface energy balance and the processes in the coma

The energy balance at the nucleus surface is given by:

$$\frac{F_{\odot}(1 - A_{\mathrm{B}})}{r_{\mathrm{h}}^2} \cos\theta = \varepsilon\sigma T^4 + Z(T)L(T) - \kappa_{\mathrm{s}}\frac{\mathrm{d}T}{\mathrm{d}z} \tag{1}$$

Here $F_{\odot}$ denotes the incident solar flux, θ the solar zenith angle, r_{h} the heliocentric distance, A_{B} the comet Bond albedo (see Sect. 7 for a definition of albedo), and ε the infrared emissivity, respectively. $Z(T)$ is the surface sublimation rate of the ices at temperature T, $L(T)$ the latent heat used for sublimation, κ_{s} is the coefficient for heat conduction into the interior along z, and $\frac{\mathrm{d}T}{\mathrm{d}z}$ the internal temperature gradient. Unfortunately, many of the critical parameters that govern the sublimation processes are only poorly known. In particular the heat conduction into the interior is uncertain because it depends on porosity, composition, conductivity of the ices, etc., of which we have only limited knowledge. Reference [174] provide an overview on heat conduction in a porous medium and the approximations used in various comet models.

A first estimate on the activity evolution of cometary nuclei with heliocentric distance can be obtained when solving the energy balance for a pure ice surface facing the Sun and neglecting internal heat conduction. On a log-log scale (Fig. 4), the sublimation of ices shows a sudden rise at the heliocentric distance, where sufficient solar energy for their sublimation becomes available. This sharp rise in Z is often referred to as the "onset of activity" for a given species. When further approaching the Sun, the sublimation increases proportional to the increasing incoming solar energy as r_{h}^{-2} because almost all energy is converted into sublimation. The heliocentric distance, at which the onset of activity is seen, depends on the volatility of the ices. Highly volatile ices,

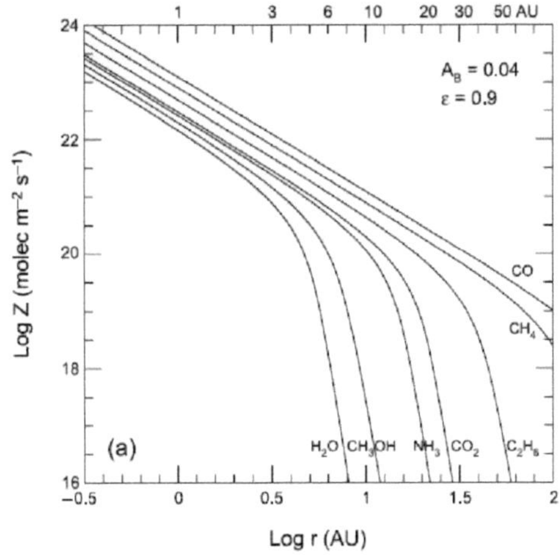

Fig. 4. Evolution of production rates for various ices versus heliocentric distance, assuming pure ice surfaces [154]

such as CO ice that sublimates already at 24 K, start to sublime at about r_h = 100–200 AU. At 3–5 AU heliocentric distance, water ice sublimation starts, when surface temperatures increase above about 150 K. Species with intermediate volatility are expected to start their activity onset at intermediate r_h in this simple pure ice model (Fig. 4).

For real cometary nuclei, the sublimation over their orbit is more complex than the pure ice case and depends on the nucleus composition and internal structure. Below, a very brief overview on the basic concepts of sublimation models and their predictions is given.

2.2 Gas Sublimation and Nucleus Differentiation

Different model approaches have been made to simulate cometary activity. However, most models agree on some general concepts of the sublimation process:

* It is generally assumed that a cometary nucleus is a porous body with dust particles mixed into the ice.
* sublimation of ices occurs at the surface as well as inside the nucleus as a result of heat conduction into the interior.
* The sublimated gases inside the nucleus will flow to the surface because a pressure gradient builds up between the site of gas sublimation and the nucleus surface.
* Gases are also flowing in opposite direction, toward the nucleus center, but they quickly re-condensate because of the lower temperature.
* When the ice sublimates, the gas flow drags along the small embedded dust particles.

* Dust particles too large to be accelerated by the gas flow can accumulate at the surface and build a crust covering the volatile icy interior. This dust crust inhibits surface sublimation, because the dust heated by solar energy re-radiates most of the energy at thermal wavelengths. It depends strongly on the porosity of the crust how much sublimated water vapor can pass from below through the crust into the coma.

We already mentioned that the details of the sublimation process depend on the internal composition and structure of the nucleus. Input parameters for model simulations are the composition of the ice mixture, the dust/gas ratio, the heat conduction, and therefore parameters such as porosity of the ice matrix, etc. (see [174] for an updated list of input parameters and their commonly used values). A main difference between models used to simulate the cometary outgassing behavior is the amount of amorphous water ice contained in the nucleus. The presence of amorphous ice affects the gas activity evolution substantially. Here we outline the two extreme model assumptions: a nucleus made of pure crystalline ices and a nucleus made of initially pure amorphous ice. Reality will be, as usual, between these two extremes.

A Nucleus made of Crystalline Ice

First, we outline the evolution of a cometary nucleus consisting of a mixture of ices in crystalline form on its way toward the Sun. Heat penetrating into the nucleus interior can sublimate volatile ices which then "escape" through the pores to the surface. Each volatile ice reacts more or less independently to the increasing heat as the comet approaches the Sun and shows individual onsets of activity with decreasing r_h. A cometary nucleus therefore evolves during subsequent perihelion passages and may chemically differentiate into a layered body after some orbits.

To illustrate this scenario, let us assume a homogeneously mixed nucleus made of crystalline ices on its first passage into the inner solar system. Highly volatile ices will start to sublimate first, followed by less volatile species and finally H_2O ice. Therefore, the top layers of the nucleus will be depleted of all minor volatiles after some perihelion passages (Fig. 5) and only the lowest layers will still contain the original composition [19, 20, 72, 74]. In an evolved porous crystalline nucleus, therefore, highly volatile ices sublimate from the inside through the pores of the ice matrix. This is different to water ice, which is at the surface, possibly covered by a dust crust.

The maximum depth from which volatile ices sublimate to the surface is determined by the penetration depth of solar energy into the nucleus. Despite the uncertainties in our knowledge of nucleus parameters, we can make some crude estimates of the skin depth for penetration of solar energy into the nucleus over the diurnal and orbital cycle. The orbital skin depth is [174]:

$$l_{\mathrm{orbit}} = \sqrt{\frac{2Ka^{\frac{3}{2}}}{\sqrt{GM_{\odot}}\rho c}} \tag{2}$$

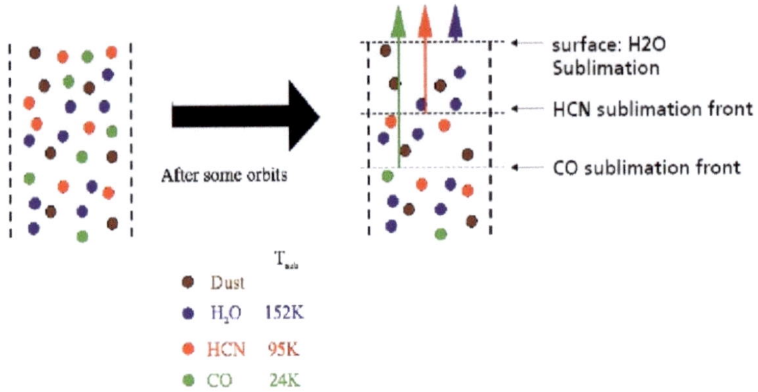

Fig. 5. Illustration of the differentiation of an originally homogeneous crystalline nucleus (left) over subsequent orbits. Circles indicate separate ice phases of volatiles (H_2O: blue, HCN: red, CO: green) and a dust component (brown). After several orbits, the surface regions are depleted of volatile ices. Their sublimation fronts have moved into the interior of the nucleus (right), and these ices sublimate through the pores of the water ice matrix to the surface

We assume a thermal conductivity of $K = 0.6\,\mathrm{J\,m^{-1}\,s^{-1}\,K^{-1}}$, a density of $\rho = 700\,\mathrm{kg\,m^{-3}}$, a specific heat $\mathrm{c} = 8 \times 10^2\,\mathrm{J\,kg^{-1}\,K^{-1}}$, and an orbital semi-major axis, a, between 5 AU for a Jupiter family comet and several hundred to thousand AU for long-period comets. Then the orbital skin depth is between 5 m up to several hundred meters. Thus, it may well be that, for example, the interior of a large Jupiter family comet is never reached by solar energy and remains at its original state (neglecting radioactive heating). The outer meters of a comet, however, are definitely modified during its subsequent orbits around the Sun.

For short-period comets, the temperature at aphelion is still sufficient to sublime very volatile ices, such as CO. Some models (e.g., [18]) predict that after many orbital revolutions, the sublimation front of this minor volatile has moved deep (several meters) into the interior. At such depths, the available energy depends only little on the orbital position of the comet. The resulting gas production rate is then expected to be almost constant along a cometary orbit in this model, whereas less volatile ices show a clear orbital variation of gas production (Fig. 6). Such an extreme scenario is most likely to occur for highly volatile species, such as CO and CO_2, which might be present only in the lowest layers of differentiated nuclei.

Cometary rotation leads to diurnal variations of the solar energy input on the surface. Again, we have a look at the skin depth:

$$l_{\mathrm{diurnal}} = \sqrt{\frac{KP}{\pi \rho c}} \tag{3}$$

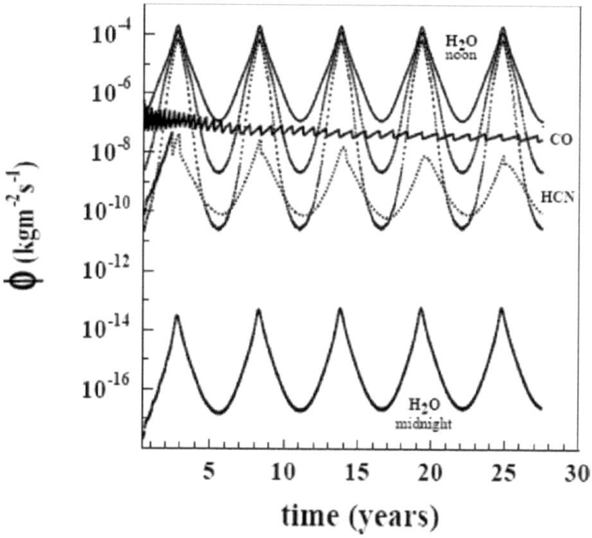

Fig. 6. The effect of differentiation of a crystalline porous nucleus on its gas production rates over several orbits for a Jupiter family comet [21]. The evolution of CO activity is almost constant, whereas water and other less volatile ices show clear variations over the orbit of the comet

With a rotation period of $P = 10\,\mathrm{h}$, the diurnal skin depth is only 0.1 m [174]. Thus, rotational modulation of solar energy influences only the top layers of a nucleus.

A Nucleus with Amorphous Ice

The level of amorphous ice present in the nucleus is an important parameter for comet sublimation. At cold temperatures and low pressure, e.g., during the formation of cometary nuclei in the outer parts of the pre-planetary disc, water ice is expected to condense in its amorphous form. However, amorphous ices crystallizes in an irreversible and exothermic process at temperatures above 136 K. Thus, to be still present in comets today, the temperature of the ice grains built into the nucleus should have never exceeded this crystallization temperature.

The energy released during the exothermic crystallization process provides an additional energy source that can lead to enhanced sublimation of the ice. Thus, when the critical crystallization temperature is reached, the comet will show gas activity. This can result, for example, in a sudden increase of activity (outbursts), which is indeed often observed in comets. The transition of amorphous to crystalline ice also changes the physical parameters of the nucleus, like heat conduction (the heat conduction of amorphous ice is about four times lower than for crystalline ice [174]), porosity, and density. This will result in a different evolution of gas activity over the orbit as compared to a nucleus made purely of crystalline ice. In addition, amorphous water ice can efficiently trap gases of more volatile ices. These are released at the moment of water ice crystallization and provide an additional source of volatiles [15, 16, 169, 170].

To illustrate the effect of amorphous ice, let us assume that at least on its first orbit into the inner solar system, a comet is made purely of amorphous water ice. If we further assume the extreme case that all volatiles are trapped in the amorphous water ice, then even highly volatile ices can be released only at the moment of water ice crystallization. When our hypothetical new comet further approaches the Sun, the sublimation front moves deeper into the nucleus and the surface layers will be crystallized. We will therefore find sublimation in the crystalline ice layer as outlined above, and this layer may eventually deplete from the highly volatile species (Fig. 7).

At present, it is unclear whether water ice is contained in the nucleus in crystalline or in amorphous form and what would be the ratio of the two. However, also comets containing amorphous ice will be porous, and they also may contain grains of frozen crystalline volatiles. Such non-trapped highly volatile ices will therefore be able to sublime through the pores in the water ice and show activity also at large r_{h}, similar to the case of a pure crystalline nucleus. Obviously, the outgassing of a cometary nucleus is a complex interplay between porosity, the presence of amorphous ice, and the abundances of volatiles. In addition, a critical factor for the internal layering of the nucleus is the rate of surface erosion by water sublimation in comparison to the time scale for penetration of the orbital heat wave. If surface erosion is fast, no equilibrium for the internal structure is reached even after many orbits.

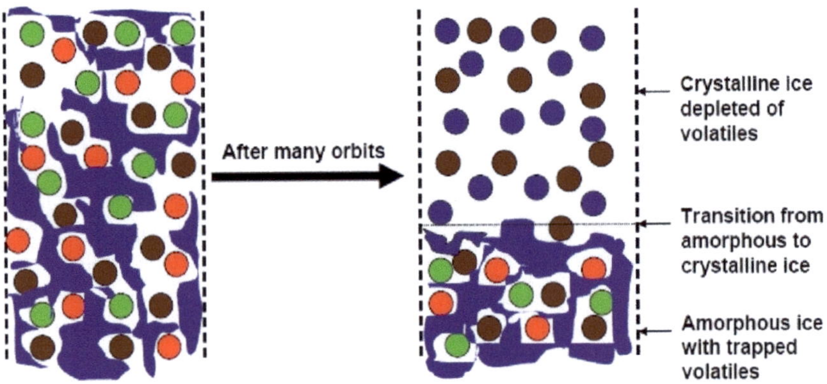

Fig. 7. Illustration of differentiation of a nucleus consisting originally of amorphous water ice and its evolution after several orbits. The meaning of colours is as in Fig. 5. Blue represents an amorphous water ice phase. Other volatiles are trapped in the amorphous ice and are also present as a separate crystalline phase (left). After several orbits, the surface layer crystallizes and is depleted from volatile ices. In a porous nucleus, the separate non-trapped crystalline ice phases may also sublime from the deeper interior

So far, we neglected the presence of dust particles in the ice in our discussion. We already mentioned that large dust particles may accumulate to form a crust at the surface. The sublimation of underlying ices then depends on the thermal conductivity and the porosity of the dust layer. The presence of dust has a strong effect on the surface temperature. Pure ice surfaces use most of their energy for sublimation. Dust-covered surfaces can heat up; for example noon temperatures with dust are expected around 360 K in comparison to about 200 K for a pure ice surface [174]. Because a dust crust may form an effective obstacle for the sublimating ices, it has been proposed that pressure built up by sublimated gases unable to penetrate through a dust crust can lead to cracks in the surface and small outbursts of gas/dust activity.

Several groups attempting to model the nucleus activity exist in addition to the examples already given. They all provide predictions on the outgassing behavior of comets (e.g., [38, 39, 114]). In general, the models predict that the evolution of gas activity seen in the coma of a comet can be quite different for a comet where the upper layers contain crystalline ice or a comet where amorphous ice is still present close to the surface. In addition, the dust content affects the activity evolution. Unfortunately, many of the parameters entering the simulations of the outgassing processes are not well known. They have to be derived by comparing model predictions to in situ and ground-based observations of the long-term activity evolution or in situ data from landers.

2.3 Observations of Gas Activity Evolution

The evolution of production rates along a cometary orbit depends on the available solar energy, the volatility of the species and the structure, and outgassing processes in the nucleus as outlined above. Key to constrain the nucleus composition and structure are observations of the cometary activity at large heliocentric distances and over a wide range of r_{h} to cover the long-term evolution. In future, in situ measurements from landers will also become available. Unfortunately, observations of gases are often possible only in the water-driven sublimation regime inside $r_{\mathrm{h}} = 3\,\mathrm{AU}$, because at large r_{h} the sublimation rates are low (Fig. 4), and the excitation of line emissions is weak. Only exceptional bright long-period comets allow us to detect gas emissions also in the CO-driven regime at large heliocentric distances. Here we discuss observations of the gas evolution in comets in view of the different model concepts for sublimation.

The unusually bright long-period comet Hale–Bopp provided us with the widest coverage of gas activity observations so far (Fig. 8). We therefore look at its activity evolution in more detail. Beyond $r_{\mathrm{h}} = 5\,\mathrm{AU}$ activity of CO has been detected in comet Hale–Bopp in the radio range [24, 25] as well as emissions of CN and HCN in the optical [177, 187] and radio [24]. The other minor volatile species were detected in the water-driven sublimation regime at less than 3–5 AU heliocentric distance. All production rates increase toward perihelion and decrease on the outbound path, following the variation in solar

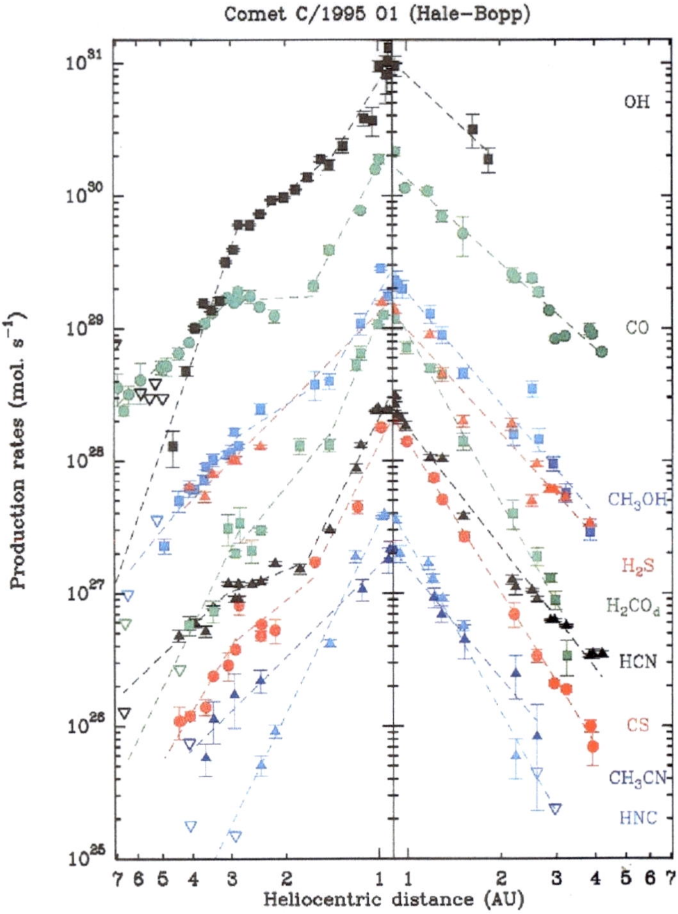

Fig. 8. Gas production rates over heliocentric distance of comet Hale–Bopp [25]

energy input. Water sublimation is first detected around 5 AU, and at r_h = 3.5–4 AU, it started to dominate over CO activity, as expected for a water ice dominated comet.

H_2CO, CS, and HNC show a steeper increase of their production rates toward perihelion than other species. This is most likely linked to their formation as a daughter product in the coma, rather than from pure nucleus sublimation (see Sect. 5).

When comparing pre- and post-perihelion observations, some systematic differences in the evolution of gas activity can be seen, like a sudden increase in activity for most species within 1.5 AU pre-perihelion and a stagnation in CO production rate near 2–3 AU pre-perihelion (Fig. 8). However, the overall activity evolution is very similar for the monitored parent species on the inbound and outbound path.

Do the measurements constrain the nucleus interior? To see how we can use observations like in Fig. 8 to study the sublimation processes, we discuss the observations in view of the two extremes of a pure amorphous and pure crystalline nucleus. First, we treat comet Hale–Bopp as an unprocessed nucleus consisting of a homogeneous mixture of amorphous ice. As the comet approaches the Sun, we expect crystallization of its surface layers that should lead to differences of the production rates between pre- and post-perihelion. This is not observed. The difference is illustrated when comparing model predictions based on amorphous water ice with some CO trapped and some CO in a separate phase (Fig. 9) to the observed activity evolution (e.g., [73,175]). We note that in the models CO is trapped by amorphous water ice only by a few percent. Most of CO is assumed to condense in a separate phase, able to outgas from the interior of the nucleus, similar to the scenario for crystalline water ice outlined above. The models predict the increase in CO production rate on the pre-perihelion path, although the simulations come to different results on where the crystallization of the amorphous ice sets in and on the detailed subsequent activity evolution. However, these models do not agree with the post-perihelion evolution of CO. Updated models, taking into account the improved knowledge on comet Hale–Bopp (Fig. 10) give significantly improved results. Nevertheless, the CO production rate post-perihelion is still high. This effect is attributed to the thermal inertia of the nucleus causing seasonal effects. For more realistic models, we need to know parameters, such as the nucleus size and the rotation axis, which are usually not available for a modeled comet. Nevertheless, there remain differences between models

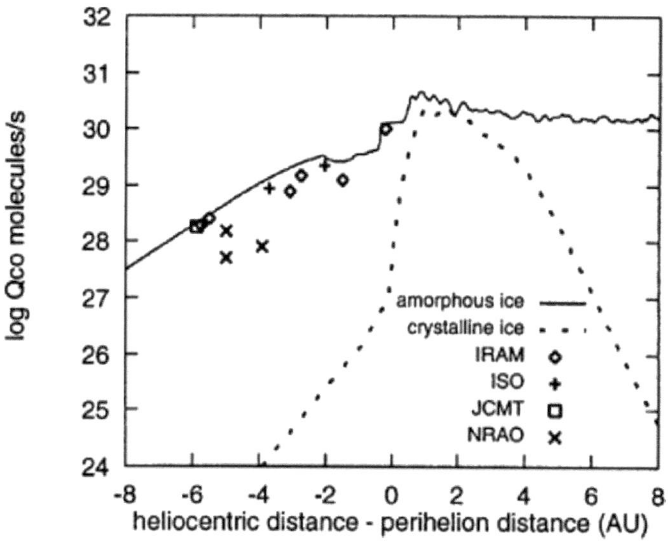

Fig. 9. Comparison of the CO production rates with measurements [73]. The modeled nucleus contained initially amorphous or crystalline H_2O ice

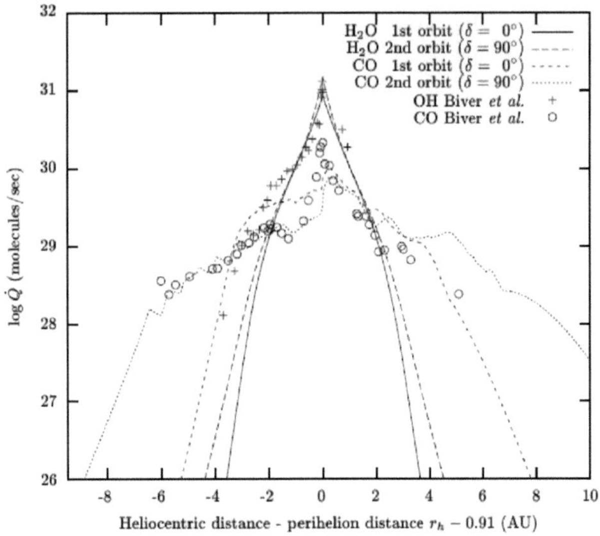

Fig. 10. Comparison of H_2O and CO production rates with measurements [74]. The modeled nucleus contained initially 40% amorphous H_2O ice, 5% CO trapped in amorphous water ice, 5% CO condensed independently, and 50% dust (by mass). The two orbits shown differ by the assumed orientation of the pole axis

and observations, even for well-studied comets such as Hale–Bopp. This is not really surprising in view of the model uncertainties. In addition, the comet was not on its first orbit into the inner solar system, and some processing probably has already happened.

We may then ask what we would expect in the other case of a purely crystalline surface layer. As outlined above, the sublimation fronts may have already moved deep into the interior of the nucleus. In the extreme case, the production rates of highly volatiles, such as CO, may not vary significantly along the orbit anymore. Modeling the comet coming into the solar system for the first time with crystalline water ice results in much lower CO gas production rates (Fig. 9), because the CO sublimation front has already moved several meters into the nucleus [73]. This is clearly not observed. Hale–Bopp's nucleus, therefore, seems not to be in a highly differentiated state. Again this is not surprising, because the orbital heat wave will penetrate to several tenths to hundreds of meters for a long-period comet such as Hale–Bopp. As a result, many revolutions are needed to reach an evolved and highly differentiated state. Possibly, this state is never reached completely, if effects such as efficient surface erosion during perihelion passage are taken into account.

When observing the evolution of gas activity, it is interesting to study the region of the onset of activity of the volatiles, because the onset depends on their volatility and in addition on the heat conduction into the nucleus interior. Unfortunately, the onset occurs at large r_{h} for most ices except H_2O (Fig. 4). However, HCN sublimation is expected to start around 7–8 AU heliocentric distance. As HCN and and its daughter product CN show strong emission bands, measurements over a wide range of r_h were possible for comet Hale–Bopp (Fig. 11), and the region of activity onset could be probed. A comparison of the observations to models solving the surface energy balance on the nucleus has been made for various heat conductivity parameters (Fig. 11). The observations are consistent with sublimation of HCN close to the surface or a very low heat conductivity of the nucleus [178]. Again, no sign of significant differentiation of the nucleus is found from observational data.

Hale–Bopp's activity evolution seems to be different from models treating the nucleus in terms of an old, evolved and differentiated body, but also from models treating it as a newcomer made of amorphous ice. More complex models to simulate the observations have been made in the meantime (see [174] for a recent overview), but it is still difficult to understand the evolution of the production rates observed. Unfortunately, most sublimation models do not include ices of intermediate volatility, such as HCN, for a comparison to observations.

The release of CO from extended coma sources (see Sect. 5) may complicate the discussion further, because it can lead to a significant increase of the CO production rate with decreasing heliocentric distance resulting from

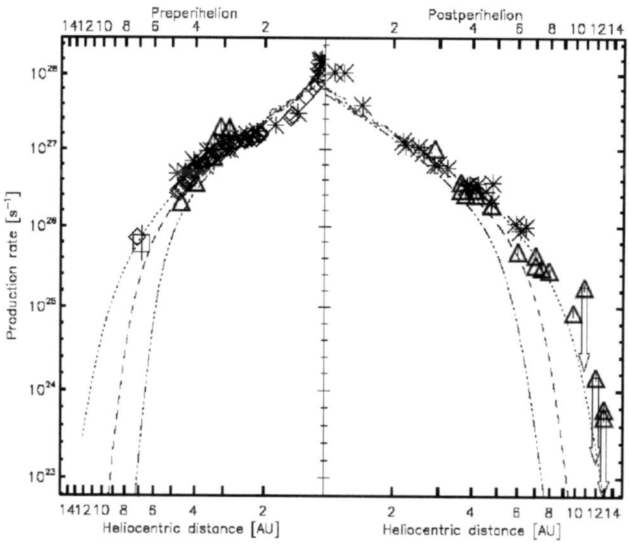

Fig. 11. Comparison of HCN and CN production rates over heliocentric distance [178]. Lines indicate sublimation models with different heat conductivities of the nucleus (dotted: k = $0\,\mathrm{W\,m^{-1}\,K^{-1}}$; dashed: k = $0.005\,\mathrm{W\,m^{-1}\,K^{-1}}$; dashed-dotted: k = $0.05\,\mathrm{W\,m^{-1}\,K^{-1}}$)

coma processes. This process is not included in the nucleus outgassing models and may explain part of the discrepancy between models and observations for molecules with coma sources. However, the significance of extended coma sources is reduced for most observed volatiles, showing again that it is important to incorporate several volatile species into the sublimation models.

The interior of short-period comets might show a higher state of differentiation than comet Hale–Bopp, because they remain closer to the Sun and therefore at higher temperatures. Unfortunately, in short-period comets, only daughter products could be monitored over a wide range of r_h so far. They provide a less stringent constraint on the differentiation processes taking place, because we add uncertainty by modeling their parent production rates with chemical models (see Sect. 5). In addition, only for few comets, the heliocentric distance range covered by observations of gas emissions is extending beyond $r_{\mathrm{h}} \approx 2.5\,\mathrm{AU}$, and we therefore only have data in the water-dominated regime.

The variation of production rate with heliocentric distance is often expressed by a power law, r_{h}^{k}, by fitting the slope of $\log(Q)$ over $\log(r_{\mathrm{h}})$. The resulting exponents, k, can vary a lot from comet to comet. For example, values of k from -0.8 to -10.1 for $Q(CN)$ have been determined [2], but the most extreme values are usually found for comets observed over only a small range in r_{h}. This already illustrates a major problem when studying a comet's activity evolution. Temporal variations (rotation, outbursts, etc.) require good coverage in r_{h} to disentangle these relatively short-term effects from the long-term orbital evolution. Extrapolation of often only poorly known model parameters, such as scale lengths (see Sect. 5), to large r_{h} can additionally introduce false distance dependencies on the production rates derived. Additional complications are found for species released by extended coma sources, as already mentioned. In view of the large number of uncertainties, one needs to be cautious when interpreting observations based only on few data points, in particular when comparing results derived with different models and parameters. Clearly, a larger statistical sample is needed.

Several surveys have been made to study the gas activity evolution with heliocentric distance so far, with different results:

- Surveys of comets made by photometric and spectroscopic measurements in the optical range [2,41,82,167] show similar production rate evolutions for CN, NH, and C_3, resulting in constant production rate ratios over distance.
- For the NH_2/CN ratio, a decrease with r_{h} has been reported [17,41]. However, this is difficult to understand, because NH does not seem to show this effect. Possibly, heliocentric distance dependencies for NH_2 are introduced by uncertainties in the excitation models used for these measurements.
- The C_2/CN ratio has been reported to decrease with heliocentric distance in some observations. While the study of five comets [167] showed a strong dependence of the C_2/CN ratio on r_{h}, only little or no variation was found

in larger samples [2, 41]. However, C_2 is likely to be a grand-daughter product from C_2H_2 and possibly additional parents, such as C_2H_6 and other organic species, which complicate the determination of production rates. The formation of C_2 needs to be clarified further before we can finally conclude to what extent the production rate of C_2 evolves different to other species or whether the differences seen at present are an expression of incomplete modeling (see Sect. 5).

- Comparing the evolution of daughter products originating from parent ices more volatile than water to the OH production rates shows in general no strong correlation. It seems OH production rates can as well vary more steeply or shallow than the minor volatiles with r_h [2]. This is difficult to understand in view of sublimation models. However, the database is sparse and may simply reflect the lack of a statistically significant number of good quality data points.

The large error bars in production rate determinations of short-period comets make the study of their orbital evolution difficult. Nevertheless, we note that all species vary over the orbit. Unfortunately, we do not have data on the CO production rate of a Jupiter family comet, which we could compare to model predictions. However, other volatile parent molecules, such as C_2H_2 and C_2H_6, would also be expected to show a different activity evolution compared with less volatile ices, such as HCN (Fig. 4) if the nuclei would be highly differentiated.

To summarize, the observations of gas production rates over heliocentric distance in comets are still insufficient to give a clear and statistically significant picture of the activity evolution of volatile species. We could turn the argument around and conclude that so far no signs for highly evolved and differentiated nuclei can be derived from the observations. However, we will need more observational constraints to further improve our understanding of the sublimation activity of cometary nuclei.

The aim of our discussion was to outline the principle of how production rate observations can be used to constrain nucleus models. Three parameters are important to provide helpful data in future:

- Observe species of different volatility, such H_2O and CO, but also ices with intermediate volatility to enlarge the data base for model comparisons.
- Observe over a wide range of heliocentric distances with good coverage to smooth out short-term temporal effects.
- Concentrate at large heliocentric distances, where the onset of activity can be observed.

The latter point is important, because the onset of activity is sensitive to the nucleus heat conduction. Unfortunately, it is usually difficult to observe, because the onset occurs at large r_h for the volatiles (Fig. 4). It has been done for H_2O and HCN/CN in comet Hale–Bopp but will, however, be extremely difficult for normal comets.

Additional clues may come from comets active at very large distances, e.g., objects such as Chiron and other Centaur objects. Their outgassing is not driven by water, but by highly volatile ices such as CO, and they provide additional clues to the sublimation of these volatiles in addition to the study of normal comets (see [154]).

Finally, the lander of ESA's Rosetta mission will study the surface layers of comet 67P/Churyumov–Gerasimenko in situ and measure the composition and structure of the top surface layers. Obviously, this will provide data input for a big step forward in refining sublimation models. A major question, also addressed by the ground-based observations outlined above, is of course whether the composition measured in the top layers is pristine or suffered from severe differentiation.

3 Coma and Tail Dynamics

In this section, we discuss the dynamical processes of gas molecules after their sublimation from the nucleus surface. We take the parameters $r_{\mathrm{h}} = 1\,\mathrm{AU}$ and $Q = 10^{30}$ molecules s^{-1} as a reference case. The conditions are similar to the values of comet Halley near the encounter of the Giotto spacecraft, which provided us with detailed in situ values of the inner coma. This is still the most comprehensive data set about a comet to date. Halley, therefore, is our reference comet in the following, unless specified explicitly. Obviously, a full and comprehensive description of gas-dust dynamics and its application is beyond the scope of this introductional text. We refer to recent reviews, such as [52] and [47], for a detailed overview.

3.1 Dynamics of the Neutral Coma

The Basic Scenario

The gas molecules sublimate from the cometary nucleus and accelerate into the coma by (adiabatic) expansion into vacuum. Figure 12 illustrates schematically the principle processes in the coma:

- The main gas acceleration occurs within the first few kilometers above the surface. After a few tens to hundreds of kilometers a mean gas velocity of the order of $1\,\mathrm{km\,s^{-1}}$ is reached. Beyond a few 10^3 km, the gas accelerates again, because it is heated in the intermediate coma by photolytic processes.
- The gas density decreases quickly as the gas expands (as $1/\mathrm{r}^2$ in case of isotropic expansion and constant velocity).
- The gas coma temperature drops from about 200 K above an active region on the surface to approximately 100 K at a nucleocentric distance of about 10^2–10^3 km. In case of pure adiabatic gas expansion, the temperatures are

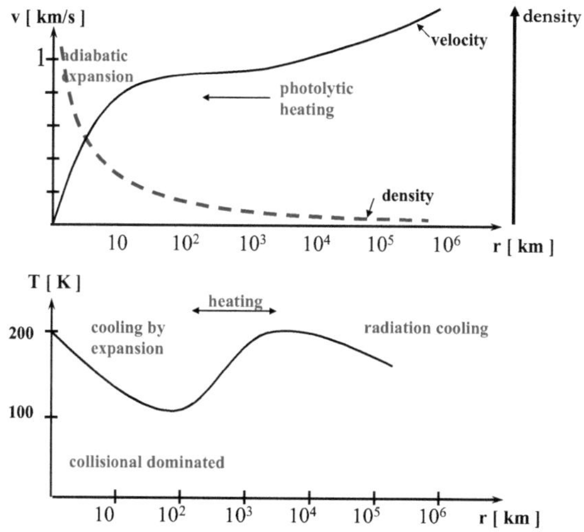

Fig. 12. Illustration of the principle processes in a cometary coma for a Halley-type comet near $r_\mathrm{h} = 1\,\mathrm{AU}$

expected to drop even lower, down to about 20 K. Heating mechanisms such as gas–dust interaction or recondensation are discussed to explain the somewhat higher observed gas coma temperatures at these distances [52].

- At larger nucleocentric distances (10^3–10^4 km), heating of the gas by photolytic processes is important resulting in an increasing gas temperature in the intermediate coma. The main heating process is photo-dissociation of water molecules into OH and H. The dissociation provides an excess energy to the daughter products, which results in molecule excess velocities in the order of $18\,\mathrm{km\,s^{-1}}$ for H and about $1.09\,\mathrm{km\,s^{-1}}$ for OH molecules (see [47] for details on excess velocities and branching ratios of H_2O photodissociation).
- At large nucleocentric distances, radiative cooling of the coma molecules decreases the temperature again.

The Choice of Mathematical Description

Above the surface, a Maxwellian velocity distribution is established in the sublimated gas after a few collisions. For a comet like Halley, the flow is dominated by collisions for the first kilometers in the coma and can be described by hydrodynamic equations. At distances beyond several 10^4 km, however, collisions are rare due to the low gas densities, and the coma can be described as a free molecular flow. At such large distances, the influence of solar gravitation and radiation pressure is important. Solar radiation pressure accelerates the gas into the anti-solar direction and leads to a deviation of the coma from spherical shape on a large scale (Fig. 13).

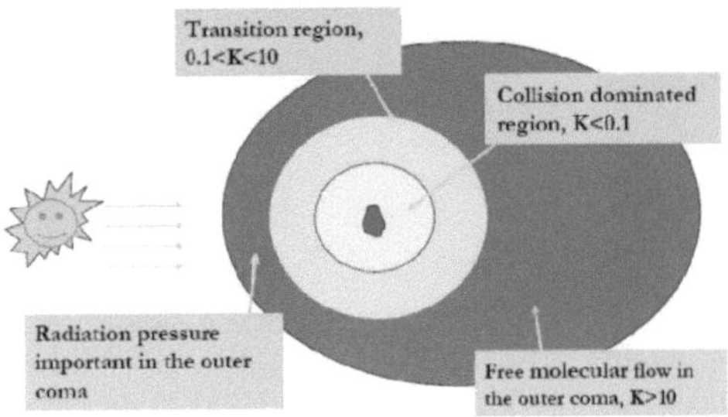

Fig. 13. Illustration of the principle flow regimes in a cometary coma for a Halley-type comet near $r_{\mathrm{h}} = 1\,\mathrm{AU}$

In general, a hydrodynamic description of the flow can be applied if the mean free path between collisions of two gas molecules, λ, is small against a characteristic length of the system, L, i.e. $\lambda < L$, with

$$\lambda = \frac{1}{\sqrt{2}\sigma n}. \tag{4}$$

On the surface we can write:

$$\lambda = \frac{u}{\sqrt{2}\sigma Z}. \tag{5}$$

Here, σ is the collisional cross-section of the gas molecules, n the gas number density, u the gas velocity at the surface, and Z the surface sublimation rate (molecules $\mathrm{s^{-1}\,cm^{-2}}$).

Often the Knudsen number $K = \frac{\lambda}{L}$ is used to characterize the flow regime (Fig. 13 and 14). Inviscid hydrodynamics can be used for K < 0.1. For K > 0.1 the flow becomes viscous but is still hydrodynamic (described by the Navier–Stokes equations), and for K > 10 we have to treat the flow as free molecular outflow.

Assuming L to be equivalent to the nucleus radius, r_{nucleus}, the size of the collisionally dominated coma is about 10^3–10^4 km. The choice of r_{nucleus} as a characteristic length L is somewhat artificial. Alternatively, the radial distance to the nucleus, r, is sometimes used.

Whenever a description by hydrodynamics is not applicable, Monte-Carlo models provide an alternative approach, although they are computationally time intensive. In a Monte-Carlo approach, the real gas is approximated by a large number of simulated molecules moving in a grid space and with time. Position, velocity, and energy exchange by collisions between particles are computed and monitored. Monte-Carlo approaches are, for example, mandatory for modeling the huge hydrogen coma of comets (e.g., [47]). They are also needed when modeling the outer regions of a coma or weakly active comets with low gas densities that never form a collisionally dominated coma region. If hydrodynamics can be applied, it is more efficient, but Monte-Carlo

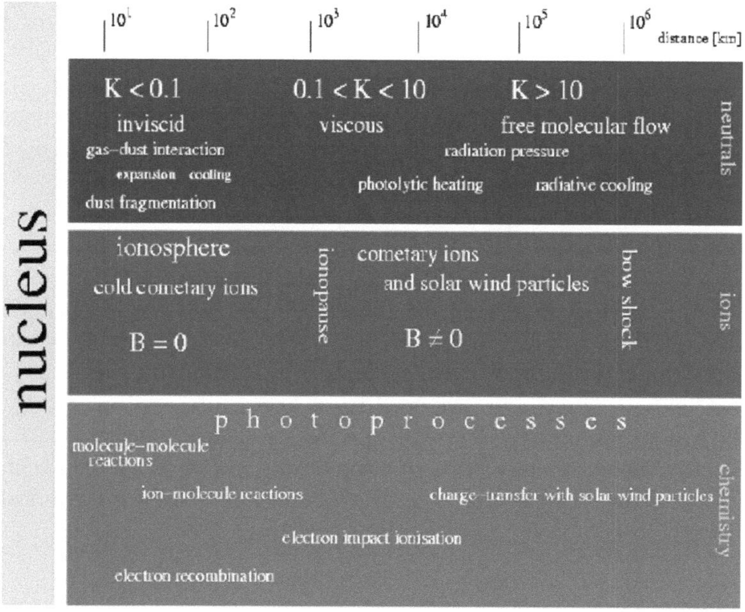

Fig. 14. Schematic diagram indicating the various flow regimes and physical and chemical processes in the cometary coma for a Halley-type comet at 1 AU heliocentric distance

approaches still form a valuable complementary method to compute the gas distribution.

In summary, the choice of mathematical method depends on the coma region and conditions studied:

- **Immediately above the surface:** Gas-kinetic approach (Boltzmann equation) include upper surface layers, Monte-Carlo models; this region is often called "Knudsen-layer."
- **Collisionally dominated coma region:** Set of hydrodynamic equations for the gas and dust components.
- **Transition region:** Description of viscous flow, gas-kinetic and/or hydrodynamic approach (check both).
- **Outer coma region:** Free molecular gas flow, Monte-Carlo models.
- **Gas and dust tails:** Free flow; molecules and dust particles move in the solar gravitational field under the influence of solar radiation pressure on Keplerian trajectories.

Hydrodynamic Description of the Gas Coma Expansion

The inner coma region is studied by space missions and therefore of special interest. Furthermore, within the inner few hundred kilometers above the cometary surface, the starting conditions for the flow observed on a large scale, e.g., from ground, are defined.

If K < 0.1, the flow can be described by the Euler equations for inviscid flow. The mass conservation equation, describing the variation of density, ρ, by volume changes of the expanding gas and by external gas sources, Q:

$$\frac{\partial \rho}{\partial t} + \nabla(\rho \boldsymbol{u}) = Q \tag{6}$$

Here, the external sources Q are simply given by the comets gas production rate, if no chemical coma reactions are taken into account. When including the production and destruction of gas molecules by chemical processes, changing number densities of each species need to be taken into account in the source term by solving in addition a chemical reaction network (see Sect. 5).

The momentum conservation equation describing the acceleration of a fluid element by pressure gradients, ∇p, or external forces, $\boldsymbol{F}$, e.g., gravitation or gas-dust interaction forces, is given by:

$$\frac{\partial \rho \boldsymbol{u}}{\partial t} + \nabla(\rho \boldsymbol{u}\boldsymbol{u}) + \nabla p = \rho \boldsymbol{F} \tag{7}$$

And finally, the energy conservation equation is:

$$\frac{\partial \rho e}{\partial t} + \nabla(h + \frac{1}{2}\boldsymbol{u}^2)\rho \boldsymbol{u} = E \tag{8}$$

with the energy density:

$$e = \varepsilon + \frac{1}{2}\rho \boldsymbol{u}^2 \tag{9}$$

and

$$h = \varepsilon + \frac{p}{\rho} \tag{10}$$

Here, ε is the inner energy and h the specific enthalpy. If gas molecules that move with distinctively different velocities compared to the dominant water molecules are considered (e.g., H atoms), a set of hydrodynamic equations for each of these species is needed, including the source terms for momentum exchange. For ions and electrons, a set of magneto-hydrodynamic equations must be used, taking into account the interaction with the solar wind (see Sect. 3.4).

In addition, there are analog equations for the dust particles, ρ_{d}, in each size interval i:

$$\frac{\partial \rho_{\mathrm{d},i}}{\partial t} + \nabla(\rho_{\mathrm{d},i}\boldsymbol{u}_{\mathbf{d},\boldsymbol{i}}) = Q_{\mathrm{d},i} \tag{11}$$

$$\frac{\partial \rho_{\mathrm{d},i} \boldsymbol{u}_{\mathbf{d},i}}{\partial t} + \nabla(\rho_{\mathrm{d},i} \boldsymbol{u}_{\mathbf{d},i} \boldsymbol{u}_{\mathbf{d},i}) = \rho_{\mathrm{d},i} \boldsymbol{F}_{\mathbf{d},i} \tag{12}$$

The hydrodynamic equations for gas and dust are complemented by the equation of state for an ideal gas:

$$\varepsilon = \frac{p}{\rho(\gamma - 1)} = \frac{R_{\mathrm{g}} T}{\gamma - 1} \tag{13}$$

The surface conditions can be expressed as

$$T_0 = T\left(1 + \frac{\gamma - 1}{2}\right)^{-1};\ u_0 = c_0 = \sqrt{\gamma R_{\mathrm{g}} T_0};\ \rho_0 = \frac{Z_0}{u_0};\ p_0 = \rho_0 R_{\mathrm{g}} T_0 \tag{14}$$

when assuming a reservoir outflow model with the reservoir temperature assumed equal to the nucleus surface temperature. These equations then apply after a few free scale lengths above the surface. We note that the knowledge of the starting conditions at the nucleus surface is an important but difficult constraint for any coma model. For the first few collisions above the surface, there is the Knudsen layer, as described above. However, detailed calculations show that the results of the reservoir model are similar to the detailed Knudsen layer calculation.

As a first estimate for the gas velocity, we can use the equation for the limiting velocity of expansion from a gas reservoir into vacuum:

$$u_{\mathrm{max}} = \sqrt{\frac{2\gamma}{\gamma - 1} R_{\mathrm{g}} T_{\mathrm{r}}} \tag{15}$$

If we set T_{r}, the temperature of the reservoir, equal to the surface temperature of an active region of a comet near 1 AU and assume $\gamma = \frac{4}{3}$ for water vapor, we obtain a terminal gas velocity of about $0.86\,\mathrm{km\,s^{-1}}$. This velocity is close to the values measured in the intermediate coma of comets near 1 AU. Figure 15 shows how the gas velocity and temperature change for different cometary gas production rates, Q.

In real comae, the gas flow depends also on heating by dust particles, mass loading by fragmenting dust and, of course, on the heliocentric distance of a comet. A detailed discussion on gas coma velocities is found in [47].

Acceleration of Dust Particles in the Inner Coma and Their Effect on the Gas Flow

The cometary dust particles are coupled to the gas flow for the first few kilometers above the surface. They are accelerated by collisions with gas molecules to velocities of a few 10–100 $\mathrm{m\,s^{-1}}$, depending on their size, shape, and density. The dynamics of dust particles on a larger scale is discussed in Sect. 3.3.

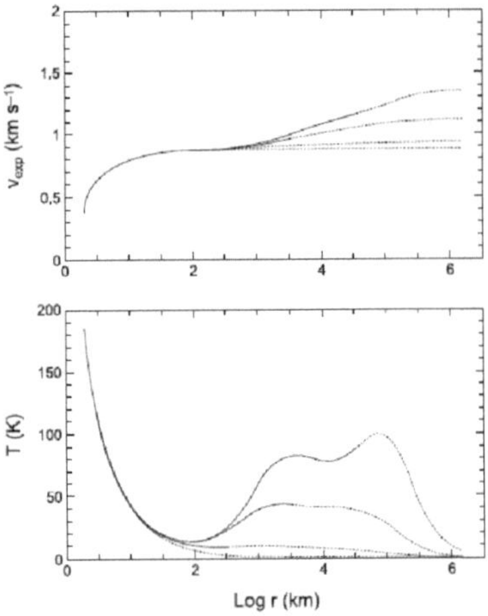

Fig. 15. Gas velocity and temperature in the coma for different gas production rates [33]. Top diagram: top curve: $Q = 10^{27}\,\mathrm{s}^{-1}$, until bottom curve: $10^{30}\,\mathrm{s}^{-1}$. Bottom diagram: again from low to high gas production rates. In the dotted regions, fluid dynamics does not apply anymore

The main factors affecting the acceleration of dust in the cometary coma can be discussed already when looking at the simple scenario of acceleration of a spherical dust particle by the gas flow in the free molecular flow approximation [104]:

$$\frac{\mathrm{d}\boldsymbol{u}_\mathbf{d}}{\mathrm{d}t} = \frac{1}{2} C_\mathrm{D} \alpha \rho_\mathrm{g} |\boldsymbol{u} - \boldsymbol{u}_\mathbf{d}| (\boldsymbol{u} - \boldsymbol{u}_\mathbf{d}) \tag{16}$$

with:

$$\alpha = \frac{\pi a^2}{m_\mathrm{d}} = \frac{3}{4} \frac{1}{\rho_\mathrm{d} a} \tag{17}$$

The acceleration depends on dust particle mass, m_d, radius, a, relative gas velocity, $\boldsymbol{u}$, and dust velocity, $\boldsymbol{u}_\mathbf{d}$. ρ_g denotes the gas density. The main factor affecting the coupling of dust particles to the gas is the cross-sectional area to mass ratio, α. The drag coefficient, C_D, depends only slightly on the shape and structure of the dust particles. So we find the following general behavior of dust particle acceleration:

- Light particles are accelerated more efficiently than massive particles.
- For a given density, small particles are accelerated more easily than large particles (Fig. 16). Therefore, small grains follow the gas flow longer than large grains.
- The gas density decreases rapidly with increasing nucleocentric distance. Furthermore, the drag coefficient, C_D, is a strong function of temperature that also drops steeply in the inner coma. Therefore, C_D decreases in the

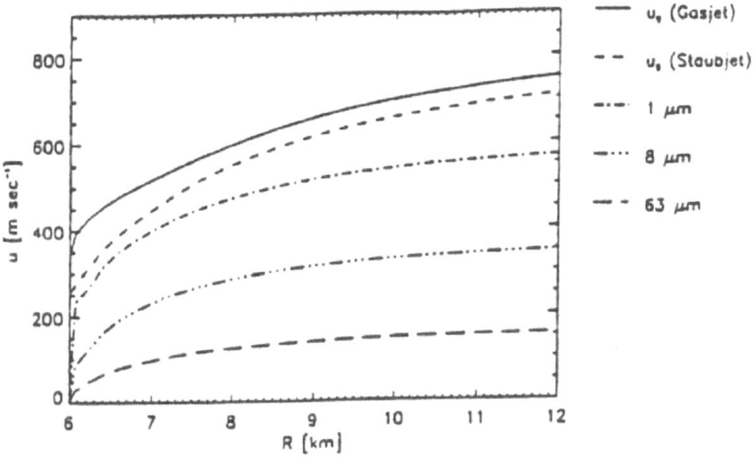

Fig. 16. Dust particle acceleration depends on the radius of the particle. The figure shows the dust particle velocity for different radii. The solid line shows for comparison the gas velocity in the dust loaded flow [132]

first kilometers above the surface and dust acceleration is efficient only in the innermost coma.

- In the intermediate coma, the dust particles decouple from the gas flow and move on trajectories according to their velocity just before decoupling. Further out, gravitation and solar radiation pressure determine their dynamics.

What happens to the gas flow when dust particles are added? The mass loading by dust reduces the initial gas outflow velocity. However, the hotter dust particles then heat the flow and lead to a faster acceleration of the gas. Finally, again gas velocities around 0.86 km s^{-1} are reached for the gas flow in the intermediate coma.

The images obtained by the HMC camera on board the Giotto spacecraft visiting comet Halley provided indications for fragmentation of dust particles in the coma within a few kilometers above the surface. Such small and hot fragmenting dust particles heat the near-nucleus coma and can therefore also modify the near-nucleus dynamics.

Measured Coma Gas Velocities and Temperatures

Figure 17 shows the velocity of H_2O gas measured in situ at comet Halley by the Giotto spacecraft [134, 139]. The acceleration of the molecules can be seen. In the inner coma, velocities are near the value predicted from adiabatic gas expansion (0.86 km s^{-1}). The velocity reaches 1.1 km s^{-1} within the first 40000 km above the nucleus.

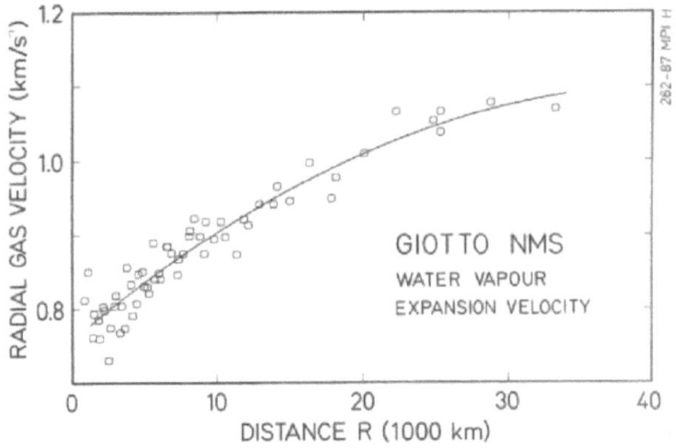

Fig. 17. Gas velocity measured in situ at comet Halley by the Giotto neutral mass spectrometer [134]

The temperature and gas velocity can also be determined by ground-based observations, although at a much lower spatial resolution and only for sufficiently bright comets. These observations are made in the infrared (IR) and at radio wavelengths range.

At IR wavelengths, recent developments of new telescopes and instruments allowed detecting molecules such as CO with long-slit spectroscopy. With high spectral and spatial resolution observations of CO lines in the near-IR range, it has been possible to study the heating processes in the coma with increasing nucleocentric distance in bright comets (e.g., [66]).

At radio wavelengths (mm and sub-mm wavelengths), velocities and temperatures of parent molecules can be measured by high-spectral resolution observations. At these wavelengths, emission lines can be fully resolved and allow us to measure the Doppler shift of the line and also to analyze the line shape. However, the beam size of radio observations is usually very large and does not allow us to spatially resolve the coma (although some spatial resolution is obtained by mosaics of pointings or interferometers). The measured emission signal results from the inner to the intermediate coma, where densities of the observed molecules are high. As illustrated in Fig. 12, the velocities and temperatures vary within this coma region. Therefore, the measurements of gas velocity in the radio range correspond to a kind of "weighted" average of the inner coma. Measurements of the gas expansion velocities of comets near 1 AU give velocities around $1\,\mathrm{km\,s^{-1}}$, as we would expect in the intermediate coma. Generally, for comets at $r_\mathrm{h} < 3\,\mathrm{AU}$ velocities around $0.6–1.8\,\mathrm{km\,s^{-1}}$ are found, and the expansion velocity increases with decreasing heliocentric distance [141,189,192]. In addition, the investigation of OH emissions in several comets indicates a dependence on the cometary gas production rate [31]. Analysis of H_2O, HCN, and OH emission profiles results in somewhat different velocities. Line asymmetries are often observed in radio lines (e.g., Fig. 18) and are interpreted as indicators for asymmetric outgassing.

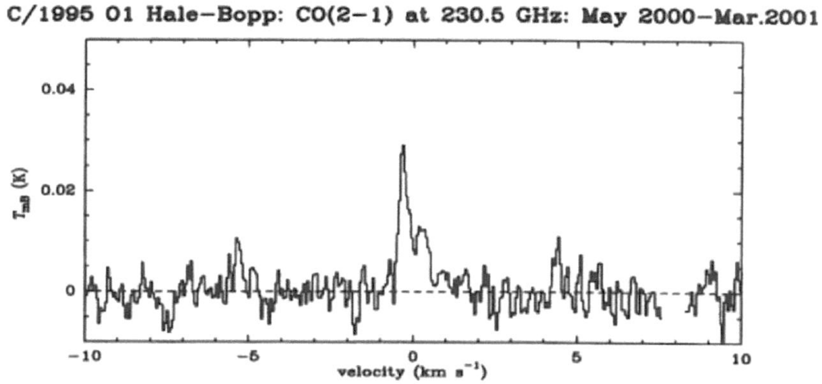

Fig. 18. Radio emission line of CO observed during the approach of comet Hale–Bopp to perihelion [24]. The line shape is asymmetric, indicative for anisotropic outgassing at the comet

Bright comets allow us to measure the variation of gas velocity with heliocentric distance. Observations of comet Hale–Bopp [24, 25] provided observational evidence for a scaling law of the expansion velocity, based on measurements extending beyond $r_\mathrm{h} = 8\,\mathrm{AU}$. Somewhat different scaling laws were found pre- and post-perihelion (Fig. 19). On average, the velocity scaled as:

$$u = 1.116(\pm 0.014) r_\mathrm{h}^{-0.40(\pm 0.01)}\, km\, s^{-1}.$$

In summary, the recent observations of u at radio wavelengths suggest that a scaling law like $u = a r_\mathrm{h}^{-b}$ (with a near $1\,\mathrm{km\,s^{-1}}$ and an exponent of $b \approx 0.5$)

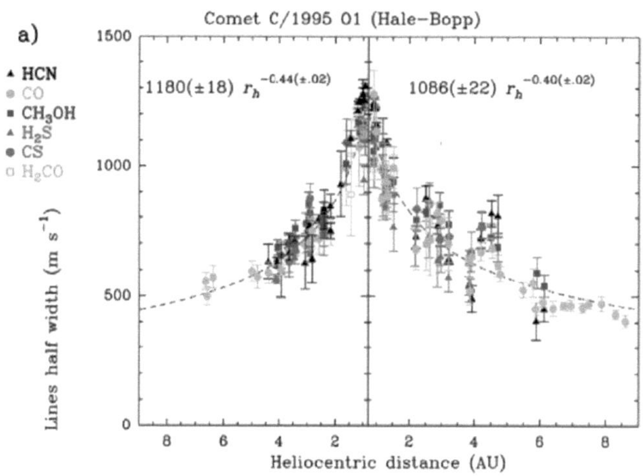

Fig. 19. Gas velocity over heliocentric distance determined from measurements at radio wavelengths of comet Hale–Bopp [25]

provides a reasonable extrapolation of the gas expansion velocity to large r_h when production rates need to be determined in comets for which no direct measurement of u can be made.

Remote observations from ground at radio wavelengths usually provide insufficient resolution to study the nucleocentric temperature profile, but allow us to derive an "average" rotational temperature (see Sect. 4) determined by the conditions in the collisional zone covered in the field-of-view (FOV) of the observations. In principle, as the line excitation is mainly caused by collisions, the derived T_{rot} corresponds to T_{kin} at the last collision of the molecules.

Again, comet Hale–Bopp was the first comet that allowed measurements of T_{rot} (e.g., [25]) over a wide range of heliocentric distances (Fig. 20). Rotational temperatures in the coma dropped from about 130 K at perihelion to about 10 K at 7–8 AU. Again, the evolution can be approximated by a power law:

$$T = 103(\pm 7) r_h^{-1.10(\pm 0.08)} \mathrm{K}.$$

Gas and Dust Jets

In many images of comae, sunward outgassing is dominating the gas flow and sunward and tailward asymmetries are often observed. Furthermore, gas and dust jet structures are present. Major gas jets are believed to originate from localized regions on the cometary surface with enhanced sublimation activity. Such increased sublimation can be caused by local differences in the ice/dust content, differences in chemical ice composition and locally different heat flow efficiencies into the nucleus interior. In addition, surface topography

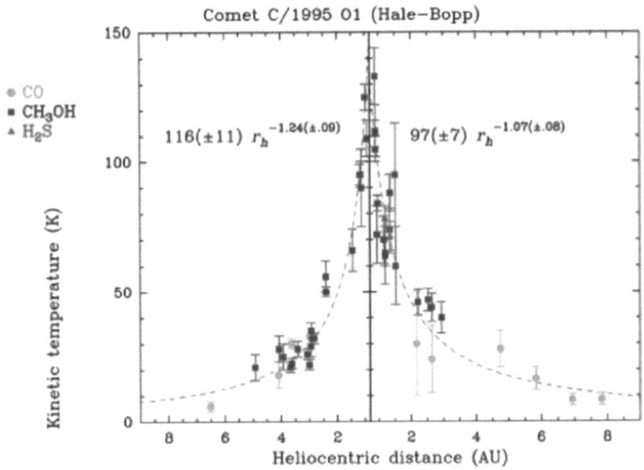

Fig. 20. Rotational temperature over heliocentric distance determined from measurements at radio wavelengths of comet Hale–Bopp [25]

leads to variations of the solar flux incident angle over the surface, resulting in differences of the energy available for sublimation.

To study the formation of jets in the inner coma, it is instructive to recall the expansion of a free jet from a nozzle into vacuum (Fig. 21). When gas leaves the nozzle, it expands laterally over a very short distance. Then, it moves along straight streamlines within the isentropic region. The maximum opening angle of the lateral expansion depends on the size of the aperture of the nozzle, the Mach number (gas velocity/sound velocity) and the adiabatic coefficient of the gas. For H_2O gas the maximum opening angle is ≈150°. Figure 22 shows the modeled gas flow field and density above an active region on a spherical nucleus [132]. The wide lateral expansion of the gas jet around the nucleus can be seen. The gas density is also shown on a somewhat larger scale in Fig. 23.

Dust particles in the gas flow decouple from the streaming gas when the density decreases, as explained above. Therefore, the lateral expansion of the dust flow is less than for the gas. The decoupling of the dust depends on the particle size and density. Small grains show wider lateral expansion than large grains. Figure 23 shows the gas and dust density distribution above an active surface region for comparison. The difference is obvious, the lateral expansion of larger (10 µm radius) dust particles is much smaller than for the gas. Nevertheless, the dust jet above an active region is still a relatively wide feature.

Filaments

In situ images of comet Halley show very narrow structures, called filaments (Fig. 30), which become clearly visible after some contrast enhancing image processing. They have narrow opening angles <10° and column density enhancements above the coma background of only a factor of two [203]. As normal gas and dust jets are expected to show much wider opening angles (see above), this observation stimulated research to reproduce narrow straight flow structures.

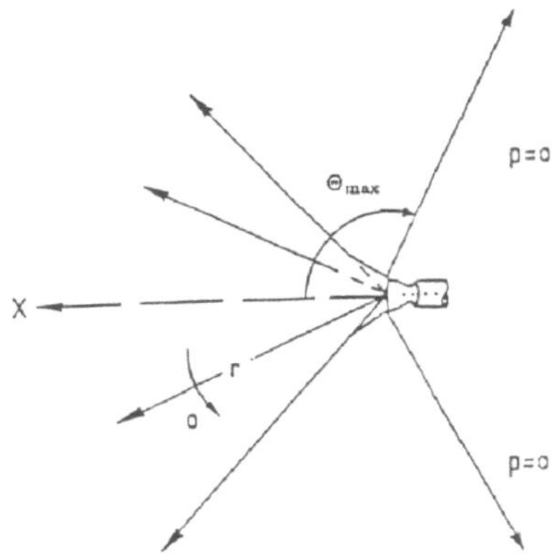

Fig. 21. Schematic view of the expansion of a free jet into vacuum [132]

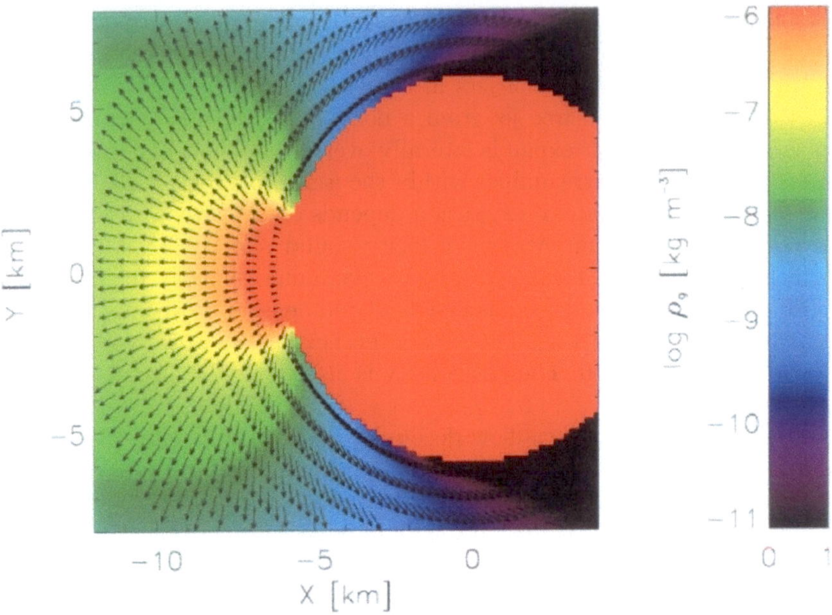

Fig. 22. Gas flow field and density distribution [132] for an active area with constant production rate on a spherical nucleus. Note the wide lateral expansion of the gas jet around the nucleus

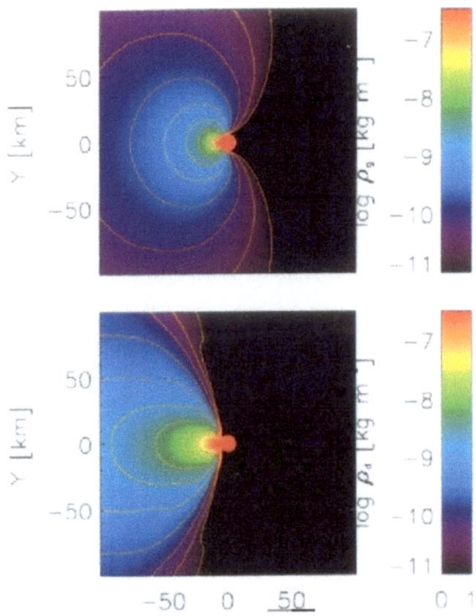

Fig. 23. Gas density distribution (top) and distribution for 10 µm dust particles (bottom) [132] in the inner coma above an active area on a spherical nucleus. Note the wide lateral expansion of the gas and the more confined distribution of the larger dust particles

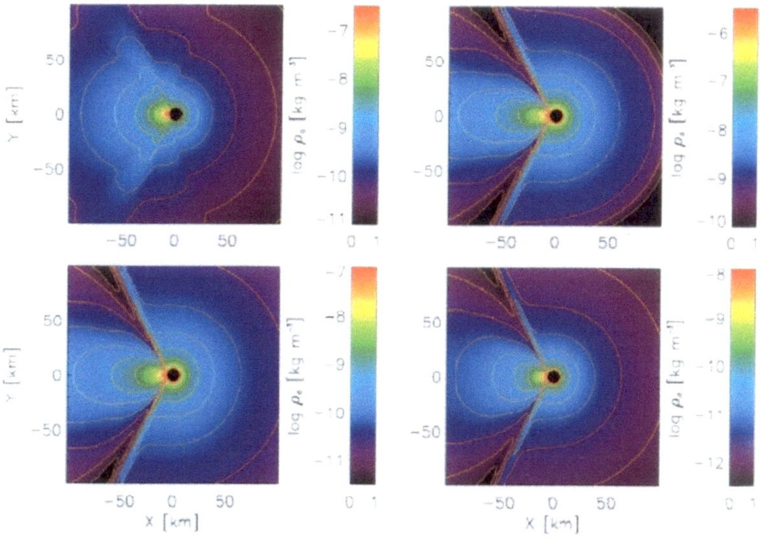

Fig. 24. Gas and dust density distribution for a jet expanding into a background gas [132]. Upper left: gas density; Upper right: dust particles with 1000 μm radius; Lower left: 30 μm radius; Lower right: 1 μm radius

To address the formation of narrow, straight filaments in the inner coma, we look at the expansion of gas into a background gas. In such a case, the gas flow is confined by the gas background, inhibiting the lateral expansion of the flow. Close to the boundaries, between outstreaming and background gas, shock systems form. Figure 24 shows the effect of the lateral expansion into a background for gas and dust particles of three sizes. At the jet boundaries, gas density enhancements form in the shock system. These enhancements would give the appearance of filaments or very confined jets in images of the near-nucleus region. The formation of narrow structures is even more pronounced for the dust. However, for comparison with observations, the integrated intensity along the line-of-sight must be computed. Looking at the coma from different aspect angles, the appearance of the interaction regions can be very different, and sometimes the narrow filament-like features produced in the interaction region are not visible (Fig. 25). Therefore, filaments formed by the interaction of gas flows may disappear as the viewing geometry changes, for example from an orbiting spacecraft or by the rotation of a nucleus as seen from Earth.

To summarize we note that

- The interaction of an outstreaming gas with a gas background produces density enhancements at the interaction region giving the appearance of filaments.
- These filaments are not located directly above an active region on the surface, but at the jet/background boundary.

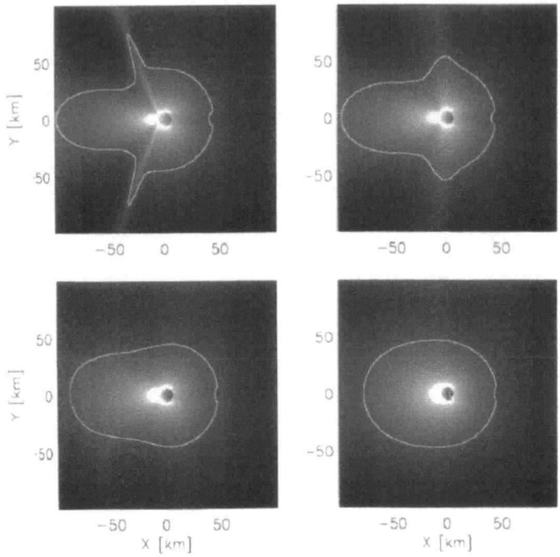

Fig. 25. Dust column density distribution for a jet expanding into a background gas for different viewing angles, Φ [132]. Upper left: $\Phi = 107°$; Upper right: $\Phi = 120°$; $\Phi = 135°$; $\Phi = 150°$. The background sublimation is set to 1% of the production rate at the active region

- These features are much narrower than expected for the free expansion of a jet into vacuum.
- Whether such filaments are indeed seen in images depends strongly on the viewing geometry (Fig. 25).

Interaction regions between streaming gas flows are not only seen for a single jet expanding into a background gas, but also for neighboring active regions. In such case, two gas jets interact, again producing shock systems with related gas and dust density enhancements (Figs. 26 and 27).

There are several possible reasons, why some areas on the surface are more active than others. An effect studied intensively in the past is the different solar illumination on an irregularly shaped nucleus. The differences in solar energy received by the various parts of the surface result in differences of surface sublimation and finally in interacting gas jet flows. Again shock systems form, giving the appearance of multiple filaments in the coma. A summary of the gas flow field around an irregular nucleus is given in [52].

Observations of Jets and Filaments

Jets

The lateral expansion of gas jets, excess energies for daughter molecules and nucleus rotation altogether lead to a relatively isotropic gas coma on a large scale (in comparison to the innermost coma), with asymmetries because

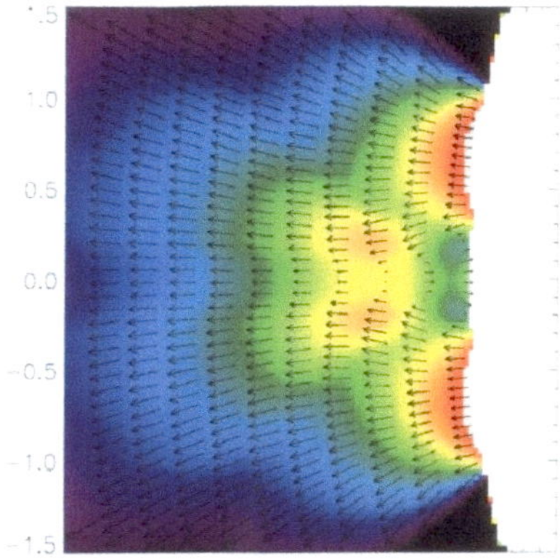

Fig. 26. Gas flow over an active region with 2 km diameter with an inactive zone in the center of 300 m radius [132]

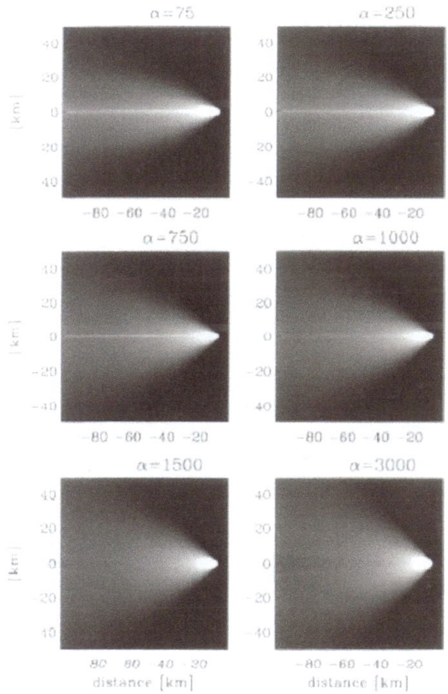

Fig. 27. Dust column densities corresponding to the flow region of Fig. 26 for different dust particle classes α. The large particles are focussed into a narrow straight filament-like structure [132]

of external forces, such as radiation pressure. Nevertheless, ground-based images of comets often show gas and dust jets. For example, in Fig. 28 two gas jets are clearly seen in an image of comet C/2004 Q2 Machholz taken through a CN filter. The continuum image shows a sun-tail asymmetry, indicating preferred sunward outgassing of the nucleus. The spatial distribution of the jets observed for gas molecules and dust particles is usually not the same. This is not surprising because gas and dust decouple after a few kilometers in the coma, as described above. Gas molecules and dust particles then move outward with different velocity. On a rotating nucleus, their spatial distribution is then expected to be different. The gas jets are often broader than the dust in agreement with a more efficient lateral expansion of the gas in comparison to the visible dust. Furthermore, jets observed in light of daughter radicals that receive isotropic excess velocities after photodissociation of their parent also appear broader than parent molecule jets. A nice illustration of the fast lateral expansion from localized surface regions is the appearance of the ejecta cloud after impact of the Deep Impact probe in comet Tempel 1. About 17 h after impact, the ejecta cloud of CN is already visible all around the nucleus, whereas the dust cloud still expands mainly in the sunward direction with a much narrower opening angle (e.g., [180]).

To distinguish jets from narrow filaments, it is helpful to plot the azimuthal intensity profile in the coma, as shown in Fig. 29. Broad jet structures with intensity enhancements of several 10%, as in the example shown here, are obviously jets related to active surface regions on the nucleus surface. Filaments produced by gas flow interaction in the coma would be much narrower than the broad features seen in the azimuthal profiles.

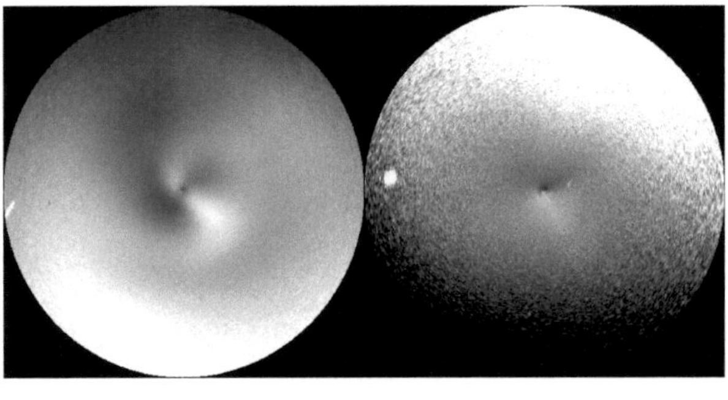

Fig. 28. Jets in comet C/2004 Q2 Machholz observed on Dez. 8/9, 2004. The comet was at $r_{\rm h} = 1.4\,\mathrm{AU}$ and $\Delta = 0.5\,\mathrm{AU}$. The field-of-view is 4.5 arcmin. The solar direction is at the bottom of each frame. At each radial distance, a mean coma intensity has been subtracted to enhance non-isotropic structures in the coma. Left: CN at 385 nm, the gray scale indicates variation of $\pm\,25\%$ from the mean value; right: dust continuum at 443 nm, the gray scale corresponds to $\pm\,15\%$ intensity deviation from the mean. Two gas jets are clearly visible in the CN frame. The dust image shows no clear jets, but enhanced intensity toward the Sun

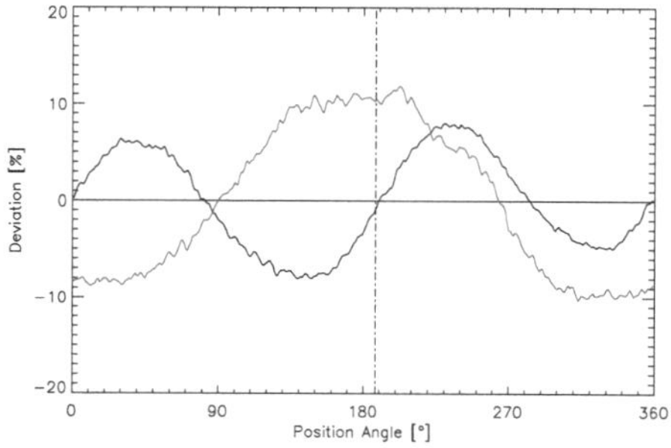

Fig. 29. Azimuthal intensity profile in the coma of comet C/2004 Q2 Machholz. The profiles correspond to the images shown in Fig. 28. The profiles have been averaged over a nucleocentric distance of 1800–3600 km. Gray line: continuum, black: CN

Filaments

Narrow straight filaments are common in the close vicinity of nuclei, as can be seen in the images of comets Halley (Fig. 30), Borrelly (Fig. 31), and Wild 2 (Fig. 32). Such filaments are narrow and faint structures, clearly different to the broad jets with relatively strong contrast to the mean coma density.

It is not obvious how to relate jets and filaments to active regions on the nucleus surface. How can we find out what we are looking at? In general, gas and dust jets produced by an isolated active region on the nucleus surface are expected to have the following appearance:

- wide opening angle, sometimes curved appearance
- relatively high contrast to the background gas and dust
- observed on a large spatial scale
- observed over long time periods

Narrow straight filaments are characterized by:

- narrow structures, straight
- low contrast to the background gas and dust
- observed in the near-nucleus region

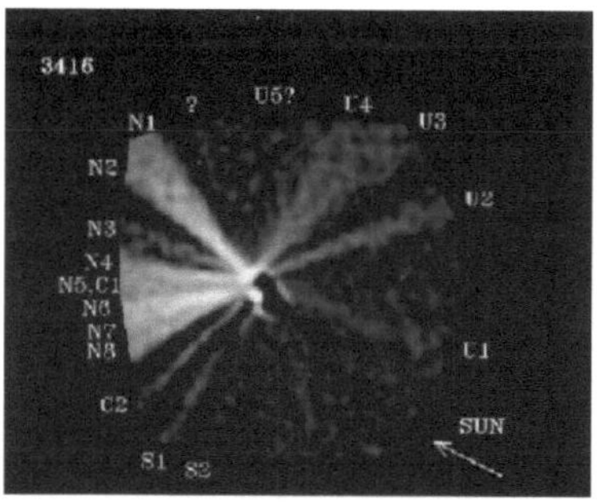

Fig. 30. Filaments seen close to the nucleus of comet Halley [203]

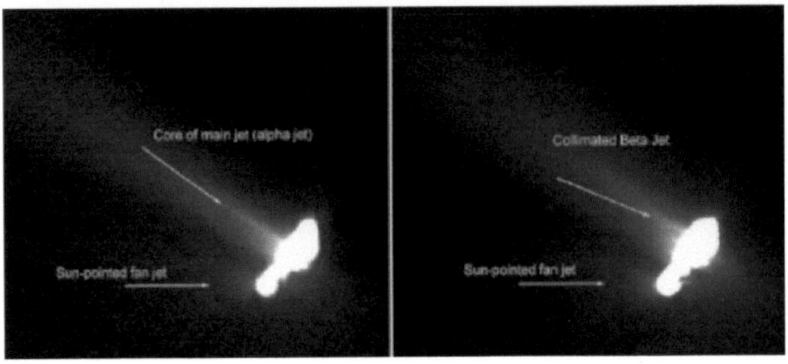

Fig. 31. Filaments seen close to the nucleus of comet Borrelly [198]

In summary, for ground-based as well as in situ images obtained in flybys, it is important to look at the contrast and opening angle of the jets and filaments seen before drawing conclusions on their relation to the location of active areas on a nucleus surface.

3.2 Dynamics in the Outer Coma and Neutral Gas Tails

On a larger scale, beyond about 10^4 km, the gas is accelerated again by heating because of photoprocesses (Fig. 12). The dominant photo reaction for heating is photodissociation of H_2O to OH and H (e.g., [47]). Further out, radiative cooling of the OH molecules begins to dominate. However, on this large scale collisions are rare and the molecules move in free flow.

In this regime, radiation pressure from the Sun is important for the gas flow. Absorption and re-emission of solar photons causes an acceleration of atoms and gas molecules in the anti-solar direction. The acceleration is small for coma molecules, but can be large for atoms with high fluorescence efficiency factors, such as hydrogen and sodium atoms. The acceleration by solar radiation pressure is given by:

$$\gamma = \frac{h}{m r_{\mathrm{h}}^2} \sum_i \frac{g_i}{\lambda_i} \tag{18}$$

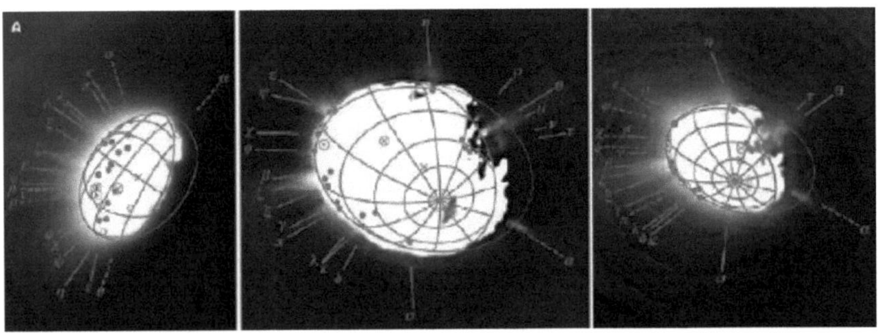

Fig. 32. Filaments seen close to the nucleus of comet Wild 2 [196]

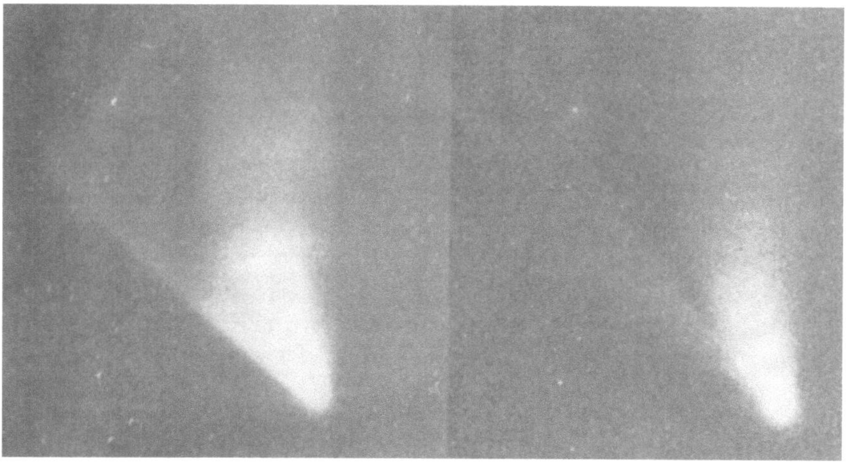

Fig. 33. The sodium tail (left) in comet Hale–Bopp and a comparison to its H_2O^+ tail (right) [51]

The effective acceleration of the molecules and atoms is then given by the balance of the solar radiation pressure force with solar gravitation. This is usually characterized by the parameter:

$$\beta_{\text{gas}} = \frac{F_{\text{rad}}}{F_{\text{grav}}} = \frac{h}{GM_\odot m} \sum_i \frac{g_i}{\lambda_i} \tag{19}$$

Here, g_i is the g-factor for fluorescence at $r_{\text{h}} = 1\,\text{AU}$ (scales with r_{h}^{-2}) for a transition at wavelength λ_i, h is the Planck constant, G the gravitational constant, $M_\odot$ the solar mass, and m the molecular mass.

Taking into account solar radiation pressure is important when computing the dynamics of cometary sodium tails (Fig. 33) and the hydrogen coma (Fig. 42) of comets. Hydrogen forms mainly by photodissociation of H_2O and OH. Its large-scale lengths and efficient acceleration leads to an extend of the coma up to several 10^7 km, turning comets into the largest objects of the solar system when taking their H-coma into account. The dynamics of H atoms is complex and needs to consider their excess energy obtained during formation in addition to solar radiation pressure. [47] provide a detailed review of the dynamics of H atoms and observations of cometary hydrogen.

3.3 Dynamics of Dust Tails

At nucleocentric distances beyond about 10^3 km, radiation pressure and solar gravity determine the motion of cometary dust particles. The solar gravity force can be expressed as:

$$F^{\text{d}}_{\text{grav}} = \frac{GM_\odot}{r_{\text{h}}^2} \left(\frac{4}{3} \pi a^3 \rho \right) \tag{20}$$

The force by solar radiation acting on a dust particles is given by:

$$F_{\rm rad}^{\rm d} = \frac{Q_{\rm pr}}{c} \left(\frac{L_\odot}{4\pi r_{\rm h}^2} \right) \pi a^2 \tag{21}$$

Here, a is the particle radius, ρ the dust particle density, $L_\odot$ the solar luminosity, c the speed of light, and $Q_{\rm pr}$ the radiation pressure efficiency given by the ratio of the radiation pressure cross-section to the geometrical cross section ($Q_{\rm pr} \approx 2$ for large particles).

Radiation and gravitational forces act in opposite directions and are both proportional to $1/r_{\rm h}^2$. Thus, the dust particles move on Keplerian orbits around the Sun with what is effectively a gravitation field reduced by $(1-\beta)$:

$$F_{\rm grav}^{\rm eff} = F_{\rm grav}^{\rm d}(1-\beta) \tag{22}$$

where the β-parameter is defined as the ratio of radiation pressure and gravitation force:

$$\beta = \frac{F_{\rm rad}^{\rm d}}{F_{\rm grav}^{\rm d}} = \frac{3L_\odot}{16\pi c G M_\odot} \frac{Q_{\rm pr}}{\rho a} \tag{23}$$

The effective acceleration of dust particles therefore depends on their size a. Generally, large dust particles remain near the nucleus for longer times, because their acceleration is low. Small particles are accelerated more efficiently. If the particles are very small and become transparent, their acceleration decreases again. Investigations of the spatial distribution of dust particles in the coma and tail can therefore provide indications on the dust particle size distribution. However, the radiation pressure efficiency, $Q_{\rm pr}$, depends on the composition of the dust and varies with its absorption and scattering properties. Therefore, it is difficult to disentangle size and material properties of dust particles from observations (see Sect. 7). Figure 34 shows the β-parameter for various materials as a function of grain radius for illustration.

The dynamics of dust particles on a large scale is determined by β and can be described by the formalism of [83]. They used the concept of synchrones and syndynes to describe the distribution of dust particles in the large-scale dust tail. Particles with any β-value ejected at the same time, t, from the nucleus are distributed along a line called "synchrone" (Fig. 35). Particles with the same β-value ejected at any time are distributed along "syndynes". For a given observing geometry, we can compute a set of synchrones and syndynes, with free parameters such as the particle size distribution and the initial velocity. The parameter set best fitting the appearance of the observed dust tail is used to derive the dust particle parameters. Usually, the optical properties and densities of dust particles are assumed to be the same, and only the size distribution is varied. Alternatively, Monte-Carlo models have been made to compute the motion of the dust particles. The inverse Monte-Carlo approach described by [87] takes into account, for example, a distribution of initial velocities and anisotropic outgassing. Reference [88] gives an overview

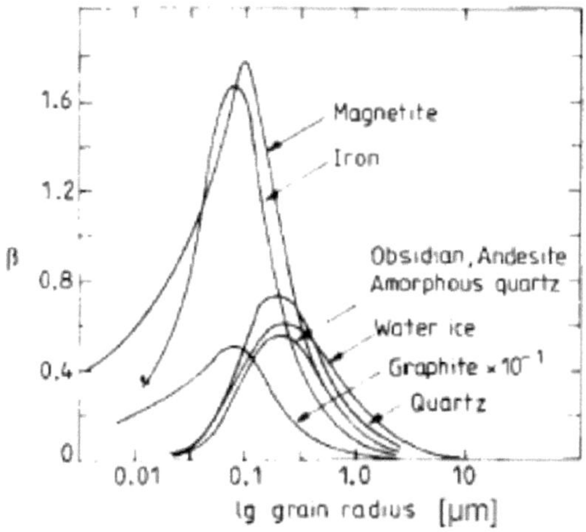

Fig. 34. The β-parameter as a function of grain radius for different materials

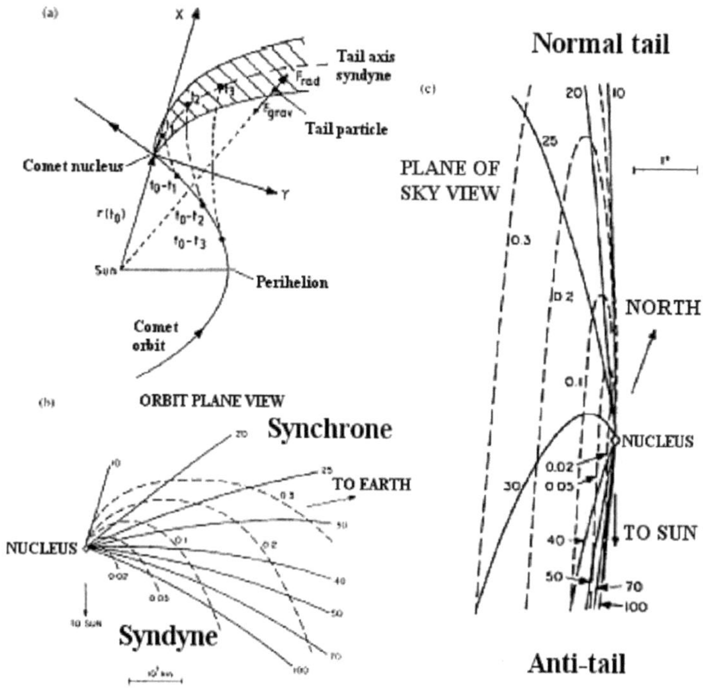

Fig. 35. The principle of synchrone and syndyne calculations as performed by Finson and Probstein [65]

on the various methods to derive information on the dust particles from studies of their distribution in the dust tail.

Sometimes, a sunward spike of the dust tail is seen when the Earth crosses the orbital plane of a comet. This phenomenon is called an "anti-tail." The anti-tail is often interpreted as "old" and large particles with small β-value, released months before the observations. Another feature sometimes observed during orbital plane crossing is called a "neck-line." Neck-lines appear as narrow, bright spikes in the dust tail aligned with the solar and anti-solar direction. As all particles ejected by a comet finally move along Keplerian orbits around the Sun, they cross the orbital plane again on the second node, 180° away from their ejection node. When the Earth passes through the orbital plane of the comet, these particles are seen lined-up in projection as a narrow spike and become bright by strong forward scattering. Possibly, most anti-tails reported in the past were actually unrecognized neck-line observations. Neck-lines are usually observed post-perihelion and are formed by dust particles ejected during the pre-perihelion path. Periodic comets could in principle also show neck-lines pre-perihelion, but this has never been observed. The interest in neck-lines arises from the potential to detect very large particles that are otherwise difficult to investigate in dust tail observations [88].

3.4 Dynamics of Ion Tails

Comet – Solar Wind Interaction

The parent and daughter molecules in the cometary coma are eventually ionized by photoionization, charge exchange or collisional ionization in the inner coma. The charged cometary particles interact with the solar wind. The solar wind consists mainly of hydrogen and helium ions. They stream approximately radially outward from the Sun with velocities of a few hundred kilometers per second, depending on heliographic latitude, solar cycle and interaction regions within the solar wind flow. The solar wind carries with it the magnetic field, which is "frozen" into the flow. This means the solar magnetic field lines are fixed to a fluid element and move outward with this flow element. Because the Sun rotates, the magnetic field lines therefore form a spiral around the Sun, the so-called Parker spiral [173]. The ionized cometary atoms and molecules form an obstacle in the solar wind, leading to a large interaction zone and the formation of a several 10^7 km long cometary ion tail.

For the following description of the comet – solar wind interaction, we choose the cometocentric frame of reference. In this frame, the solar wind streams at the comet. For simplicity, we assume the frozen magnetic field at an angle of 90° to the flow field, which streams straight at the comet (see Fig. 36). Figure 37 shows the physical parameters along the path of the Giotto spacecraft through the coma of comet Halley computed by a

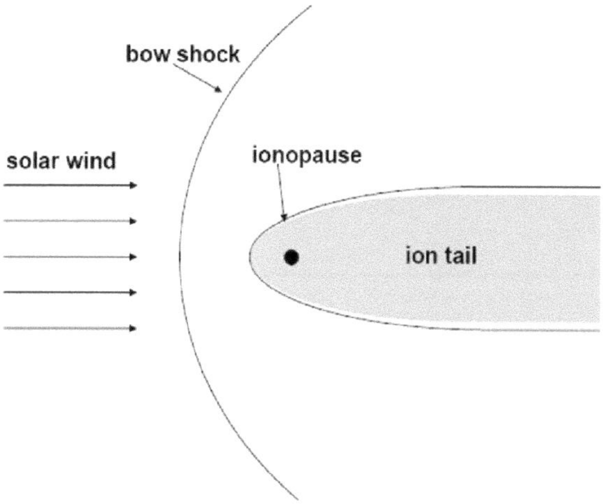

Fig. 36. The principle comet - solar wind interaction

magneto-hydrodynamical model. We use these figures to guide us through the main interaction zones of an active comet with the solar wind:

- As the solar wind approaches the comet, its magnetic field picks up an increasing number of cometary ions. This mass loading leads to a reduction of the solar wind speed with decreasing distance to the comet. Eventually, a bow shock forms in front of the comet separating the supersonic solar wind flow from the subsonic ion flow around the nucleus. In Fig. 37, a jump in velocity, temperatures, and magnetic field around 10^6 km marks the location of the bow shock.
- Behind the shock, the solar wind flow is increasingly mass loaded by cometary ions and the velocity further reduced. At large distances, sideways from the comet, however, the solar wind passes undisturbed. The interplanetary magnetic field, which is frozen into the solar wind flow, therefore folds (Fig. 38) around the comet [5].
- At the pressure boundary of the outstreaming cometary ions with the onstreaming mass loaded solar wind, an ionopause forms. Inside the ionopause, a magnetic field-free cavity forms (around 3.5×10^3 km in Fig. 37). Here, we find purely cometary plasma.
- In front of the ionopause, the magnetic field piles-up and the magnetic field strength increases.
- The temperatures of ions and electrons also drop from solar wind values to relatively cool conditions in the inner coma (Figs. 37 and 39).
- Magnetic curvature and pressure forces quickly accelerate the cometary ions up to velocities of a few hundred kilometers per second. The cometary ion tail forms.

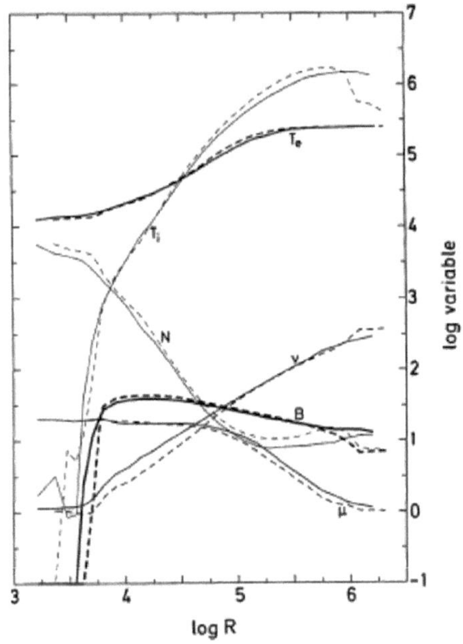

Fig. 37. Results of a 3D-MHD model for the ion tail of comet Halley during Giotto encounter [210]. Ion, T_i, and electron temperatures, T_e, magnetic field, B, ion velocity, v, ion density, N and mean molecular weight, μ, are shown along the inbound and outbound paths of the spacecraft

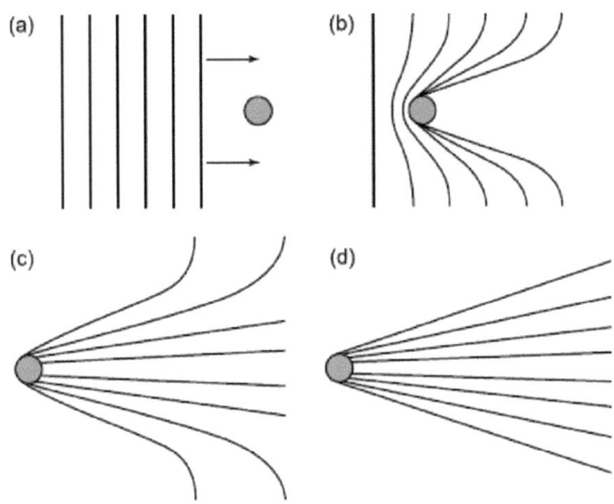

Fig. 38. Schematic sketch of the folding of the solar wind magnetic field lines around a comet [5]

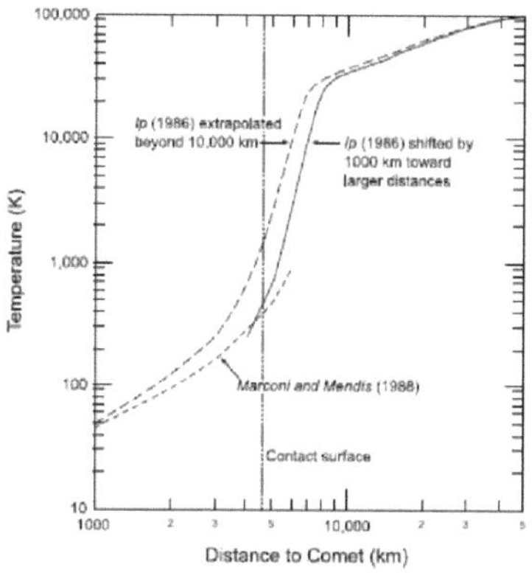

Fig. 39. Profile of the electron temperature used in various model calculations [97]

The size of the interaction region of comets with the solar wind can be approximated by the stand-off distance of the bow shock, R_{I}. It depends on the solar wind flux, $\rho_{\odot} u_{\odot}$, the cometary gas production rate, Q, the average particle mass, m_{C}, the neutral gas speed, u, and the ionization rate, κ:

$$R_{\mathrm{I}} = \frac{\kappa m_{\mathrm{C}} Q}{4\pi u \rho_{\odot} u_{\odot}} \tag{24}$$

R_{I}, therefore, scales with the gas production rate for given solar wind conditions. We note, however, that in weakly active comets or at large heliocentric distances, no bow shock will form. Weak comets may also lack a diamagnetic cavity around the nucleus, and the solar wind magnetic field and solar wind particles may even penetrate to the nucleus surface.

Good knowledge of the temperatures is crucial to understand the ion chemistry in the inner coma (see Sect. 5). Unfortunately, the temperature of the electrons, $\mathrm{T_e}$, in the energy range relevant for electron recombination – which is an important loss process for ions in the inner coma – could not be measured in situ so far. Several attempts have been made to derive the temperature distribution from measurements of the ions in Halley's coma by various models (Fig. 39) with different results. They all agree on a steep decrease of electron temperature in the inner coma, because in the diamagnetic cavity region, electrons are cooled efficiently by collisions with water molecules. However, where the steep decrease in $\mathrm{T_e}$ occurs is difficult to determine. [98] derived a distribution for $\mathrm{T_e}$ (Fig. 39) that could match well the measurements of H_3O^+ ions in comet Halley (see Sect. 5). Spatial mapping of HCO^+ ions in comet Hale–Bopp (e.g., [147, 148]) showed a reduced column density in the inner coma, which could also be explained by low electron temperatures leading to increased loss processes [177]. Except for such indirect evidence, the low electron temperature range is difficult to measure, and we have to wait for future space missions for in situ data.

Observations of Ion Tails

Observations of ion tails have shown that they always point almost radially away from the Sun, with only a slight abberation angle of a few degrees. On the basis of the appearance of cometary ion tails, Biermann concluded in 1951 [22] that a flow of charged particles must exist streaming radially away from the Sun, the solar wind. This may have been the most important implication of observations of cometary ion tails for our understanding of the solar system. In addition, cometary ion tails serve as a laboratory for plasma phenomena, which are difficult to simulate in a laboratory on Earth.

Our today's general picture of the comet–solar wind interaction has been confirmed by the ICE and Giotto spacecrafts visiting comets P/Giacobini-Zinner, P/Halley, and P/Grigg-Skjellerup (e.g., [12, 152, 165]). For example, magnetometer measurements showed the folding of the magnetic field lines around comet Halley and the existence of a diamagnetic cavity [165]. The results of the plasma experiments on Giotto are summarized in numerous reviews and books, for example [151], and it would require too much space to summarize even the most important measurements here.

On a large scale, ion tails show many highly time variable phenomena, such as rays folding toward the main tail, disconnections of the whole tail from the nucleus region, formation of clouds moving down the tail, etc. It has often been proposed that the response of the cometary ion tails to changes in the solar wind can be used as a tracer of the solar wind conditions. Observations of ion tails and model simulations to understand structure formation (e.g., [168, 179, 193, 211]) have provided considerable improvements in our understanding on the comet–solar wind interaction. However, it has also become evident, that a one-to-one correspondence of ion tail structures to solar wind features may be an oversimplifying assumption. Different conditions in the solar wind can lead to similar appearances of the ion tails, e.g., the formation of tail rays or disconnecting clouds, making the identification of the origin of ion tail variations in the solar wind difficult.

4 Emission Excitation in the Gas Coma

The molecules, atoms, and ions in the cometary coma are visible through their emitted radiation. Higher energy levels are excited by, for example, solar energy or collisional excitation. Transitions can occur among rotational levels within a vibronic band (pure rotational transitions), among rotational levels of different vibrational bands (ro-vibrational transitions) and ro-vibrational levels in different electronic bands (electronic transitions) (Fig. 40). Allowed transitions between energy levels with wave number ν depend on the relevant transition levels:

$$\nu \propto (E'_{\mathrm{el}} - E''_{\mathrm{el}}) + (E'_{\mathrm{vib}} - E''_{\mathrm{vib}}) + (E'_{\mathrm{rot}} - E''_{\mathrm{rot}}) \tag{25}$$

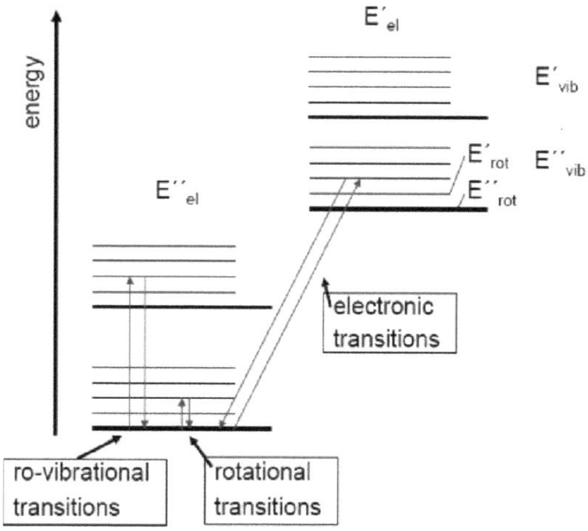

Fig. 40. Schematics of energy levels of a hypothetical molecule. Two electronic energy levels are shown with two vibrational energy levels each. For the vibrational levels, several rotational energy levels, are indicated

Emissions of electronic transitions ($E'_{\mathrm{el}} - E''_{\mathrm{el}}$) occur in the UV, optical, and the near-infrared range. Emissions in the UV are not transmitted through the Earth atmosphere and can be detected only by rockets and from spacecrafts. This is the case for most atoms and atomic ions in cometary comae. Vibrational transitions ($E'_{\mathrm{vib}} - E''_{\mathrm{vib}}$) are observed in the infrared wavelengths range. In the Earth atmosphere, water molecules are efficient absorbers of IR radiation, and therefore, detections of cometary emissions from the ground are possible only in atmospheric windows free of water absorption bands. However, infrared space telescopes allow us to observe emissions over a wide wavelengths range. Pure rotational transitions ($E'_{\mathrm{rot}} - E''_{\mathrm{rot}}$) are observed at radio wavelengths. It is beyond the scope of this introduction text to explain in detail molecular excitation, and we refer to the standard literature [106–108].

The wavelengths range at which emissions are primarily observed differs for parent molecules and daughter radicals. The lifetime of electronic transitions, $\tau \propto 1/A_{\mathrm{ul}}$ (A_{ul}: Einstein A coefficient of the transition) is in the order of 10^{-8} s. This is much shorter than the lifetime of vibrational ($\tau \cong 10^{-3}$ s) or rotational ($\tau \cong 1$ s) transitions. Therefore, depending on the energy levels excited, the observed emission of a molecule is found in different wavelengths ranges:

- Atoms: They emit by electronic transitions at UV and optical wavelengths.
- Daughter radicals are observed mainly at optical wavelengths, because solar photons excite their upper electronic bands, which have very short lifetimes.
- Parent molecules: High energy photons in the optical to UV range lead to fast photo-dissociation of the molecule. Therefore, emission excitation

occurs instead by collisional and radiative excitation of the lowest rotational and ro-vibrational energy levels. At radio wavelengths, for example, we observe rotational transitions in the lowest vibrational bands.

- Symmetric parent molecules: Symmetric molecules without permanent dipole moment are a special case because for them pure rotational transitions are not allowed, and they can therefore only be observed by their ro-vibrational transitions at IR wavelengths, but not in the radio range (examples: CH_4, C_2H_2, C_2H_6).

In the UV range atoms such as H, O, C, and S are detected (see overview by [75]), as well as parent molecules such as CO (the so-called Fourth positive system at 1450 Å and the CO Cameron bands at 2050 Å indicating CO_2 photodissociation) and the S_2 molecule (see overview by [32]).

Observations of comets in the optical wavelength range have the longest history and statistical baseline of cometary observations. Therefore, this wavelengths range is important when statistical comparisons between comets are made, although observations at longer wavelengths ranges (IR, radio) increase in modern times with improved instruments. Furthermore, the high fluorescence efficiency of some molecular emissions, for example CN, make optical emissions an ideal tracer of gaseous activity in faint comets and comets at large heliocentric distances.

Transitions of several daughter radicals are well known in comets, as shown in Fig. 41. The transitions appear as band sequences in low-resolution spectra, because emission lines with the same Δv between upper and lower vibrational

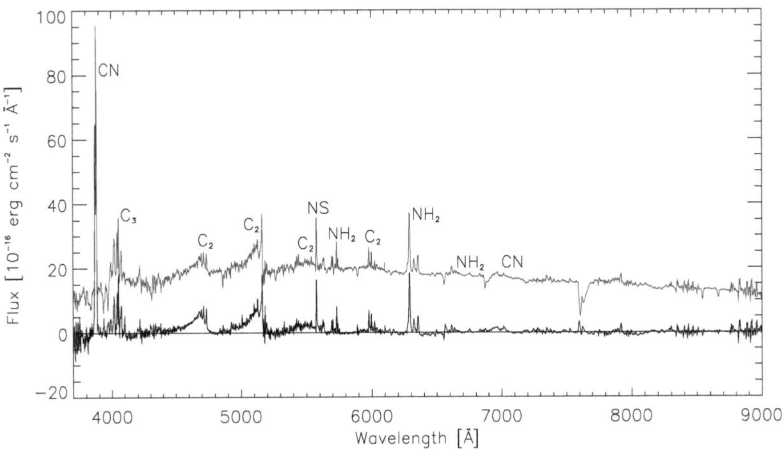

Fig. 41. Optical spectrum of comet 9P/Tempel 1. The observations were made on July 3/4, 2005, at ESO. The comet was at $r_{\mathrm{h}} = 1.6\,\mathrm{AU}$ and $\Delta = 0.9\,\mathrm{AU}$, respectively. Gray: Spectrum with night sky subtracted. The underlying continuum caused by scattered solar light on cometary dust particles can be seen. Black: the same spectrum, but continuum subtracted

band, v, are very close in wavelengths and can be resolved only with high-resolution spectroscopy. The most prominent emission bands in the optical range arise from CN, CH, C_3, C_2, NH, and NH_2 molecules.

Cometary parent molecules are observed mainly by their rotational and ro-vibrational transitions in the radio and IR-domain. The most important species to be observed include the main cometary ices: H_2O, CO, and CO_2, in addition to a large number of minor species, such as C_2H_2, C_2H_6, CH_4.

Because of the complex nature of some emission bands because of overlapping emissions of different species, an identification is sometimes difficult. For example, around 3.4 μm a broad emission feature is seen in many comets, e.g., in comet Halley [45]. The main emission is usually attributed to the C-H stretching vibration mode. It is, however, still unclear which molecules contribute to this emission. Parts are believed to result from methanol and formaldehyde, but other organic molecules may contribute. Recently, the CH-feature has been observed also by the Deep Impact spacecraft in comet Tempel 1 [1] in the ejecta material after the impact. The relatively strong feature after impact was preliminarily interpreted as evaporating organic material. Clearly, identifying the contributions to this emission band is an interesting future task, but deserves more modeling efforts of the excitation conditions, including optical depths effects and non-LTE (LTE: local thermodynamical equilibrium) excitation conditions.

Altogether, a large number of observations of cometary emissions exist from the UV up to the radio range. These emissions are used to spectroscopically identify the species present in the coma and provide us with an inventory of cometary constituents. [75] provides a list of the spectroscopically observed daughter species in comets up to now, together with the wavelength/frequency of their main emissions. [32] discuss the detected parent molecules.

At present, 24 parent molecules have been detected by spectroscopic emission features. The presence of two parent molecules, CS_2 and N_2, is, however, inferred only by their daughter products CS and N_2^+. In addition, some species believed to be parent molecules may instead be daughter products of more complex organic species. Such complex parent molecules have been suggested, for example, for CO and H_2CO (see Sect. 5).

Emissions of the refractory component of comets (Ca, Co, Cr, Cu, Fe, K, Mn, Ni) can be detected mainly in sun-grazing comets, with the exception of Na, which could be observed in several comets up to now (e.g., [51, 109, 166, 176]). Therefore, sodium is one of the rare species that allows us to study the non-refractory component of cometary nuclei.

Many more molecular species have been searched for, but remained undetected. In particular, complex parent molecules are difficult to detect spectroscopically because the intensity of the individual rotational and vibrational lines of these molecules is very low. A list of the undetected parent molecules searched for at radio wavelengths can be found in [55]. Line catalogs of high-resolution spectra at optical wavelengths (e.g., [42]) show the well-known daughter radicals, but also contain a large number of unidentified emission lines. Although most of them are probably part of the band systems of the already well-known species, the possibility of yet unidentified molecules in comets contributing to these spectra cannot be ruled out. Comets are studied over a wide range of wavelengths. However, not the whole wavelength range is fully exploited yet by observations, and our inventory of cometary parent molecules is clearly not yet complete.

Indications for complex organic species resulted from mass spectrometer data obtained during the Giotto fly-by on comet Halley. The neutral and ion mass spectrometers on board Giotto gave the first in situ measurements of the volatiles in a cometary coma. Unfortunately, a clear identification of these high mass ranges is difficult. Polyoxymethylene (POM) molecules have been proposed to explain the regular mass peaks observed [111]. An overview of the results of the ion mass spectrometer can be found, for example, in [14].

4.1 Resonance Fluorescence

The dominant excitation mechanism for the electronic transitions observed in the optical, near-UV, and near-IR range is resonance fluorescence by the solar flux. The strength of an observed emission is calculated using the fluorescence efficiencies or g-factors (in units of [photon s^{-1}]). The g-factors are calculated from the absorption oscillator-strength, f, the Einstein A coefficients, A_{ik}, and the solar flux, F_λ at the observed wavelength, λ [40]:

$$g_{ik} = \frac{\pi e^2}{m_e c^2} \lambda^2 (f F_\lambda)_{ik} \frac{A_{ik}}{\Sigma_k A_{ik}} \tag{26}$$

The g-factor calculated for the solar flux at $r_\mathrm{h} = 1\,\mathrm{AU}$ scales with distance as r_h^{-2}.

If an observed emission band is caused by pure resonance fluorescence and the relevant g-factors are known, the conversion from observed fluxes to molecular column densities, N, is straightforward:

$$N = \frac{4\pi}{g} \frac{1}{\Omega} F \tag{27}$$

Here, F denotes the observed emission flux in the aperture and Ω the aperture size used. Converting from column densities to molecular gas production rates of the comet then usually requires a model of the spatial distribution of the molecules in the coma (see Sect. 6).

The solar spectrum as seen by a coma molecule is Doppler-shifted because of the velocity of the molecule. The fluorescent excitation of emission bands near solar Fraunhofer absorption lines, therefore, can be a strong function of the heliocentric velocity component of a molecule. The Doppler shift has two components: the heliocentric velocity component of the comet's orbital velocity (Swings effect [201]), and the motion of the molecule with respect to the nucleus (Greenstein effect [92]). The latter is important only for fast moving atoms, such as Na, or ions, moving with velocities of several tenth to hundred km/s. Bands most strongly affected by the Swings and Greenstein effects are lying near strong solar Fraunhofer absorption lines. An emission band significantly affected by the Swings effect is, for example, the CN (0-0) band at 389 nm.

In addition, variations of the incoming solar radiation affect the excitation of emission bands. The solar flux arriving at a cometary molecule depends on the 11-year solar cycle. It may also vary with solar rotation and the occurrence of sunspots.

Fluorescence excitation models are usually made for observations near 1 AU, and their application to large heliocentric distances needs to be verified. The relative importance of collisional and radiative excitation processes in the coma can vary with heliocentric distance and cometary activity and can lead to deviations of the excitation observed from model predictions. For example, the relative band strengths of NH_2 detected in comet Hale–Bopp beyond $r_h = 3$ AU did not agree with the fluorescence excitation models known at that time [177]. New g-factors [128–130] gave agreement with the observations and should now be used.

4.2 Prompt Emission

Prompt emission means a daughter molecule or atom is formed in an excited state and then performs a radiative transition into the ground state. A well-known example for prompt emission in the optical range is the formation of oxygen by:

$$H_2O + h\nu \rightarrow O^* + H_2$$

In this reaction, the oxygen atom is formed in an excited state (indicated by *). Observed emissions of oxygen are O 1D: 6300 Å, 6364 Å, and 1S: 5577 Å. In case of prompt emission, every transition occurs only once, immediately after formation of the oxygen atom. This makes prompt emission an ideal tracer of the parent molecule H_2O. Unfortunately, oxygen atoms are produced by the same process also in the Earth atmosphere. Therefore, high spectral resolution and a sufficiently large Doppler shift of the comet geocentric velocity component are needed to separate the cometary emission from Earth atmosphere contamination. Further complications arise, for example, when collisional quenching of the upper levels plays a role in very active comets.

Other examples for prompt emission are the CO "Cameron bands" at 2050 Å, which are excited after formation by photodissociation of CO_2, and H Ly_α emission. Images and spectra of H Ly_α have shown the enormous extent of the cometary neutral hydrogen coma (Fig. 42). Recent modeling of the excitation and dynamics of cometary hydrogen can be found in [46, 47] and [182].

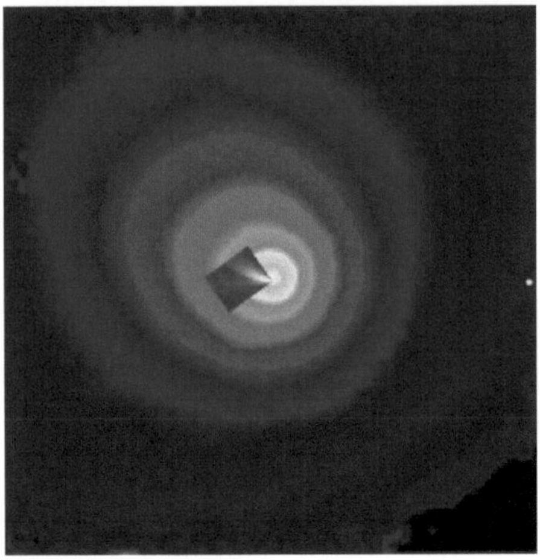

Fig. 42. Image of the hydrogen coma of comet Hale–Bopp observed by Ly_α emission [48]. The white dot to the right gives the solar disc to scale. Inserted is a picture of the comet taken at optical wavelengths for comparison with the elongation of the ion and dust tails

4.3 Optical Depth Effects

Obviously, sunlight reaches the nucleus surface and sublimates the volatile ices. The cometary coma, therefore, is generally optically thin. However, optically thin conditions may not be fulfilled for strong resonance lines, such as H Ly_α. High optical depths in the coma will affect the resonance excitation of radiation in the coma as well as photochemical processes.

The optical depth is a function of wavelengths and position in the coma, R:

$$\tau(\lambda, R) = \sum_{i=1}^{s} \int_{\mathrm{R}}^{\infty} N_i(R)\sigma_i(\lambda)\mathrm{d}R \tag{28}$$

Here, N_i is the number density of molecules in the line-of-sight to the Sun, σ_i is the absorption cross section at wavelengths λ.

An additional note is added here on optical thickness effects at visual and infrared wavelengths by light scattering on coma dust particles. This effect may be important for active dusty comets, as discussed in [161] and references therein. These models compute the ambient solar flux on the nucleus as input for surface sublimation models. We remark, however, that such effects can also influence the excitation of molecules very near the surface.

4.4 Excitation of Rotational and Vibrational Transitions

To calculate the population of the rotational levels, fluorescence equilibrium is often assumed throughout the outer coma. However, this assumption is not fulfilled for short-lived molecules and in the inner coma where collisional processes are important. Depending on the molecules observed, both excitation mechanisms must therefore be considered. An overview of the excitation of rotational and vibrational transitions can be found in for example [56, 59] and [58].

Calculating the excitation by collisions requires knowledge of the collisional cross-sections of the molecules and the coma temperature. The thermal excitation of a molecular energy level is determined by the Boltzmann distribution:

$$N^{\mathrm{t}} = \frac{N_{\mathrm{u}}}{N_0} = \frac{g_{\mathrm{u}}}{g_0} e^{-\frac{E}{kT}} \tag{29}$$

Here, N_{u} and N_0 are the number of molecules in the excited and ground level, respectively. E is the excitation energy of the upper level. $g_{\mathrm{u},0}$ are the statistical weights of the upper and lower levels. In general, the temperature profile in the coma is a function of nucleocentric distance, r, and depends on the radiative cooling and photolytic heating processes (see Sect. 3). The population of the ro-vibrational levels by collisions is therefore also a function of nucleocentric distance.

To determine the excitation of the molecules, we assume optically thin conditions and neglect the influence of dust particles. We can also assume that collisions are important only among neutrals because ion densities in the innermost coma are low. Furthermore, we assume radiative excitation occurs only between vibrational bands. The rate of collisions is given by:

$$\kappa_{\mathrm{C}} = \sigma_{\mathrm{C}} n_{\mathrm{H_2O}} u \tag{30}$$

Here, u is the mean velocity between two collisions. The relative excitation of a rotational level i in case of a non-steady state calculation is then given by:

$$\frac{dN_i}{dt} = -N_i \sum_{j \neq i} p_{ij} + \sum_{j \neq i} N_j p_{ji} \tag{31}$$

Here, p_{ij} is a transition rate from level i to j, and $p_{ij} = C_{ij} + g_{ij}$ for $E_i < E_j$ or $p_{ij} = C_{ij} + A_{ij}$ for $E_i > E_j$ [32].

In addition to this set of stiff differential equations describing the population of the rotational and vibrational levels, one needs a model of the spatial distribution of the molecules (see Sect. 3).

To derive the rotational temperature, T_{rot}, from the population of excited rotational levels, several emission lines of a molecule (e.g., Fig. 43) need to be observed simultaneously (e.g., [30, 164, 209]). If the observed transitions are relaxing only slowly to fluorescence equilibrium, their population is determined by the collisions occurring in the inner coma, and T_{rot} is close to the kinetic temperature, T_{kin}.

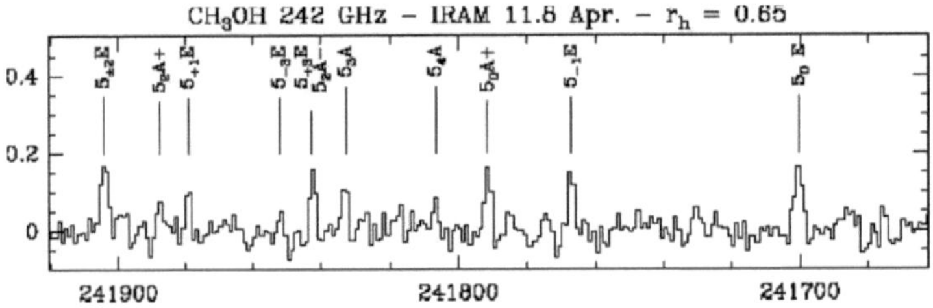

Fig. 43. Methanol lines observed in comet Hyakutake [28].

4.5 OH Maser Emission

The excitation of OH radio lines is the only case of stimulated emission in the cometary coma. The 18-cm lines arise from excitation of the Λ-doublet in the molecular ground state by solar UV radiation, building up an inversion population. Because transitions between the Λ-doublets are inhibited, the resulting maser transitions are very sensitive to quenching by collisions of OH with neutrals and ions in the coma. The radius up to which quenching needs to be considered in calculations of the OH excitation depends on the activity of a comet. It is around 10^5 km for comets with production rates near $10^{29}\,\mathrm{s}^{-1}$. Proper treatment of collisional quenching is difficult because collisional cross-sections and ion densities are often not sufficiently known in the coma. The quenching of the OH population is therefore usually approximated by appropriate scaling laws, unless it can be measured by spatial mapping the OH distribution in the coma [44, 90, 91, 190].

4.6 X-ray Emission

The first observations of a comet with an X-ray telescope lead to the discovery of X-ray radiation of comets [144]. Since then, many model attempts and further observations of several comets have been made (e.g., [6, 96, 135, 136, 142]) to explain the cause of the emission, and it has become a new field of cometary science. See [143] for an overview.

5 Chemical Processes in the Coma

The composition of the cometary coma changes with increasing nucleocentric distance because of a number of chemical reactions. Reactions occur between neutral gas molecules, neutral and ionic species, with the solar wind particles, and the solar radiation field (Fig. 14). Table 2 provides an overview of the different types of reactions considered in chemical models of cometary comae and gives an example for each reaction type. More detailed discussions also including reactions of minor importance can be found in [194], [110], and [184].

To model the composition of the coma, a set of equations including the relevant chemical reactions has to be solved. The density of a given species, n_i,

Table 2. Types of chemical reactions in cometary comae

Reaction type	Example
Photodissociation	$H_2O + h\nu \rightarrow OH + H$
Photoionisation	$H_2O + h\nu \rightarrow H_2O^+$
Photodissociative ionisation	$CO_2 \rightarrow O + CO^+ + e$
Electron impact dissociation	$C_2H_2 + e \rightarrow C_2 + H_2 + e$
Electron impact ionisation	$CO + e \rightarrow CO^+ + 2e$
Dissociative electron recombination	$C_2H^+ + e \rightarrow C_2 + H$
Charge exchange	$CO^+ + H_2O \rightarrow CO + H_2O^+$
Neutral-neutral reactions	
3-Body reactions	

varies by the sum of its chemical formation and loss processes. In the simple case of a reaction of two species, A + B → C + D, the change of their number densities by this reaction is:

$$\frac{\partial n_{\mathrm{A}}}{\partial t} = \frac{\partial n_{\mathrm{B}}}{\partial t} = -k n_{\mathrm{A}} n_{\mathrm{B}}; \quad \frac{\partial n_{\mathrm{C}}}{\partial t} = \frac{\partial n_{\mathrm{D}}}{\partial t} = k n_{\mathrm{A}} n_{\mathrm{B}} \tag{32}$$

Here, k is the reaction rate. In general form, for a total number of s species in a reaction network, this can be written as (see also [194]):

$$\frac{\mathrm{d}n_i}{\mathrm{d}t} = \sum_{j=1}^{q} \nu_{ij} k_j \prod_{l=1}^{s} n_l^{m_{ij}}, \qquad i = 1, ..., s \tag{33}$$

Here, q is the number of chemical reactions and ν_{ij} is the stoichiometric coefficient of species i in reaction j, which is positive for products and negative for reactants. The reaction order m_{ij} is equal to $\mid \nu_{ij} \mid$ if negative and zero otherwise.

The rate coefficients for collisional reactions, k_j, are usually expressed by the Arrhenius-law:

$$k_j = A_j \left(\frac{T}{300}\right)^{B_j} \exp\frac{-C_j}{T} \tag{34}$$

Here A, B, C are coefficients derived from laboratory experiments or theoretical calculations. Tabulated coefficients can be found, for example, in [194]. T denotes the gas or ion temperature, depending on the reaction considered. Note that the Arrhenius-law is valid only for LTE-conditions (LTE: local thermodynamic equilibrium).

The neutral gas molecules are also dissociated and ionized by solar UV radiation. The photolytic rate coefficients are usually given for $r_{\mathrm{h}} = 1\,\mathrm{AU}$ and scale with the solar flux as the inverse square of the heliocentric distance:

$$k_j^{\mathrm{ph}} = \frac{k_{j,1\,\mathrm{AU}}^{\mathrm{ph}}}{(r_{\mathrm{h}}\,[\mathrm{AU}])^2}. \tag{35}$$

For active comets inner molecules are shielded from sunlight by the gas molecules towards the sun and the effective optical depth of the coma has to be taken into account. The photolytic rate coefficients are therefore given by:

$$k_j^{\mathrm{ph}}(r) = \int_0^\infty F_{\mathrm{sun}}(\lambda)\sigma_i(\lambda)e^{-\tau(\lambda,r)}\mathrm{d}\lambda \qquad (36)$$

Here, r is the nucleocentric distance of the molecules considered. $\sigma_i(\lambda)$ denotes the photo cross section for a specific reaction which is computed from the absorption cross section of the molecule or ion, $\sigma_i'(\lambda)$, and the quantumn yield, $\phi(\lambda)$, for the reaction considered:

$$\sigma_i(\lambda) = \sigma_i'(\lambda)\phi(\lambda) \qquad (37)$$

The optical depth, $\tau(\lambda)$, is given by:

$$\tau(\lambda, r) = \sum_{i=1}^{s} \int_r^\infty n_i(R)\sigma_i{\prime}(\lambda)\mathrm{d}R \qquad (38)$$

and is integrated along the line-of-sight from the location of the molecule considered towards the Sun, R. Cross sections and photo rates relevant for cometary chemistry can be found in [61,112,194] and at amop.space.swri.edu (interactive web-page provided by Huebner et al.). Note that photo rates depend on the solar cycle and are often given for quiet and active Sun conditions.

It is instructive to look at an order-of-magnitude overview of the reaction rate coefficients for the most important reactions (see Table 3). Photo reactions have by far the highest rate coefficients and are therefore important throughout the coma (see also Fig. 14).

The most important photoreaction for the dynamics of the neutral molecules in the gas coma is the dissociation of water molecules. The excess energy given to the daughter products, mainly to H, heats the coma. This leads to an acceleration of the gas, as discussion in Sect. 3.

Rate coefficients for reactions involving two or three neutral molecules are generally much less efficient. In addition, the reactions also depend on the number density of the species involved (32), which drops as $\approx$r^{-2} in the inner

Table 3. Order-of-magnitude overview of reaction rates to compare the various reaction types. Assumptions: $r_{\mathrm{h}} = 1\,\mathrm{AU}$, $T_{\mathrm{g}} = 300\,\mathrm{K}$ and $T_{\mathrm{e}} = 10^4\,\mathrm{K}$

photo reactions	10^{-3} - $10^{-7}\,\mathrm{s}^{-1}$
electron impact reactions	10^{-10} - $10^{-13}\,\mathrm{cm}^{-3}\,\mathrm{s}^{-1}$
neutral-neutral collisions	10^{-7} - $10^{-11}\,\mathrm{cm}^{-3}\,\mathrm{s}^{-1}$
neutral-ion collisions	10^{-9} - $10^{-10}\,\mathrm{cm}^{-3}\,\mathrm{s}^{-1}$
3-body collisions	10^{-29} - $10^{-32}\,\mathrm{cm}^{-3}\,\mathrm{s}^{-1}$

coma. Collisional reactions between neutrals are therefore important only very close to the nucleus.

The ionic chemistry is more complex. Ions are mainly produced by photoionisation, photodissociative ionisation, electron impact and charge transfer. Photodissociative ionisation by extreme UV photons produces ions in an excited state, which then dissociate quickly. This process can be efficient in the inner coma, because extreme UV photons can penetrate deeply into the coma. Electrons produced by ionisation processes in the inner coma are involved in a number of reactions, like electron ionisation, electron dissociation and electron dissociative recombination. Positive ion charge transfer reactions are also important in the inner coma. Solar wind particles, however, do not penetrate into the cometopause (see Sect. 3), and the reactions involving these species are therefore *most* important at large nucleocentric distances, where solar wind particle densities are the highest. An overview of the relative importance of the various ionic reactions can be found in [194].

5.1 Chemistry of Some Frequently Observed Species

NH, NH_2, NH_3

The formation of NH and NH_2 by ammonia (NH_3) photodissociation is well established now. However, in earlier measurements (e.g., [78,216]) of NH_2 and NH, their production rates were consistently below the mass-spectrometer data of NH_3 for comet Halley (see detailed discussion in [11]). This was still the case in first comparisons of NH and NH_2 production rates to direct radio observations of NH_3 in comet Hale–Bopp [23, 187]. With the recently revised NH_2 g-factors [128–130], the agreement is however improved.

C_2, C_3, C_2H_2, C_2H_6

The chemistry of the carbon-bearing radicals is complex (Fig. 44). The main parent molecule of the C_2 radical is C_2H_2 (ethane). C_2H_2 had been proposed to form the parent molecule of the well-known C_2 radical [120] by the reaction $C_2H_2 + h\nu \rightarrow C_2H$, followed by $C_2H + h\nu \rightarrow C_2$ (see also discussion in [46]). Abundance ratios of C_2 and of C_2H_2 in comet Hale–Bopp [187, 207] are consistent with acetylene playing a major role as parent molecule for C_2. However, additional parents are likely. For example, C_2H_6 (acetylene) can lead to the formation of C_2 through decay to C_2H_4, which subsequently dissociates to C_2H_2, again resulting in C_2. Another parent species detected is HC_3N [35, 146], which dissociates into $C_2H + CN$, again leading to the C_2 radical. It has been shown [105] that the production rates of the main C_2 parent species can be derived from observations using a chemistry scheme involving not only photoreactions, but also electron impact reactions (Fig. 44).

No parent molecule of C_3 has been detected yet. The radical is most likely formed by C-bearing molecules like C_3H_4, or more complex species (see discussion in [105]). Because of the expected weak spectral emission lines of such

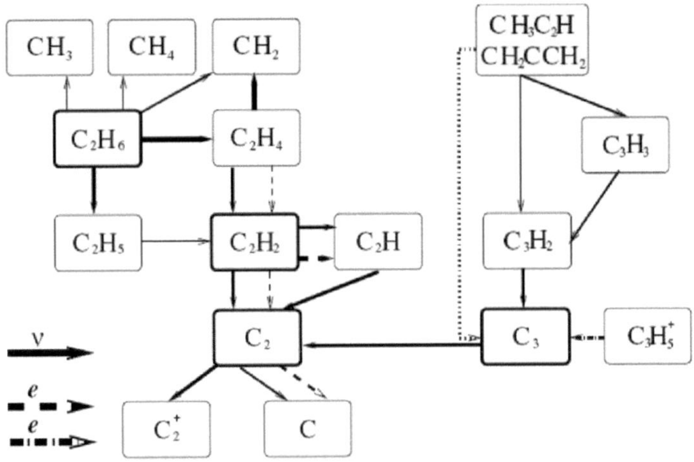

Fig. 44. Overview of the main reaction pathways for the formation and destruction of C_2 and C_3 radicals [105]

complex organic molecules, a direct detection of C_3 parent molecules probably has to wait for future in situ observations.

CN, HCN, HNC, H_2CO, CO, Extended Coma Sources

A question still under debate is whether photodissociation of HCN is the only process forming CN radicals. While it was shown that HCN is sufficient to explain the observations of CN in comet Hale–Bopp beyond $r_h = 3$ AU, it is still unclear whether additional sources of CN exist in comets near perihelion. The arguments for and against HCN being the only CN source steam from disagreements in their production rates, analysis of spatial CN profiles and comparisons on the spatial CN distribution in optical images with 2D maps of the HCN radio signal. The result is inconclusive so far. Recent evidence for an additional parent of the CN radical also comes from measurements of the $^{15}N/^{14}N$ isotopic ratio in high-resolution optical spectra [10, 121]. This ratio is significantly different to the value measured for HCN in comet Hale–Bopp (see also Sect. 6.4). Speculations about the nature of the additional CN source range from sublimation of icy-dust-particles in the coma to HCN polymers on dust grains. Recently, laboratory experiments have been set-up to investigate the formation of CN from HCN polymers experimentally (e.g., [85]). For a recent review on the problem of CN sources, see [84].

CO is a relatively abundant parent ice in cometary nuclei. However, in-situ and ground-based measurements [68] of comet P/Halley have shown an extended source for CO (or distributed source, depending on naming convention) in the coma in addition to its nucleus source (Fig. 45). The coma source peaked at about 10^4 km from the nucleus and accounted for about half of the CO observed in comet Halley.

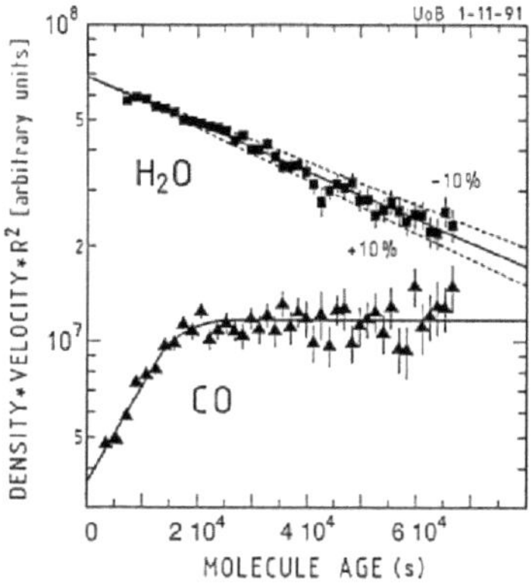

Fig. 45. Indication for an extended source for CO by the NMS spectrometer on board Giotto [70]

By high-resolution near-IR long-slit spectra, it has been possible to spatially separate the nucleus CO source in several long-period comets since the mid-1990s. The results of observations of six comets are summarized in [163]. Studies of the variation of CO abundances with heliocentric distance in comet Hale–Bopp [66] showed that beyond about 2 AU only the nucleus CO source seemed to be present, whereas closer to the Sun about half of CO came from the nucleus and extended coma sources. Part of the extended coma source of CO molecules in the coma is photodissociation of formaldehyde. However, the abundance of H_2CO in comets is too low to fully explain the extended CO coma source. Several mechanisms have been proposed, including CHON grains, sublimation of the outer mantle of unaltered interstellar grains, or polymerized formaldehyde (POM: polyoxymethylene) (e.g., [66, 68]). A discussion of possible mechanisms to explain the Halley measurements and an overview of the proposed ideas to explain extended coma sources can be found, for example, in [134].

Formaldehyde, H_2CO, has been observed by its radio transitions, but also in the IR-range, in many comets with abundances of 0.13–4% [32, 43, 60, 191, 197]. In situ measurements of H_2CO in comet Halley [89, 134] show the release by an extended coma source. Possibly, formaldehyde is completely released from a source in the coma and not a parent ice molecule [155]. However, the relatively easy synthesis of formaldehyde in ice mixtures containing H_2O and CO makes it unlikely that no H_2CO at all should be present in the nucleus [50]. H_2CO molecules can polymerize and form polyoxymethylene (POM: $(-CH_2-O-)_n$). It has therefore been suggested that the distributed H_2CO source consists of formaldehyde polymers on grains releasing single H_2CO molecules during sublimation [111, 113, 155]. Laboratory studies [86] showed that a combination of photodegradation and thermal degradation of POMs could provide sufficient H_2CO in the coma, in agreement with the measurements in comet Halley [50]. Unfortunately, our knowledge of the organic

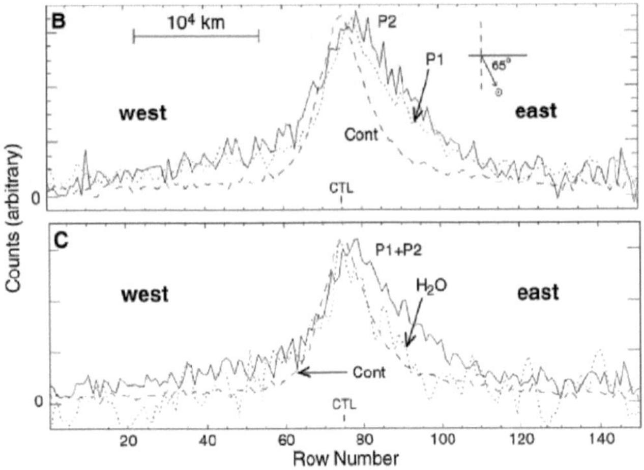

Fig. 46. Indication for an extended source for CO in comet Hale–Bopp [66]

refractory component of cometary grains is still poor and allows us only to set constraints on the extended gas sources. Improvements are expected from further laboratory measurements and the direct analysis of cometary grains, for example of probes from the Stardust mission.

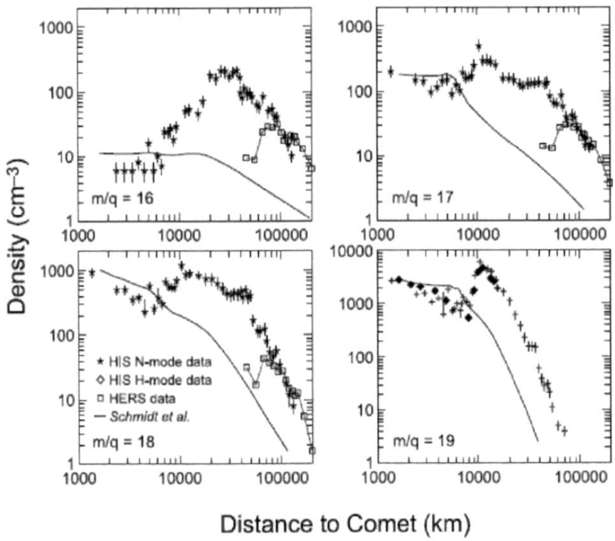

Fig. 47. Distribution of ion densities for mass/charge 16–19 amu e^{-1} in the inner coma of comet Halley as measured by Giotto IMS [8]. The solid line is a fit of a model by [194]

Ions

Observations of cometary plasma tails show H_2O^+ and CO^+ ions to be their dominant constituents. These ions are formed by photoionisation:

$$H_2O + h\nu \rightarrow H_2O^+ + e \tag{39}$$

$$CO + h\nu \rightarrow CO^+ + e \tag{40}$$

In the inner coma, however, NH_4^+ and H_3O^+ dominate inside a few 10^4 km. We take the formation and destruction of H_3O^+ as an example. H_3O^+ is formed by ion–neutral reactions of water ions with neutral water molecules and is destroyed by dissociative recombination.

$$H_2O^+ + H_2O \rightarrow H_3O^+ + OH \tag{41}$$

$$H_3O^+ + e \rightarrow H_2O + H \tag{42}$$

In general, ion destruction by dissociative electron recombination is a major loss process for the dominant ions in the inner cometary coma. Figure 47 shows the spatial distribution of ions of the mass/charge range 16–19 amu e^{-1} as measured by Giotto IMS [8] in the inner coma of comet Halley. Surprisingly, the peak number density was not near the nucleus, but at about 10^4 km distance (called "ion pile-up region" by [12]). The formation of the enhanced ion region was a puzzle for some time. The most likely explanation is that no ion enhancement is seen, but instead a decreased ion density in the innermost coma region. The decrease in ion density is caused by a sudden drop in electron temperature, T_e, in the inner coma because of efficient cooling of electrons by collisions with water molecules ($T_e \sim T_{gas}$ inside the cometopause) (Fig. 39). The rate coefficient for dissociative electron recombination is a strong function of T_e, and decreasing T_e therefore results in efficient ion dissociation in the inner coma. The major difficulty of modeling the observed distributions of H_3O^+ and other ions was caused by the fact that T_e in the energy range relevant for these reactions could not be measured by Giotto. As a result, many attempts have been made to determine the correct position of the sudden change in T_e (see summary in [95], Fig. 47). However, a good agreement of modeled and measured H_3O^+ ion density was finally obtained [98].

Radio observations of HCO^+ ions in comet Hale–Bopp from the ground also showed a region of decreased density around the nucleus [215]. Again, destruction of HCO^+ ions by electron dissociative recombination seems to play a major role [213]. Radio observations therefore allow us to image from ground the region where electron dissociation is important (see also discussion in [177]), at least for exceptionally bright comets. However, determining the low-energy electron temperature to model the inner coma ion densities remains difficult if no in situ data are available.

6 Gas Production Rates

Our knowledge of the chemical composition of cometary nuclei, its variation among comets and its variation with time and heliocentric distance is based on the measurements of production rates of molecules observed in the coma. When accurate coma production rates are available, models of the outgassing processes in the nucleus help to translate coma observations to nucleus abundances. Parent molecule production rates are best measured by direct observations of their emission lines, mainly in the radio and IR domain. Observations of daughter products in the optical range complement these observations. They significantly extend the range of heliocentric distances covered for many species and allow to include observations of faint comets.

The first step to derive gas production rates is to convert the measured emission fluxes into column densities or the number of molecules in the field-of-view (FOV) of the observation. This requires knowledge of the excitation mechanism of the observed emission as described in Sect. 4.

The detailed physical and chemical coma processes are often not taken into account when interpreting ground-based observations of comets. Only simple models of the cometary coma are used because fundamental parameters such as velocities and temperatures can often not be measured, in particular at large r_h. When discussing cometary gas activity and composition, it is therefore important to have in mind the various difficulties encountered.

6.1 Simple Coma Models

If the whole coma can be covered in the FOV of an observation and resonance fluorescence is the main excitation mechanism, the production rate, Q, can be derived from the measured flux, F, by the simple relation

$$4\pi \Delta^2 F = gQ\tau, \tag{43}$$

equating the flux at geocentric distance, Δ, with the fluorescence efficiency, g, multiplied by the mean production rate of the molecules with lifetime τ.

However, remote observations usually do not cover the whole coma in their FOV, and to correct for the molecules missed in the observations, a coma model needs to be applied. The most commonly used model is the Haser model [103], because it provides a simple analytic formula for the dependence of column density on nucleocentric distance. Exponential decay of a parent molecule and its daughter product is assumed. The number density of the daughter product versus nucleocentric distance r is then given as:

$$n(r) = \frac{Q}{4\pi u r^2} \frac{l_d}{l_p - l_d} \left(\exp\left(-\frac{r}{l_p}\right) - \exp\left(-\frac{r}{l_d}\right) \right) \tag{44}$$

The column density is then derived by integrating the computed profiles along the line-of-sight to the comet. This results in an analytical expression of the column density with a series of Bessel functions [103].

The model assumes isotropic radial outflow of parent and daughter molecules. The velocities of parents and daughters are assumed to be constant and in most cases set equal to $1\,\mathrm{km\,s^{-1}}$ at 1 AU. This is not in agreement with the real velocity profile in the coma, which depends on nucleocentric distance as a result of gas expansion, collisions, and photolytic heating (Sect. 3.1). However, constant velocities can reasonably well approximate high-resolution line profiles of the parent molecules observed at radio wavelengths and are a satisfying first approximation of the coma flow on the scales measured in remote ground-based observations. See Sect. 3 for a more detailed discussion on gas velocities.

To apply the Haser model, the scale lengths of the parent, l_p, and daughter products, l_d, need to be known. In principal, the scale lengths are given by the product of life time versus photodestruction and the gas velocity. However, as already mentioned, the gas velocity is not constant in the coma. In addition, temporal production rate variations, coma inhomogeneities, and extended sources can modify the spatial column density profiles. Furthermore, the observed daughter product may result from multiple parent molecules. In practice, therefore, scale lengths of daughter species and their parents are determined by approximating their observed spatial distribution with a column density profile derived from a Haser model.

We also add a word of caution when computing production rates at large heliocentric distances by simply extrapolating scale lengths determined near 1 AU. Such scale lengths might not correctly account for the changes in chemistry and gas flow occurring in the coma. Clearly, the Haser model serves only as a rough model of the coma flow, and it is often oversimplified.

Somewhat more sophisticated models of the coma flow used to compute production rates allow us to account for temporal variations of Q, for example due to nucleus rotation, and excess energies given to daughter products after formation from their parent molecules [49,81]. The more realistic modeling of the coma, however, is paid by the additional need of accurate values for the outflow velocities and the excess energies from photodissociation.

Realistic coma models need to take into account non-isotropic outgassing, temporal variability such as rotation, orbital variations, outbursts, as well as extended coma sources, and additional chemical processes other than simple two-step photoreactions. However, such models reach a state of complexity that makes them inconvenient to be used for fast production rate determinations. In addition, because of the large number of unknowns entering the modeling the reliability of the derived Q probably does not increase. The simple models, on the other hand, allow us to derive production rates relatively easy for a large number of comets. However, one needs to have their shortcomings in mind.

6.2 Abundance Ratios and Compositional Differences among Comets

Abundance Ratios

The main volatile constituent of the nucleus is water ice, followed by carbon monoxide. Although the activity of cometary nuclei is dominated by these two main ices, the minor species with abundances of at most a few percent give important clues for the understanding of the origin and formation of comets. Their abundance ratios reflect the processes of condensation to cometary grains in the presolar nebula or preplanetary disc and the amount of re-processing during later phases of solar system evolution.

Until the first direct drilling experiments on cometary nuclei will be made, the composition of comets can only be derived from measurements of their coma composition. The abundance of minor volatiles is usually characterized by providing the production rate ratio to the main activity driver, H_2O. In case of pure ices and in the regime where sublimation is controlled solely by the variation of solar insolation, all production rates vary with r_h^{-2} and the abundance ratios are constant. In a real comet, however, sublimation of the minor species also depends on the physical parameters of the nucleus (see Sect. 2), and the variation with heliocentric distance may not be the same as for water. In addition, the evolution of production rates obviously differs at larger distances where species are in the regime where their onset of activity occurs. Consequently, abundance ratios are distance dependent and comets can be compared only when observed at the same r_h, or if the variation of their abundance ratio is known to be small over the distance range considered. Most of the parent abundance ratios presented below were determined near 1 AU. Many reviews discussing volatile abundances can be found, the most recent in [32].

Water ice, H_2O, is the most abundant volatile constituent in cometary nuclei (80% by number in comet Halley [134]) and dominates the gaseous activity within about 3 AU heliocentric distance. The determination of H_2O production rates is therefore of vital importance to characterize cometary activity. H_2O emission bands in the infrared range were first observed from the Kuiper-airborne observatory [164] and from Earth orbit using infrared satellites, e.g. ISO and Spitzer. Unfortunately, observations from space can provide only poor coverage of comet activity because of the limited availability of IR-space telescopes and their restrictions by the solar elongation angle. Improvements in infrared technology allowed to detect water bands also from ground-based telescopes (e.g., [64, 162]). These observations require bright comets, and a sufficient Doppler shift if the observed transitions are affected by the telluric water absorption bands. The water production rate is therefore in most comets derived from its daughter products, mainly OH and O.

Abundance ratios of the minor parent species with respect to water are shown in Table 4.

Table 4. Production rate ratios relative to water. The minimum and maximum reported values are given for parent molecules from radio and infrared observations as summarized in [26] and [32]. For optical daughter radicals, the ratios for typical and depleted comets are taken from [2]

Molecule	Parent molecules	Daughter radicals	
		Typical	Depleted
H_2O	100		
CO	2–30		
CO_2	3–6		
CH_4	0.8–1.5		
C_2H_2	0.1–0.5		
C_2H_6	0.11–0.67		
CH_3OH	1.8–6.2		
H_2CO	0.13–4		
HCOOH	0.09		
$HCOOCH_3$	0.08		
CH_3CHO	0.02		
C_2		0.13–0.79	0.0074–0.1
C_3		0.0055–0.081	0.0014–0.02
NH_2CHO	0.015		
NH_3	0.5–1.5		
HCN	0.08–0.25		
HNCO	0.04–0.1		
HNC	0.005–0.04		
CH_3CN	0.01–0.035		
HC_3N	0.02		
NH		0.17–1.6	0.11–1.2
CN		0.15–0.68	0.11–0.32
H_2S	0.12–1.5		
OCS	0.1–0.4		
SO_2	0.2		
CS_2	0.06–0.2		
H_2CS	0.05		
S_2	0.0012–0.005		

C-bearing Species

The second most abundant volatile in comets is CO (2–30% relative to H_2O). Whether a possible extended CO coma source is included in ground-based measurements depends on the field-of-view (FOV), or beam size, of the observation. Small FOVs tend to cover mainly the nucleus source, whereas large FOVs, include the extended coma source and overestimate the nucleus production rates if a parent Haser model distribution is used (see Sect. 5). Therefore, part of the spread in CO/H_2O ratios is caused by the presence of the additional coma sources.

Remote observations of CO_2 and the in situ data in comet Halley near 1 AU provided abundances of 2–6% relative to water. Observations performed with ISO in comet Hale–Bopp showed an abundance of 22% at $r_h = 2.9$ AU [57]. Prompt emission of the CO Cameron band near 200 nm after photodissociation of CO_2 can be used to trace carbondioxide. However, because of only poorly known excitation parameters, reliable production rates are difficult to derive. Observations of the CO Cameron bands in comets 103P/Hartley 2 and C/1991 T2 (Shoemaker-Levy) with HST [208] and three additional comets using the IAU-satellite [77] are in agreement with CO_2/H_2O abundances of 2–6% in comets near 1 AU.

CH_3OH abundances range from 1.8–6.2%. Because the excitation of CH_3OH can be well determined, reliable abundances have been derived. The number of known carbon-bearing molecules has increased significantly by many new detections (HCOOH, $HCOOCH_3$, CH_4, C_2H_2, C_2H_6, CH_3CHO and C_2H_5OH) with abundances <1% in comets Hyakutake and Hale–Bopp [32,34,36,55,162]. Methane was first inferred from analysis of mass spectrometry data in comet Halley. However, the derived abundances were uncertain and model dependent, ranging from 0.5 to 2% [7,99]. The detection of CH_4 emission lines in the IR-range provides abundances of 0.8–1.5% [162,207]. The presence of C_2H_6 and C_2H_2 with abundances comparable to methane indicates that cometary material is not formed by processes in thermochemical equilibrium. The carbon chemistry might be a result of hydrogenation processes of CH_4 bearing ices on grain surfaces. However, radiation processing can be an alternative formation mechanism for C_2H_2 and C_2H_6 (see discussion in [36,162]). The most complex organic molecule detected by spectroscopy in comets is ethylene glycol (C_2H_5OH) observed in comet Hale–Bopp [54].

Nitrogen-bearing Species

The most abundant N-bearing molecule in the gas phase is ammonia. The abundance ratio of NH_3 derived from radio observations to water is of the order of 1% ([23], and references therein), in agreement with the revised value from mass-spectrometry measurements by the Giotto spacecraft in comet Halley [156]. Radio emissions of the NH_3 molecule are weak and can be detected only in bright comets. However, its daughter products NH_2 and NH can be observed in the optical range, and the sample of ammonia abundance measurements increases significantly if they are used to determine production rates of their parent.

HCN has been observed in several comets and abundances range from 0.08 to 0.25%. Evidence for additional N-bearing parent molecules came with the detection of HC_3N, CH_3CN, NH_2CHO, HNCO, and HNC in comets Hyakutake and Hale–Bopp with abundances of 0.01–0.03% relative to water [32,34,119,146]. The HCN isomer, HNC, was first detected in comet Hyakutake and subsequently observed in comet Hale–Bopp [25,119,146]. The variable HCN/HNC ratio in Hale–Bopp suggests its formation mainly as a daughter product by chemical reactions in the coma, rather than nucleus sublimation, which was also confirmed later in several comets [117, 118, 183, 185, 186].

Molecular nitrogen, N_2, has no favorable transitions to be observed. Its presence is inferred from detections of its ion, N_2^+. Contaminations of the N_2^+ emission at 319 nm with CO^+ and CO_2^+ emission features in addition to atmospheric lines result in large uncertainties of the derived column densities. Mass-spectrometry on board the Giotto spacecraft visiting comet Halley faces the problem of additional contamination of mass-channels by other species. The uncertainty in N_2 measurements is therefore very large. In comet Halley, N_2/H_2O ranged from <0.15% from IMS data [12] to 0.02% from measurements of N_2^+ [216].

Sulfur-bearing species

Six sulfur-bearing parent species are currently known in comets, H_2S, OCS, S_2, SO_2, H_2CS, and CS_2. The most abundant sulfur parent is H_2S (0.12–1.5%). CS_2 is inferred by its daughter product CS [120]. A second parent molecule of CS, OCS, has been detected [64, 146, 214] with abundances of 0.1–0.4%. Possibly, OCS is released by an extended source in Hale–Bopp [64]. Molecular sulfur, S_2, was detected in comets IRAS-Araki-Alcock and probably Hyakutake [3, 76, 138]. However, the abundances are very uncertain because of the complex excitation conditions. An updated model for S_2 excitation has been proposed by [181]. Similar uncertainties are found for observations of SO and SO_2, for which abundances up to 0.2% have been derived [146]. Both molecules require improved excitation models to better constrain their abundance ratios.

Atomic Abundances

The elemental C/O ratio is mainly determined by the abundance ratio of CO and CO_2 relative to water. In comet Halley, the ratio is in agreement with the solar system value of 0.42 [9] for a dust/gas ratio of about 1.7 and with $(C/O)_{gas} \approx 0.17$ and $(C/O)_{grain} \approx 0.91$ [124].

The N/O ratios derived from N-bearing molecules in the gas phase of comet Halley was $(N/O)_{gas} \approx 0.03$ [123]. Nitrogen is, therefore, depleted by a factor of about 4 relative to the solar system value $(N/O)_{solar}=0.13$ [9]. The N/O ratio in comet Halley's dust particles was 0.05 and, thus, depleted by a factor of about 3 [124] in comparison to the solar value. The recently detected N-bearing molecules CH_3CN, HC_3N, HNCO, and NH_2CHO have abundance of less than 0.1% and therefore do not contribute significantly to the N-content in comets. Despite the large uncertainty in deriving reliable N/O ratios, the measured depletion with respect to solar system values seems to be real, unless a significant amount of additional N-bearing molecules has remained undiscovered, either in the gas or dust phase.

The elemental (S/O) ratio is about 0.03 ± 0.02 in the gas phase. This is somewhat higher but still in agreement with the solar value of 0.02 [9]. Analysis of grains in comet Halley gave $(S/O)_{grain} \approx 0.08$ [124].

6.3 Compositional Differences Among Comets

Even though the abundance ratios measured by remote sensing in the coma of comets might not equal the actual nucleus composition, we expect comets with the same nucleus composition and internal structure to show the same coma abundances when observed at similar r_h. By comparing abundances in samples of comets, we therefore obtain information on the diversity among comets that may be related to their formation and/or later processing in the solar system. Obviously, we have to be careful in our interpretation and check to what extent different results in abundances reflect real differences of comets or just measurement uncertainty.

Abundances are computed by the ratio of gas production rates of minor species relative to H_2O or OH. When observations are made in the optical or radio range, ratios are also given relative to CN or HCN, because due to their strong emission lines, CN and HCN can be observed in faint comets.

The largest samples of cometary observations exist for the radicals observed in the near-UV and optical range (OH, NH, CN, C_3, C_2, NH_2). Several comparative studies of comets have been made with different sample sizes ranging from 17 to 85 comets [2,41,82,167]. From the evaluation of the largest sample [2], the existence of a class of comets depleted in carbon (mainly C_2) was found (Fig. 48). All C-depleted comets belong to the Jupiter family, but not all Jupiter family comets are depleted.

The number of comets in which parent molecules could be observed directly is still relatively small, but increasing strongly these days as sensitive

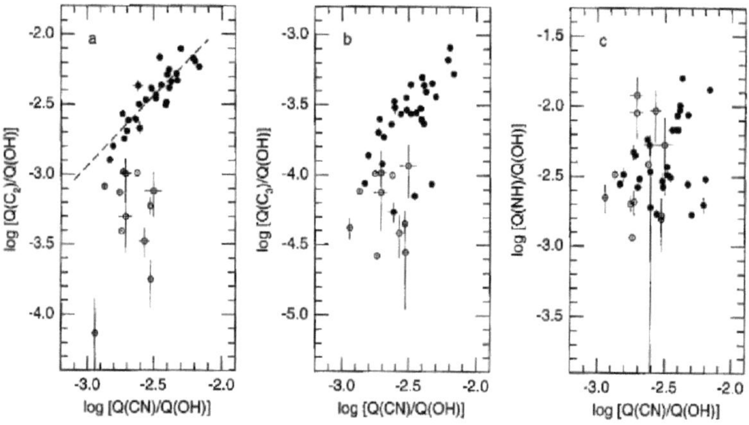

Fig. 48. Differences between Jupiter family and long-period comets [2]

observations can be made at radio and IR wavelengths. A study of HCN, HNC, CH_3CN, CH_3OH, H_2CO, CO, CS, and H_2S of comets in the radio range (4 Jupiter -family, 2 Halley-family, 13 long-period) has shown no significant differences among Jupiter family and long-period comets [27]. This result is not in contradiction with the finding of C-depleted short-period comets because the C_2 and C_3 bearing parent molecules (C_2H_2, C_2H_6, and C_3H_4) were not included in the radio observations. Methanol has been determined at radio wavelengths in more than 17 comets and has therefore one of the largest data base of parent species. Variations of CH_3OH/H_2O of 2–6% among comets and 0.5–5% among Oort-cloud comets seem to reflect real compositional differences [27, 60]. Comets rich in methanol also seem rich in formaldehyde [27].

Cometary parent molecules can also be observed at IR wavelengths, although only in bright objects. A study of six long-period comets [163] included CO, CH_4, C_2H_2, C_2H_6, HCN, and CH_3OH. Out of the six comets in this study, five comets and comet P/Halley show relatively similar abundances. Nevertheless, their nucleus source of CO still varied by a factor of 10 from comet to comet. The other species showed variations less than a factor of two. One comet however, C/1999 S4, is very different compared with the other long-period comets. It is depleted in almost all volatiles observed. Possibly C/1999 S4 is indicating different formation conditions with reduced abundances of minor species relative to H_2O and therefore variations among Oort-cloud comets. However, we have to wait for a larger statistical sample for a definite conclusion. In view of the importance of the possible existence of classes of comets with different chemical composition, further observations of large samples need to be made.

In summary, among the parent molecules nucleus CO seems to show large variations among comets up to a factor of ten. Such large differences among Oort cloud comets and the existence of a truly different comet (C/1999 S4) indicate compositional variations among long-period comets. Comparing long-period to short period comets, a class of Jupiter family comets depleted in C_2 and C_3 exists. This difference between short- and long-period comets could not yet be confirmed for parent molecules because of the insufficient detection limit for these parents for weak comets. We have to await further technical improvements and a larger statistical data base for a more detailed study of the compositional differences among comets.

6.4 Isotopic Ratios

Isotopic ratios provide severe constraints on the origin of comets. Table 5 provides a list of the abundances determined in comets in comparison to the values in the solar system and the interstellar medium (see also [206]). The excellent correlation of isotopic ratios of cometary parent molecules with the solar system values is consistent with formation of cometary nuclei in the solar nebula.

Table 5. Isotopic ratios

	ratio	comet	solar system	ISM	reference
$H^{12}CN/H^{13}C$	111±15	HB			[126]
	90±15	HB	89.9	77±7	[146]
	109±22	HB			[217]
	34±12*	Hya			[145]
$^{12}CN/^{13}CN$	95±12	Halley			[131]
	90±10	dV			[121]
	90±25	IZ			[121]
	115±20	WM1			[10]
	93±30	HB			[149]
	90±15	Q4			[149]
	90±15	K4			[149]
	100±30	S4			[116]
	90±10	Q1			[116]
	95±15	T1			[122]
$^{12}C_2/^{13}C_2$	70±15	Ik			[199]
	100±20	TSK			[172]
	115^{+30}_{-20}	K			[62]
	135^{+65}_{-45}	K			[62]
	100^{+20}_{-30}	KBM			[205]
$HC^{14}N/HC^{15}N$	323±46	HB	272	450±100	[126]
	330±98	HB			[217]
$C^{14}N/C^{15}N$	140±20	dV			[121]
	170±50	IZ			[121]
	150±40	HB			[149]
	135±20	Q4			[149]
	135±20	K4			[149]
	140±30	WM1			[10]
	150±40	S4			[116]
	140±15	Q1			[116]
	145±20	T1			[122]
$C^{32}S/C^{34}S$	27±3	HB	22.6	32±5	[126]

comets: HB: Hale–Bopp; Hya: Hyakutake; Ik: Ikeya 1963I; K: Kohoutek 1973XII; TSK: Tago–Sato–Kosaka 1969IX; KBM: Kobayashi–Berger–Milon 1975IX; dV: 122P/1995 S1 de Vico; IZ: 153P/2002 C1 Ikeya–Zhang; Q4: C/2001 Q4 (NEAT); K4: C/2003 K4 (LINEAR; WM1: C/2000 WM1 (LINEAR); T1: Tempel 1)
*: possibly contaminated by SO_2 emission

The difference of the $^{14}N/^{15}N$ ratio derived from radio observations of HCN and optical observations of CN is still a puzzle. Possibly, the difference is caused by an additional parent molecule of CN with an isotopic nitrogen abundance different to HCN [10, 121, 149].

Table 6. D/H ratio in cometary water and HCN

Comet	Species	D/H $[10^{-4}]$	Reference
Halley	H_2O	$3.08^{+0.38}_{-0.53}$	[13]
	H_2O	3.16 ± 0.34	[69]
Hyakutake	H_2O	2.9 ± 1.0	[34]
Hale–Bopp	H_2O	3.3 ± 0.8	[157]
	HCN	23.0 ± 4.0	[157]

The deuterium abundance of water derived by in situ measurements in comet Halley [13, 69] and by remote sensing at radio wavelengths in comet Hyakutake and Hale–Bopp is given in Table 6. The values are significantly higher than the standard mean ocean water values (SMOW) on Earth (Fig. 49). The abundances of D/H in comets suggest incorporation of interstellar material during comet formation. The difference to SMOW implies, that probably only a minor fraction of water on Earth comes from impacting comets. A detailed discussion on possible formation scenarios can be found in [34] and [157], and references given in Table 6.

7 Dust Particles

The cometary dust component is visible as a prominent dust tail in many comets (Fig. 1). The dust particles can be seen because they scatter the solar light efficiently. In Sect. 3, we already discussed the dynamics of dust particles in the inner coma and in the dust tail. Here, an overview of the nature of the dust particles is given.

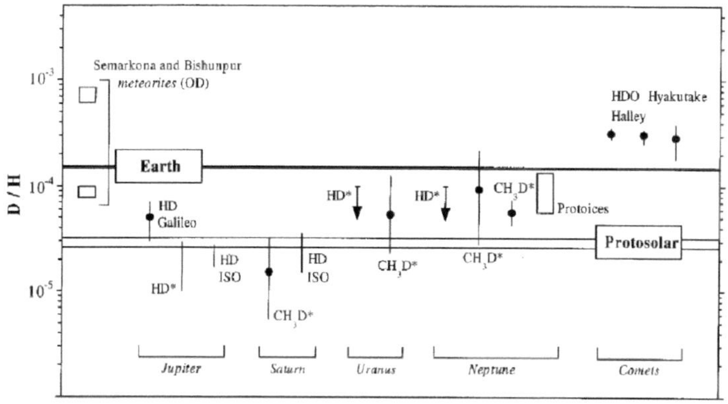

Fig. 49. The D/H ratio in comets and solar system objects [34]

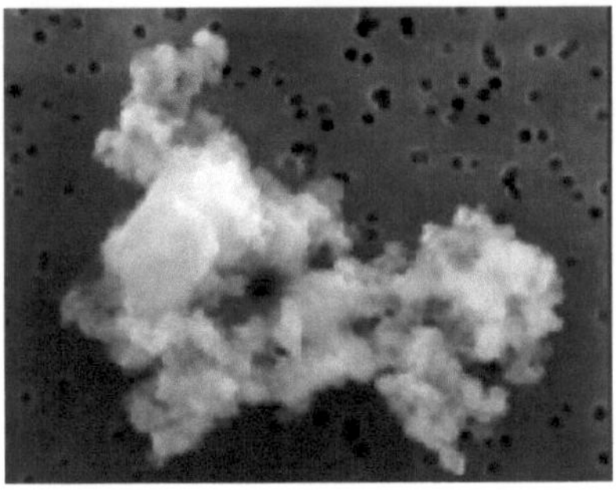

Fig. 50. Interplanetary dust particles have irregular shape and low density. This particle is about 10 μm in length. (NASA/JPL)

It is generally believed that dust particles are irregular and porous particles, very much like the interplanetary dust particles collected at the top of the Earth atmosphere (Fig. 50). At cold temperatures, under the conditions found in the interstellar medium, silicate minerals form in the amorphous state. Formation at higher temperatures or heating of the dust grains to temperatures above ~1000 K will lead to crystalline silicates. The proportion of amorphous to crystalline silicates in dust grains will therefore give us some information on whether interstellar grains may have survived the conditions in the pre-planetary disc and about the temperature regime in the disc where the grains formed which have then been incorporated into comets. The chemical composition of the dust grains will show the homogeneity of the refractory cometary material. The study of dust grains is therefore an important component in the puzzle that needs to be completed to understand the formation of comets.

7.1 Composition

In situ data for the chemical composition of cometary dust grains were available only from the Giotto spacecraft (ESA) until recently. New information is expected from the analysis of the grain samples returned to Earth by the Stardust mission (NASA). The in situ analysis of dust grains in the coma of comet Halley showed in general three types of grains with about equal relative particle abundance [140]:

- Particles similar to CO chondrites: Na, Mg, Si, Ca, Fe (~35%).
- Particles consisting mainly of light atoms: C, H, O, N (so-called CHON-particles) (~30%).
- Particles consisting of silicates, but also light elements (mixture of the two cases) (~35%).

For individual particles, however, large variations in composition have been found [94]. The density of dust particles was estimated to $\sim 1\,\mathrm{g\,cm^{-3}}$, consistent with their expected fluffy nature [94].

CHON particles are believed to consist mainly of organic material. However, a clear identification has been difficult so far. It is often proposed that these organic particles may form part of the extended coma sources discussed for some of the gas species (see Sect. 5).

Information on the chemical and mineralogical composition of cometary dust grains can also be obtained from spectral observations. Spectra in the infrared wavelengths range show prominent emission features (Fig. 51). Around 10 μm stretching vibrations of Si–O bonds in silicates produce a well-known emission feature. Additional emissions are present at longer wavelengths, e.g. at 16 μm and 35 μm caused by bending modes. Fortunately, the atmospheric window around 10 μm allows to study this feature in many comets in ground-based observations. At longer wavelengths, we need space telescopes.

The silicate emission features in Hale–Bopp and many other comets have been compared to spectral models and laboratory spectra of silicates to identify the emission peaks in terms of mineralogical composition. A combination of crystalline and amorphous grains, consisting mainly of pyroxene and olivine, is found to match best with the observed spectra (e.g., [159]). Thus, cometary dust grains are a mixture of material condensed at low temperatures (amorphous) and of material processed by high temperatures (crystalline). This composition probably results from mixing processes in the proto-planetary disc.

It is interesting to note that no strong 10 μm silicate emission feature has been detected in short-period comets yet [102]. Whether this is caused by compositional differences of comets belonging to different dynamical classes remains, however, unclear so far. Nevertheless, this finding is, among others, a strong motivation for future spectral observations of cometary silicate features.

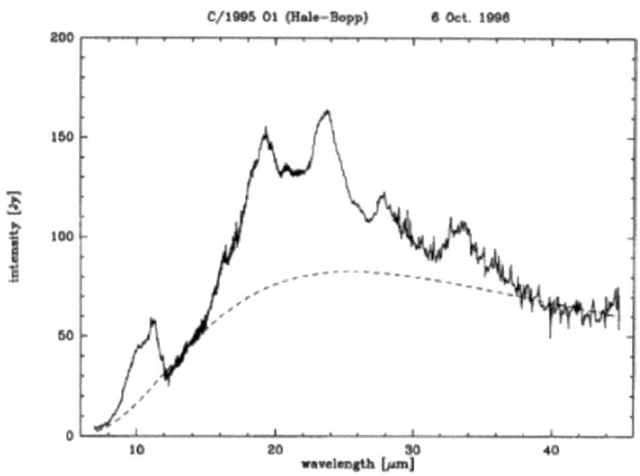

Fig. 51. ISO SWS spectrum of comet Hale–Bopp [57]. Several silicate emission features are seen superimposed on a black body spectrum

7.2 Size Distribution

To derive the size distribution of dust particles again in situ measurements are ideal. Giotto measured the dust particles at comet Halley. Figure 52 shows the resulting size distribution [153]. The diagram shows the number of detected dust particles versus their radius and mass after assuming a mean density of $1\,\mathrm{g\,cm^{-3}}$ for the dust particles. We note that:

- Most particles have small radii.
- The dust mass is concentrated in a few large particles.

At large radii, only very few particles were detected by Giotto, and beyond 10^{-3} m radius, the distribution is just a simple extrapolation. The dominance of small dust particles in number is also supported by data of the Stardust mission obtained in the coma of comet Wild 2 [93, 204].

If the exact size distribution of dust particles in a cometary coma can not be determined observationally, as it is usually the case, it is typically expressed by a function like [101]:

$$f(a) = \bar{N}\left(1 - \frac{a_0}{a}\right)^M \left(\frac{a_0}{a}\right)^N \tag{45}$$

Here, $\bar{N}$ is a scaling factor, a the particle radius, and a_0 the lower size limit. Figure 53 shows an example of size distributions for different constants M and N, which determine the position of the maximum and the slope of the distribution.

As in situ observations of comets are rare, many attempts are made to derive the dust size distribution from ground-based observations. This can be made by using infrared observations of the thermal emission of cometary dust grains in comparison to light scattered at optical wavelengths, by polarization

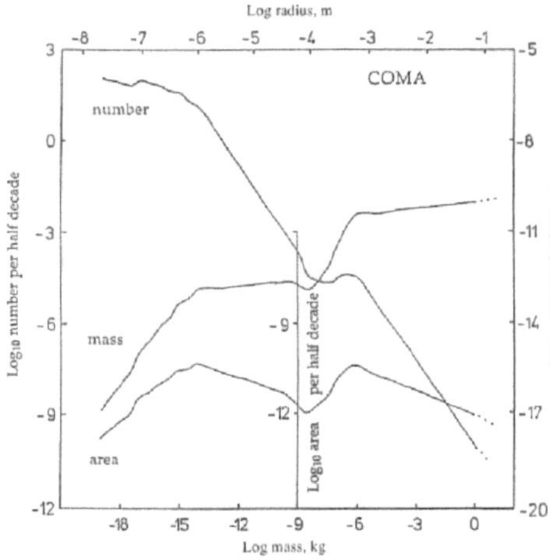

Fig. 52. Dust size distribution measured at comet Halley [153]

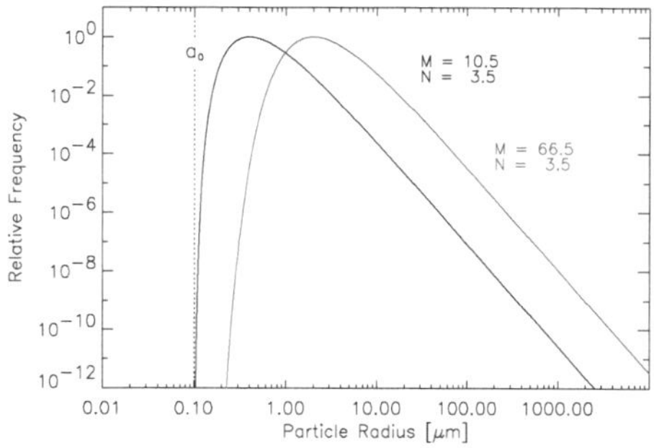

Fig. 53. Hypothetical dust size distributions computed from (45) for different constants M and N

measurements and by studying the dynamical distribution of grains in the dust tail .

A method to estimate the size distribution using IR measurements is to determine the so-called "super-heat" parameter. At thermal equilibrium the grain temperature is balanced by the absorbed flux at UV and visible wavelengths and the re-radiated flux in the infrared, and for a spherical particle:

$$\int_0^\infty \frac{L_\odot}{4\pi r_\mathrm{h}^2} Q_\mathrm{abs}(\lambda, a)\pi a^2 \mathrm{d}\lambda = \int_0^\infty \pi B(\lambda, T(a,r)) Q_\mathrm{abs}(\lambda, a) 4\pi a^2 \mathrm{d}\lambda \qquad (46)$$

To obtain the emitted flux, we need to integrate over the size distribution of the grains.

The temperature of very small grains can deviate from the equilibrium temperature, because small grains absorb solar light very efficiently at optical wavelengths with absorption coefficient $\mathrm{Q}_\mathrm{abs}^\mathrm{vis}$, but because of their low infrared emissivity, $\mathrm{Q}_\mathrm{e}^\mathrm{IR}$, it is difficult to reradiate this energy again. They therefore can heat up to relatively high temperatures. For particles larger than a few micron, $\mathrm{Q}_\mathrm{abs}^\mathrm{vis}$ and $\mathrm{Q}_\mathrm{e}^\mathrm{IR}$ are about equal and the particle temperature is closer to the black body temperature [79]. The so-called "superheat" parameter has been defined as the ratio of the grain temperature, $\mathrm{T_g}$, to the equilibrium black body temperature, $\mathrm{T_b}$, at the same heliocentric distance. A high superheat parameter, therefore, is an indication for small grains.

For the spatial distribution of dust grains in the tail and the relation to their size and other material parameters, see Sect. 3 and the references therein. A detailed overview on how to derive the material properties of dust grains including the use of polarization measurements is given in [133].

In general, observations suffer from the fact that very large dust particles represent only a small total cross-section because they are very rare. Therefore we gain little information about them, resulting in large uncertainty of the total dust mass. Very small dust particles are not efficient light scatterers and also remain undetected. If they are very numerous, they could also significantly increase the error of the dust mass determination. Therefore, the determination of a good dust mass production rate is a difficult task.

7.3 The Dust Production Rate

To be able to derive the dust-to-gas ratio of comets, we need to determine the dust mass production rate. To quantitatively determine the dust production rate from observations, we need to know the particle size distribution, their density and mass, their optical properties, as well as their dynamics. These quantities are difficult to derive observationally, as outlined above. Therefore, often a simplified approach is made based on observational quantities, such as the spatial dust distribution or the scattered optical light.

Reference [4] introduced a quantity called "Afρ"-parameter to determine the dust content of a cometary coma. We discuss this parameter here, because it is often used as an estimate of the cometary dust content.

The Afρ parameter is given by:

$$Af\rho = \frac{(2\Delta r)^2}{\rho}\left(\frac{F}{F_\odot}\right) \tag{47}$$

Here, A is the average grain albedo, f is the filling factor of the grains in the FOV of the observations, ρ is the projected aperture radius used for the photometry at the comet, and F is the observed photometric continuum flux. $F_\odot$ is the solar flux measured in the same filter bandpass used to measure F. Therefore, the parameter can be directly derived from photometric observations. For isotropic outflow with constant velocity (and without dust fragmentation), the Afρ parameter is independent of the aperture size ρ and the geocentric distance, Δ, of the comet. It is therefore an ideal quantity if different comet observations are to be compared.

The Afρ parameter is a measure for the effective dust scattering area in the FOV. This can be seen when looking at the filling factor which is given as:

$$f = \frac{N(\rho)\sigma}{\pi\rho^2} \tag{48}$$

$N(\rho)$ is the number of grains in the FOV and σ denotes the grain scattering cross-section.

Although the Afρ parameter is related to the amount of dust in the aperture and scattered solar light, it is not a direct measure of the dust production rate. To obtain a production rate, the dust expansion velocity, v, the scattering phase function, $D(\theta)$, the dust size distribution, $f(a)$, and the density of dust particles, $\rho_{\rm dust}$, must be taken into account. A method to derive the dust production rate, $Q_{\rm dust}$, from Afρ was given, for example, by [127]:

$$Q_{\rm dust} = \frac{2}{3\pi}\frac{Af\rho}{A_{\rm B}D(\theta)}\left(\int_{a_1}^{a_{\max}}\frac{f(a)a^2}{v(a)}{\rm d}a\right)^{-1}\int_{a_1}^{a_{\max}}\rho_{\rm dust}(a)a^3 f(a){\rm d}a \tag{49}$$

The dust velocity is often simply scaled with heliocentric distance as $1/\sqrt{(r_{\rm h})}$. However, more realistic results are obtained when computing the dust terminal velocity from a model of the dust coma dynamics [212]. Another important parameter is the upper limit of the integration over the dust size distribution. As most of the mass is usually contained in the large particles, it is important to derive the upper size limit, $a_{\max}$ as accurately as possible. How to derive the upper limit of particles that can just be lifted from the nucleus is explained in the chapter by Dave Jewitt [125]. But the uncertainty remains whether such large dust particles really exist, leading to a large error in dust production rates.

Another often unknown parameter is the scattering phase function, $D(\theta)$, of the cometary grains. The geometry is illustrated in Fig. 54. The scattered light intensity is not isotropic. Instead, a strong forward and backward intensity peak is usually observed. The exact scattering properties depend on the particle size, shape, and composition. Often, the particles are approximated by a sphere because then Mie scattering theory [63, 158] can be applied. To describe the scattering properties and related theory of dust grains in detail is beyond the very limited space of this chapter. We refer to [115] for scattering theory.

Here, we also provide a note on the meaning of albedo. In general, the single scattering albedo, A, is defined as the ratio of the energy scattered into all directions to the energy removed from the incident beam by extinction. Extinction includes reflection, absorption, diffraction, and refraction. Therefore, the albedo of a particle is, of course, a function of its material properties.

The Bond albedo, A_{B}, provides the incident light scattered in all directions. It does not take into account diffraction. Thus, it is defined as the ratio of [100]:

$$A_{\mathrm{B}} = \int \sigma(\theta) d\omega / G \tag{50}$$

Here, $\sigma(\theta)$ denotes the differential scattering cross-section and G the geometrical cross-section of the particle.

The geometrical albedo, A_{p}, corresponds to the light scattered relative to the light scattered by a Lambertian surface of the same geometric cross-section [100]. It is given by:

$$A_{\mathrm{p}}(\theta) = \frac{\pi}{G} \sigma(\theta) \mathrm{d}\omega \tag{51}$$

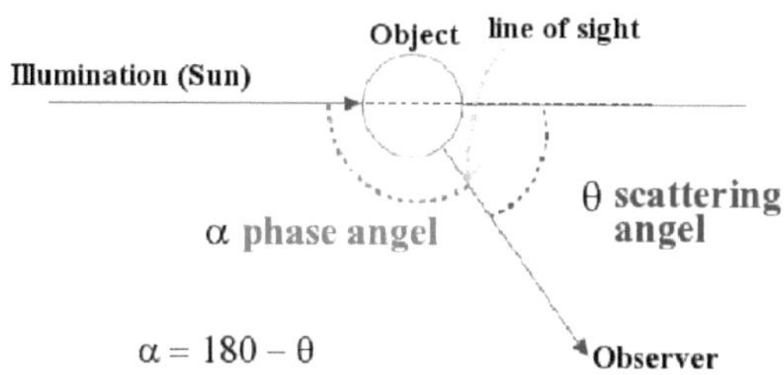

Fig. 54. Scattering angle and phase angle

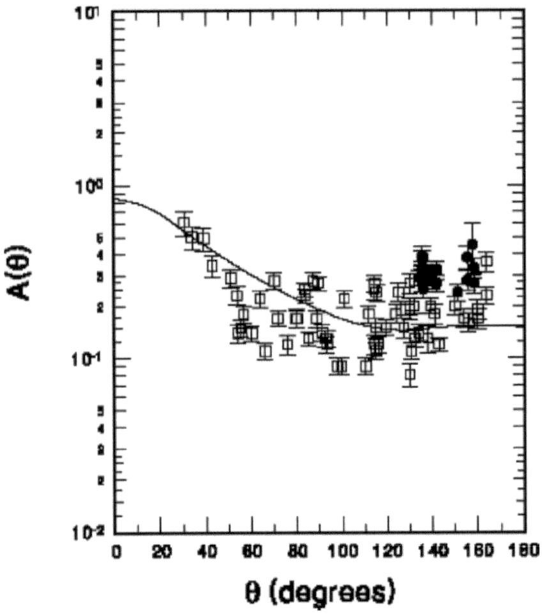

Fig. 55. albedo versus scattering angle measured in 12 comets [150]

Both albedos are related by: $A_{\mathrm{B}} = A_{\mathrm{P}} q$. Here, q is the phase integral and is equal to 4 in case of isotropic scattering where $\mathrm{A_B} = 1$ and $\mathrm{A_p} = 0.25$. In general q is given as:

$$q = 2 \int_0^{\pi} j(\theta) sin\theta \mathrm{d}\theta \tag{52}$$

Where $j(\theta) = \sigma(\theta)/\sigma(180°)$.

Figure 55 shows the albedo versus scattering angle measured in 12 comets [150]. The albedo is derived from measurements made quasi simultaneously at thermal IR and optical wavelengths. The scattering function can be derived when comets are observed over a wide range of phase angles. This is however, only rarely possible. Therefore, often a “standard” phase function is assumed based on [67]. We note, however, that the phase function for individual comets can be quite different to the standard phase curve (e.g., [188]).

8 Outlook

At the beginning of this chapter, we recalled the main motivation for cometary studies: How did comets form and what do we learn from them about the formation of our solar system? Answering this fundamental questions requires knowledge of many different fields of physics, such as gas dynamics, plasma physics, chemistry, scattering theory, molecular physics, etc., including their applications to the conditions found in comets. In the previous sections, we looked at some of the basic concepts scientists use to address the problems in the different areas of comet research. Of course, students specialize in their field of interest to obtain a much deeper understanding in that area. However, I encourage students to have a broader view on the context of their specialist research fields to be able to see interrelations and avoid misconceptions.

An enormous progress has been made in the past 20 years in our understanding of comets, although important key processes are still far from being understood (e.g., activity). Space missions have been sent to observe comets directly and provide in situ data. Ground-based technology progressed with new detector technologies and telescopes allowing us to observe comets not only in the optical but also at radio and near-IR wavelengths with high sensitivity. In addition, space telescopes provide access to wavelengths regions we can not investigate from the ground. The technical progress resulted in an enlarged volume of high quality data, which stimulated the development of theory to explain what had been observed. Nevertheless, there is still no generally accepted theory on comet formation. We need to proceed and obtain further data and improve our theory in the future. Future investigations need to include

- Direct investigations of nucleus material by landers.
- Sample return of cometary material (nucleus and coma).
- Improved statistical database of cometary gas and dust composition.
- Measurements of gas and dust activity over wide ranges of heliocentric distances.
- Laboratory comet simulation experiments (e.g., outgassing processes, ice formation).
- Numerical models of cometary activity.
- Understanding of the relation to Kuiper belt objects.

Some of these point will be already addressed in the near future by space missions. The Stardust mission has already transported in situ samples of coma dust particles to Earth [37]. ESA's Rosetta [195] mission will follow a comet over a substantial part of its orbit and will drop a lander on its nucleus to investigate the upper surface layers directly. In parallel, ground-based facilities, such as the radio telescope array ALMA [29, 202] will provide a new wealth of data on parent gas molecules. The field of cometary science therefore needs young motivated students educated in cometary science to explore the information to come.

Solving the problem of comet formation is like solving a big puzzle. We have to ask the right questions to find the right puzzle pieces and then find the right places for them to be able to solve "the big picture" of comet and planet formation. Each student works on his/her own small or larger "puzzle piece." It is the combination of all of them that will help us to obtain the whole picture.

References

1. M. F. A'Hearn, M. J. S. Belton, W. A. Delamere, J. Kissel, K. P. Klaasen, L. A. McFadden, K. J. Meech, H. J. Melosh, P. H. Schultz, J. M. Sunshine, P. C. Thomas, J. Veverka, D. K. Yeomans, M. W. Baca, I. Busko, C. J. Crockett, S. M. Collins, M. Desnoyer, C. A. Eberhardy, C. M. Ernst, T. L. Farnham, L. Feaga, O. Groussin, D. Hampton, S. I. Ipatov, J.-Y. Li, D. Lindler, C. M. Lisse, N. Mastrodemos, W. M. Owen, J. E. Richardson, D. D. Wellnitz, and R. L. White. Deep impact: Excavating comet tempel 1. *Science*, 310:258–264, October 2005.
2. M. F. A'Hearn, R. L. Millis, D. G. Schleicher, D. J. Osip, and P. V. Birch. The ensemble properties of comets: Results from narrowband photometry of 85 comets, 1976–1992. *Icarus*, 118:223–270, December 1995.
3. M. F. A'Hearn, D. G. Schleicher, and P. D. Feldman. The discovery of S2 in comet IRAS-Araki-Alcock 1983d. *ApJL*, 274:L99–L103, November 1983.
4. M. F. A'Hearn, D. G. Schleicher, R. L. Millis, P. D. Feldman, and D. T. Thompson. Comet Bowell 1980b. *AJ*, 89:579–591, April 1984.
5. H. Alfven. On the theory of comet tails. *Tellus*, 9:92–96, 1957.
6. R. Ali, P. A. Neill, P. Beiersdorfer, C. L. Harris, M. J. Raković, J. G. Wang, D. R. Schultz, and P. C. Stancil. On the significance of the contribution of multiple-electron capture processes to cometary X-Ray emission. *ApjL*, 629:L125–L128, August 2005.
7. M. Allen, M. Delitsky, W. Huntress, Y. Yung, and W.-H. Ip. Evidence for methane and ammonia in the coma of comet P/Halley. *A&A*, 187:502–512, November 1987.
8. K. Altwegg, H. Balsiger, J. Geiss, R. Goldstein, W.-H. Ip, A. Meier, M. Neugebauer, H. Rosenbauer, and E. Shelley. The ion population between 1300 KM and 230000 KM in the coma of comet P/Halley. *A&A*, 279:260–266, November 1993.
9. E. Anders and N. Grevesse. Abundances of the elements – Meteoritic and solar. *GCA*, 53:197–214, January 1989.
10. C. Arpigny, E. Jehin, J. Manfroid, D. Hutsemékers, R. Schulz, J. A. Stüwe, J.-M. Zucconi, and I. Ilyin. Anomalous Nitrogen Isotope Ratio in Comets. *Science*, 301:1522–1525, September 2003.
11. C. Arpigny. Physical Chemistry of Comets: Models, Uncertainties, Data Needs. In I. Nenner, editor, *AIP Conf. Proc. 312: Molecules and Grains in Space*, page 205, 1994.
12. H. Balsiger, K. Altwegg, F. Buhler, J. Geiss, A. G. Ghielmetti, B. E. Goldstein, R. Goldstein, W. T. Huntress, W.-H. Ip, A. J. Lazarus, A. Meier, M. Neugebauer, U. Rettenmund, H. Rosenbauer, R. Schwenn, R. D. Sharp, E. G. Shelly, E. Ungstrup, and D. T. Young. Ion composition and dynamics at comet Halley. *Nature*, 321:330–334, May 1986.
13. H. Balsiger, K. Altwegg, and J. Geiss. D/H and O-18/O-16 ratio in the hydronium ion and in neutral water from in situ ion measurements in comet Halley. *JGR*, 100:5827–5834, April 1995.
14. H. Balsiger. *Measurements of ion species within the coma of comet Halley from Giotto*, page 129. Comet Halley: Investigations, Results, Interpretations. Vol. 1: Organization, Plasma, Gas, 1990.
15. A. Bar-Nun and D. Laufer. First experimental studies of large samples of gas-laden amorphous "cometary" ices. *Icarus*, 161:157–163, January 2003.

16. A. Bar-Nun, D. Prialnik, D. Laufer, and E. Kochavi. Trapping of gases by water ice and implications for icy bodies. *Advances in Space Research*, 7:45–47, 1987.
17. J. E. Beaver, R. M. Wagner, D. G. Schleicher, and B. L. Lutz. Anomalous molecular abundances and the depletion of NH_2 in Comet P/Giacobini-Zinner. *Astrophysical Journal*, 360:696–701, September 1990.
18. J. Benkhoff and D. C. Boice. Modeling the thermal properties and the gas flux from a porous, ice-dust body in the orbit of P/Wirtanen. *Planetary and Space Science*, 44:665–673, July 1996.
19. J. Benkhoff and W. F. Huebner. Influence of the vapor flux on temperature, density, and abundance distributions in a multicomponent, porous, icy body. *Icarus*, 114:348–354, April 1995.
20. J. Benkhoff and W. F. Huebner. Modeling the gas flux from a Jupiter-family comet nucleus. *Planetary and Space Science*, 44:1005–1013, September 1996.
21. J. Benkhoff. On the flux of water and minor volatiles from the Surface of Comet Nuclei. *Space Science Reviews*, 90:141–148, 1999.
22. L. Biermann. Kometenschweife und solare Korpuskularstrahlung. *Zeitschrift für Astrophysik*, 29:274, 1951.
23. M. K. Bird, P. Janardhan, T. L. Wilson, W. K. Huchtmeier, P. Gensheimer, and C. Lemme. K-band radio observations of comet Hale–Bopp: Detections of ammonia and (possibly) water. *Earth Moon and Planets*, 78:21–28, 1999.
24. N. Biver, D. Bockelée-Morvan, P. Colom, J. Crovisier, J. K. Davies, W. R. F. Dent, D. Despois, E. Gerard, E. Lellouch, H. Rauer, R. Moreno, and G. Paubert. Evolution of the outgassing of comet Hale–Bopp (C/1995 O1) from radio observations. *Science*, 275:1915–1918, 1997.
25. N. Biver, D. Bockelée-Morvan, P. Colom, J. Crovisier, B. Germain, E. Lellouch, J. K. Davies, W. R. F. Dent, R. Moreno, G. Paubert, J. Wink, D. Despois, D. C. Lis, D. Mehringer, D. Benford, M. Gardner, T. G. Phillips, M. Gunnarsson, H. Rickman, A. Winnberg, P. Bergman, L. E. B. Johansson, and H. Rauer. Long-term evolution of the outgassing of comet Hale–Bopp from radio observations. *Earth Moon and Planets*, 78:5–11, 1999.
26. N. Biver, D. Bockelée-Morvan, P. Colom, J. Crovisier, F. Henry, E. Lellouch, A. Winnberg, L. E. B. Johansson, M. Gunnarsson, H. Rickman, F. Rantakyrö, J. K. Davies, W. R. F. Dent, G. Paubert, R. Moreno, J. Wink, D. Despois, D. J. Benford, M. Gardner, D. C. Lis, D. Mehringer, T. G. Phillips, and H. Rauer. The 1995 2002 Long-term monitoring of Comet C/1995 O1 (HALE BOPP) at radio wavelength. *Earth Moon and Planets*, 90:5–14, June 2002.
27. N. Biver, D. Bockelée-Morvan, J. Crovisier, P. Colom, F. Henry, R. Moreno, G. Paubert, D. Despois, and D. C. Lis. Chemical composition diversity among 24 comets observed at radio wavelengths. *Earth Moon and Planets*, 90: 323–333, March 2002.
28. N. Biver, D. Bockelée-Morvan, J. Crovisier, J. K. Davies, H. E. Matthews, J. E. Wink, H. Rauer, P. Colom, W. R. F. Dent, D. Despois, R. Moreno, G. Paubert, D. Jewitt, and M. Senay. Spectroscopic monitoring of comet C/1996 B2 (Hyakutake) with the JCMT and IRAM Radio Telescopes. *AJ*, 118:1850–1872, October 1999.
29. N. Biver. Comets with ALMA. In A. Wilson, editor, *The Dusty and Molecular Universe: A Prelude to Herschel and ALMA*, pages 151–156, January 2005.
30. D. Bockelée-Morvan, J. Crovisier, P. Colom, and D. Despois. The rotational lines of methanol in comets Austin 1990 V and Levy 1990 XX. *A&A*, 287: 647–665, July 1994.

31. D. Bockelée-Morvan, J. Crovisier, and E. Gerard. Retrieving the coma gas expansion velocity in P/Halley, Wilson (1987 VII) and several other comets from the 18-cm OH line shapes. *A&A*, 238:382–400, November 1990.
32. D. Bockelée-Morvan, J. Crovisier, M. J. Mumma, and H. A. Weaver. *Comets II*, Chapter The composition of cometary volatiles. University of Arizona Press, 2005.
33. D. Bockelée-Morvan and J. Crovisier. The role of water in the thermal balance of the coma. In E. J. Rolfe and B. Battrick, editors, *ESA SP-278: Diversity and Similarity of Comets*, pages 235–240, September 1987.
34. D. Bockelée-Morvan, D. Gautier, D. C. Lis, K. Young, J. Keene, T. Phillips, T. Owen, J. Crovisier, P. F. Goldsmith, E. A. Bergin, D. Despois, and A. Wootten. Deuterated water in Comet C/1996 B2 (Hyakutake) and its implications for the origin of comets. *Icarus*, 133:147–162, May 1998.
35. D. Bockelée-Morvan, J. Wink, D. Despois, P. Colom, N. Biver, J. Crovisier, D. Gautier, E. Gérard, E. Lellouch, R. Moreno, G. Paubert, H. Rauer, J. K. Davies, and W. R. F. Dent. A molecular survey of Comet C/1995 O1 (Hale–Bopp) at the IRAM telescopes. *Earth Moon and Planets*, 78:67–67, 1999.
36. T. Y. Brooke, A. T. Tokunaga, H. A. Weaver, J. Crovisier, D. Bockelée-Morvan, and D. Crisp. Detection of acetylene in the infrared spectrum of Comet Hyakutake. *Nature*, 383:606–608, 1996.
37. D. E. Brownlee, G. Flynn, F. Hörz, L. Keller, K. McKeegan, S. Sandford, P. Tsou, and M. E. Zolensky. Comet Samples Returned by the Stardust Mission. In S. Mackwell and E. Stansbery, editors, *37th Annual Lunar and Planetary Science Conference*, page 2286, March 2006.
38. M. T. Capria, A. Coradini, and M. C. de Sanctis. Modelling of cometary nuclei: Planetary missions preparation. *Advances in Space Research*, 31: 2543–2553, June 2003.
39. M. T. Capria. Sublimation mechanisms of comet nuclei. *Earth Moon and Planets*, 89:161–178, 2002.
40. J. W. Chamberlain and D. M. Hunten. Theory of planetary atmospheres: an introduction to their physics and chemistry . *Orlando FL Academic Press Inc International Geophysics Series*, 36, 1987.
41. A. L. Cochran, E. S. Barker, T. F. Ramseyer, and A. D. Storrs. The McDonald observatory faint comet survey – gas production in 17 comets. *Icarus*, 98:151–162, August 1992.
42. A. L. Cochran and W. D. Cochran. A high spectral resolution Atlas of Comet 122P/de Vico. *Icarus*, 157:297–308, June 2002.
43. P. Colom, J. Crovisier, D. Bockelée-Morvan, D. Despois, and G. Paubert. Formaldehyde in comets. I - Microwave observations of P/Brorsen-Metcalf (1989 X), Austin (1990 V) and Levy (1990 XX). *A&A*, 264:270–281, October 1992.
44. P. Colom, E. Gérard, J. Crovisier, D. Bockelé-Morvan, N. Biver, and H. Rauer. Observations of the OH Radical in Comet C/1995 O1 (Hale–Bopp) with the Nançay Radio Telescope. *Earth Moon and Planets*, 78:37–43, 1999.
45. M. Combes, J. Crovisier, T. Encrenaz, V. I. Moroz, and J.-P. Bibring. The 2.5-12 micron spectrum of Comet Halley from the IKS-VEGA Experiment. *Icarus*, 76:404–436, December 1988.
46. M. R. Combi, M. E. Brown, P. D. Feldman, H. U. Keller, R. R. Meier, and W. H. Smyth. Hubble space telescope ultraviolet imaging and high-resolution

spectroscopy of water photodissociation products in comet hyakutake (C/1996 B2). *ApJ*, 494:816, February 1998.
47. M. R. Combi, W. M. Harris, and W. H. Smyth. *Comets II*, Chapter Gas dynamics and kinetics in the cometary coma: Theory and observations. University of Arizona Press, 2005.
48. M. R. Combi, A. A. Reinard, J.-L. Bertaux, E. Quemerais, and T. Mäkinen. SOHO/SWAN observations of the structure and evolution of the hydrogen Lyman-α coma of comet Hale–Bopp (1995 O1). *Icarus*, 144:191–202, March 2000.
49. M. R. Combi and W. H. Smyth. Monte Carlo particle-trajectory models for neutral cometary gases. I - Models and equations. II - The spatial morphology of the Lyman-alpha coma. *ApJ*, 327:1026–1059, April 1988.
50. H. Cottin, Y. Bénilan, M.-C. Gazeau, and F. Raulin. Origin of cometary extended sources from degradation of refractory organics on grains: polyoxymethylene as formaldehyde parent molecule. *Icarus*, 167:397–416, February 2004.
51. G. Cremonese, H. Boehnhardt, J. Crovisier, H. Rauer, A. Fitzsimmons, M. Fulle, J. Licandro, D. Pollacco, G. P. Tozzi, and R. M. West. Neutral sodium from comet Hale–Bopp: A third type of tail. *ApJL*, 490:L199+, December 1997.
52. J. F. Crifo, M. Fulle, N. I. Kömle, and K. Szego. *Comets II*, Chapter Nucleus-coma structural relationships: Lessons from physical models. University of Arizona Press, 2005.
53. T. Encrenaz, J. Crovisier, *Les Cometes*. CNRS Editions, 1995.
54. J. Crovisier, D. Bockelée-Morvan, N. Biver, P. Colom, D. Despois, and D. C. Lis. Ethylene glycol in comet C/1995 O1 (Hale–Bopp). *A&A*, 418: L35–L38, April 2004.
55. J. Crovisier, D. Bockelée-Morvan, P. Colom, N. Biver, D. Despois, D. C. Lis, and the Team for target-of-opportunity radio observations of comets. The composition of ices in comet C/1995 O1 (Hale–Bopp) from radio spectroscopy. Further results and upper limits on undetected species. *A&A*, 418:1141–1157, May 2004.
56. J. Crovisier and T. Encrenaz. Infrared fluorescence of molecules in comets – The general synthetic spectrum. *A&A*, 126:170–182, September 1983.
57. J. Crovisier, K. Leech, D. Bockelée-Morvan, T. Y. Brooke, M. S. Hanner, B. Altieri, H. U. Keller, and E. Lellouch. The spectrum of Comet Hale–Bopp (C/1995 01) observed with the Infrared Space Observatory at 2.9 AU from the Sun. *Science*, 275:1904–1907, 1997.
58. J. Crovisier and F. P. Schloerb. The study of comets at radio wavelengths. In R. L. Newburn, M. Neugebauer, and J. Rahe, editors, *ASSL Vol. 167: IAU Colloq. 116: Comets in the post-Halley era*, pages 149–173, 1991.
59. J. Crovisier. Rotational and vibrational synthetic spectra of linear parent molecules in comets. *A&AS*, 68:223–258, March 1987.
60. J. Crovisier. Molecular Abundances in Comets. In A. Milani, M. di Martino, and A. Cellino, editors, *IAU Symp. 160: Asteroids, Comets, Meteors 1993*, page 313, 1994.
61. J. Crovisier. Photodestruction rates for cometary parent molecules. *JGR*, 99:3777–3781, February 1994.
62. A. C. Danks, D. L. Lambert, and C. Arpigny. The C-12/C-13 ratio in comet Kohoutek /1973f/. *ApJ*, 194:745–751, December 1974.

63. P. Debye. *Ann. Physik*, 30:59, 1909.
64. N. dello Russo, M. A. Disanti, M. J. Mumma, K. Magee-Sauer, and T. W. Rettig. Carbonyl sulfide in Comets C/1996 B2 (Hyakutake) and C/1995 O1 (Hale–Bopp): Evidence for an extended source in Hale–Bopp. *Icarus*, 135: 377–388, October 1998.
65. I. de Pater and J. J. Lissauer. *Planetary Sciences.* Planetary Sciences, by Imke de Pater and Jack J. Lissauer, Page. 544. ISBN 0521482194. Cambridge, UK: Cambridge University Press, December 2001.
66. M. A. Disanti, M. J. Mumma, N. Dello Russo, and K. Magee-Sauer. Carbon monoxide production and excitation in Comet C/1995 O1 (Hale–Bopp): Isolation of native and Distributed CO sources. *Icarus*, 153:361–390, October 2001.
67. N. Divine. *A Simple Radiation Model of the Cometary Dust for P/Halley.* ESA-SP 174, 1981.
68. P. Eberhardt, D. Krankowsky, W. Schulte, U. Dolder, P. Lammerzahl, J. J. Berthelier, J. Woweries, U. Stubbemann, R. R. Hodges, J. H. Hoffman, and J. M. Illiano. To CO and N_2 abundance in Comet P/Halley. *A&A*, 187:481, November 1987.
69. P. Eberhardt, M. Reber, D. Krankowsky, and R. R. Hodges. The D/H and $^{18}O/^{16}O$ ratios in water from comet P/Halley. *A&A*, 302:301, October 1995.
70. P. Eberhardt. Comet Halley's gas composition and extended sources: Results from the neutral mass spectrometer on Giotto. *Space Science Reviews*, 90: 45–52, 1999.
71. K. E. Edgeworth. The evolution of our planetary system. *Journal of the British Astronomical Association*, 53:181–188, 1943.
72. A. Enzian, H. Cabot, and J. Klinger. A 2 1/2 D thermodynamic model of cometary nuclei. I. Application to the activity of comet 29P/Schwassmann-Wachmann 1. *A&A*, 319:995–1006, March 1997.
73. A. Enzian, H. Cabot, and J. Klinger. Simulation of the water and carbon monoxide production rates of comet Hale–Bopp using a quasi 3-D nucleus model. *Planetary and Space Science*, 46:851–858, August 1998.
74. A. Enzian. On the prediction of CO outgassing from comets Hale–Bopp and Wirtanen. *Space Science Reviews*, 90:131–139, 1999.
75. P. D. Feldman, A. L. Cochran, and M. R. Combi. *Comets II*, Chapter Spectroscopic investigations of fragment species in the coma. University of Arizona Press, 2005.
76. P. D. Feldman, M. F. A'Hearn, and R. L. Millis. Temporal and spatial behavior of the ultraviolet emissions of comet IRAS-Araki-Alcock 1983d. *ApJ*, 282: 799–802, July 1984.
77. P. D. Feldman, M. C. Festou, G. P. Tozzi, and H. A. Weaver. The CO_2/CO abundance ratio in 1P/Halley and several other comets observed by IUE and HST. *ApJ*, 475:829, February 1997.
78. P. D. Feldman, K. B. Fournier, V. P. Grinin, and A. M. Zvereva. The abundance of ammonia in Comet P/Halley derived from ultraviolet spectrophotometry of NH by ASTRON and IUE. *ApJ*, 404:348–355, February 1993.
79. J. A. Fernández. *Comets; Nature, Dynamics, Origin and their Cosmogonical Relevance.* Springer, 2005.
80. Keller H. U. Weaver H. A. Festou, M., editor. *Comets II.* University of Arizona Press, 2005.

81. M. C. Festou. The density distribution of neutral compounds in cometary atmospheres. I - Models and equations. *A&A*, 95:69–79, February 1981.
82. U. Fink and M. D. Hicks. A survey of 39 comets using CCD spectroscopy. *Astrophysical Journal*, 459:729–743, March 1996.
83. M. L. Finson and R. F. Probstein. A theory of dust comets. I. Model and equations. *ApJ*, 154:353–380, October 1968.
84. N. Fray, Y. Bénilan, H. Cottin, M.-C. Gazeau, and J. Crovisier. The origin of the CN radical in comets: A review from observations and models. *PSS*, 53:1243–1262, October 2005.
85. N. Fray, Y. Bénilan, H. Cottin, M.-C. Gazeau, R. D. Minard, and F. Raulin. Experimental study of the degradation of polymers: Application to the origin of extended sources in cometary atmospheres. *Meteoritics and Planetary Science*, 39:581–587, April 2004.
86. N. Fray, Y. Bénilan, H. Cottin, and M.-C. Gazeau. New experimental results on the degradation of polyoxymethylene: Application to the origin of the formaldehyde extended source in comets. *Journal of Geophysical Research (Planets)*, 109:7, June 2004.
87. M. Fulle. A dust-tail model based on Maxwellian velocity distribution. *A&A*, 265:817–824, November 1992.
88. M. Fulle. *Comets II*, Chapter Motion of cometary dust. University of Arizona Press, 2005.
89. J. Geiss, K. Altwegg, E. Anders, H. Balsiger, A. Meier, E. G. Shelley, W.-H. Ip, H. Rosenbauer, and M. Neugebauer. Interpretation of the ion mass spectra in the mass per charge range 25–35 amu/e obtained in the inner coma of Halley's comet by the HIS-sensor of the Giotto IMS experiment. *A&A*, 247:226–234, July 1991.
90. E. Gérard, J. Crovisier, P. Colom, N. Biver, D. Bockelée-Morvan, and H. Rauer. Observations of the OH radical in comet C/1996 B2 (Hyakutake) with the Nançay radio telescope. *PSS*, 46:569–577, May 1998.
91. E. Gerard. The discrepancy between OH production rates deduced from radio and ultraviolet observations of comets. I – A comparative study of OH radio and UV observations of P/Halley 1986 III in late November and early December 1985. *A&A*, 230:489–503, April 1990.
92. J. L. Greenstein. High-resolution spectra of Comet MRKOS (1957d). *ApJ*, 128:106–106, July 1958.
93. S. F. Green, N. McBride, M. T. S. H. Colwell, J. A. M. McDonnell, A. J. Tuzzolino, T. E. Economou, B. C. Clark, Z. Sekanina, P. Tsou, and D. E. Brownlee. Stardust wild 2 dust measurements. *LPI Contributions*, 1280:59, September 2005.
94. E. Gruen and E. Jessberger. *Physics and chemistry of comets*, chapter Dust, page 113. Springer-Verlag.
95. R. M. Häberli, K. Altwegg, H. Balsiger, and J. Geiss. Heating of the thermal electrons in the coma of comet P/Halley. *JGR*, 101:15579–15590, July 1996.
96. R. M. Häberli, T. I. Gombosi, D. L. DeZeeuw, M. R. Combi, and K. G. Powell. Modeling of cometary X-rays caused by solar wind minor ions. *Science*, 276:939–942, May 1997.
97. R. M. Häberli. The temperature of the thermal electrons in the coma of comet P/Halley. *Advances in Space Research*, 18:215–215, 1996.
98. R. M. Häberli, K. Altwegg, H. Balsiger, and J. Geiss. Physics and chemistry of ions in the pile-up region of comet P/Halley. *A&A*, 297:881, May 1995.

99. S. A. Haider, A. Bhardwaj, and R. P. Singhal. Role of auroral and photoelectrons on the abundances of methane and ammonia in the coma of Comet Halley. *Icarus*, 101:234–243, February 1993.
100. M. S. Hanner, R. H. Giese, K. Weiss, and R. Zerull. On the definition of albedo and application to irregular particles. *A&A*, 104:42–46, December 1981.
101. M. S. Hanner. Thermal emission from cometary dust. *Advances in Space Research*, 2:157–157, 1982.
102. M. Hanner and J. P. Bradley. *Comets II*, Chapter Composition and mineralogy of cometary dust. University of Arizona Press, 2005.
103. L. Haser. Distribution d'intensite dans la tete d'une comete. *Bulletin de la Societe Royale des Sciences de Liege*, 43:740–750, 1957.
104. Probstein R. F. Hayes, W. D. *Hypersonic flow theory.* Academic Press, New York, USA, 1959.
105. J. Helbert, H. Rauer, D. C. Boice, and W. F. Huebner. The chemistry of C_2 and C_3 in the coma of Comet C/1995 O1 (Hale–Bopp) at heliocentric distances $r_h \geq 2.9$ AU. *A&A*, 442:1107–1120, November 2005.
106. G. Herzberg. *Molecular Spectra and Molecular Structure II. Infrared and Raman Spectra of Polyatomic Molecules.* Van Nordstrand Reinhold Company, 1945.
107. G. Herzberg. *Molecular Spectra and Molecular Structure I. Spectra of Diatomic Molecules.* Van Nordstrand Reinhold Company, 1950.
108. G. Herzberg. *Molecular Spectra and Molecular Structure III. Electric Spectra of Polyatomic Molecules.* Van Nordstrand Reinhold Company, 1966.
109. M. D. Hicks and U. Fink. Spectrophotometry and the Development of Emissions for C/1996 B2 (Comet Hyakutake). *Icarus*, 127:307–318, June 1997.
110. W. F. Huebner, D. C. Boice, H. U. Schmidt, and R. Wegmann. *Comets in the Post-Halley Era*, Chapter Structure of the coma: Chemistry and solar wind interaction, pages 907–936. Kluwer, Dordrecht, 1991.
111. W. F. Huebner, D. C. Boice, and C. M. Sharp. Polyoxymethylene in Comet Halley. *ApJL*, 320:L149–L152, September 1987.
112. W. F. Huebner, J. J. Keady, and S. P. Lyon. Solar photo rates for planetary atmospheres and atmospheric pollutants. *ApSS*, 195:1–289, September 1992.
113. W. F. Huebner. First polymer in space identified in Comet Halley. *Science*, 237:628–630, August 1987.
114. W. F. Huebner. Composition of comets: Observations and models. *Earth Moon and Planets*, 89:179–195, 2002.
115. van de Hulst. *Light Scattering by Small Particles.* New York: John Wiley and Sons, 1957.
116. D. Hutsemékers, J. Manfroid, E. Jehin, C. Arpigny, A. Cochran, R. Schulz, J. A. Stüwe, and J.-M. Zucconi. Isotopic abundances of carbon and nitrogen in Jupiter-family and Oort Cloud comets. *A&A*, 440:L21–L24, September 2005.
117. W. M. Irvine, E. A. Bergin, J. E. Dickens, D. Jewitt, A. J. Lovell, H. E. Matthews, F. P. Schloerb, and M. Senay. Chemical processing in the coma as the source of cometary HNC. *Nature*, 393:547, 1998.
118. W. M. Irvine, J. E. Dickens, A. J. Lovell, F. P. Schloerb, M. Senay, E. A. Bergin, D. Jewitt, and H. E. Matthews. The HNC/HCN ratio in comets. *Earth Moon and Planets*, 78:29–35, 1999.
119. W. M. Irvine. Spectroscopic evidence for interstellar ices in Comet Hyakutake. *Nature*, 383:418–420, 1996.

120. W. M. Jackson. Laboratory studies of photochemical and spectroscopic phenomena related to comets. In L. L. Wilkening, editor, *IAU Colloq. 61: Comet Discoveries, Statistics, and Observational Selection*, pages 480–495, 1982.
121. E. Jehin, J. Manfroid, A. L. Cochran, C. Arpigny, J.-M. Zucconi, D. Hutsemékers, W. D. Cochran, M. Endl, and R. Schulz. The Anomalous $^{14}N/^{15}N$ ratio in comets 122P/1995 S1 (de Vico) and 153P/2002 C1 (Ikeya-Zhang). *ApJL*, 613:L161–L164, October 2004.
122. E. Jehin, J. Manfroid, D. Hutsemékers, A. L. Cochran, C. Arpigny, W. M. Jackson, H. Rauer, R. Schulz, and J.-M. Zucconi. Deep Impact: High-resolution optical spectroscopy with the ESO VLT and the Keck I Telescope. *ApJL*, 641:L145–L148, April 2006.
123. E. K. Jessberger, A. Christoforidis, and J. Kissel. Aspects of the major element composition of Halley's dust. *Nature*, 332:691–695, April 1988.
124. E. K. Jessberger and J. Kissel. Chemical properties of cometary dust and a note on carbon isotopes. In R. L. Newburn, M. Neugebauer, and J. Rahe, editors, *ASSL Vol. 167: IAU Colloq. 116: Comets in the Post-Halley Era*, pages 1075–1092, 1991.
125. D. Jewitt. Kuiper Belt and Comets: An Observational Perspective. In K. Altwegg, W. Benz and N. Thomas, editors, *Trans-Neptuninan Objects and Comets*, Vol. 35, pages 1–78, 2008.
126. D. Jewitt, H. E. Matthews, T. Owen, and R. Meier. The 12C/13C, 14N/15N and 32S/ 34S Isotope ratios in comet Hale–Bopp (C/1995 O1). *Science*, 278:90–93, October 1997.
127. L. Jorda. *Atmospheres Cometaires: Interpretation des Observations dans le visible et Comparasion avec les Observations Radio.* PhD thesis, Observatoire de Paris Meudon, 1995.
128. H. Kawakita, J.-I. Watanabe, D. Kinoshita, S. Abe, R. Furusho, H. Izumiura, K. Yanagisawa, and S. Masuda. High-dispersion spectra of NH_2 in the Comet C/1999S4 (LINEAR): Excitation mechanism of the NH_2 molecule. *PASJ*, 53:L5–L8, February 2001.
129. H. Kawakita and J.-I. Watanabe. Revised fluorescence efficiencies of cometary NH_2: Ammonia abundance in comets. *ApJL*, 572:L177–L180, June 2002.
130. H. Kawakita and J.-I. Watanabe. *Revised Fluorescence Efficiencies of Cometary NH_2: Ammonia Abundance in Comets*, page 32. Annual Report of the National Astronomical Observatory of Japan, Volume 5, Fiscal 2002. Editors: Kiyotaka Tanikawa, Masatoshi Imanishi, Makoto Miyoshi, Toshiya Muramatsu, Takashi Sekii, Mitsuru Sôma, Akitoshi Ueda, Yoshiko Yamashita, Naoki Yasuda, National Astronomical Observatory of Japan, Osawa, Mitakashi, Tokyo, Japan. ISSN 1346-1192, 2004, page 32, 2004.
131. M. Kleine, S. Wyckoff, P. A. Wehinger, and B. A. Peterson. The carbon isotope abundance ratio in comet Halley. *ApJ*, 439:1021–1033, February 1995.
132. J. Knollenberg. *Modellrechnungen zur Staubverteilung in der inneren Koma von Kometen unter spezieller Bercksichtigung der HMC Daten der Giotto-Mission.* PhD thesis, University Gttingen, 1994.
133. L. Kolokolova, M. S. Hanner, A.-Ch. Levasseur-Regourd, and B. A. S. Gustafson *Comets II*, Chapter Global solar wind interaction and ionospheric dynamics. University of Arizona Press, 2005.
134. Eberhardt P. Krankowsky, D. *Comet Halley*, Chapter Evidence for the composition of ices in the nucleus of comet Halley, page 273. Ellis Horwood, 1990.

135. V. A. Krasnopolsky, M. J. Mumma, M. Abbott, B. C. Flynn, K. J. Meech, D. K. Yeomans, P. D. Feldman, and C. B. Cosmovici. Detection of Soft X-rays and a Sensitive Search for Noble Gases in Comet Hale–Bopp (C/1995 O1). *Science*, 277:1488–1491, September 1997.
136. V. A. Krasnopolsky. On the nature of soft X-Ray radiation in comets. *Icarus*, 128:368–385, August 1997.
137. G. P. Kuiper. O the Origin of the Solar System. In J. A. Hynek, editor, *Proceedings of a Topical Symposium, Commemorating the 50th Anniversary of the Yerkes Observatory and Half a Century of Progress in Astrophysics.* New York: McGraw-Hill, page 357, 1951.
138. C. Laffont, D. C. Boice, G. Moreels, J. Clairemidi, P. Rousselot, and H. Andernach. Tentative identification of S_2 in the IUE spectra of Comet Hyakutake (C/1996 B2). *GRL*, 25:2749–2752, July 1998.
139. P. Lammerzahl, D. Krankowsky, R. R. Hodges, U. Stubbemann, J. Woweries, I. Herrwerth, J. J. Berthelier, J. M. Illiano, P. Eberhardt, U. Dolder, W. Shulte, and J. H. Hoffman. Expansion Velocity and Temperatures of Gas and Ions Measured in the Coma of Comet p/ Halley. *A&A*, 187:169, November 1987.
140. Y. Langevin, J. Kissel, J.-L. Bertaux, and E. Chassefiere. First statistical analysis of 5000 mass spectra of cometary grains obtained by PUMA 1 (Vega 1) and PIA (Giotto) impact ionization mass spectrometers in the compressed modes. *A&A*, 187:761–766, November 1987.
141. H. P. Larson, M. J. Mumma, and H. A. Weaver. Kinematic properties of the neutral gas outflow from comet P/Halley. *A&A*, 187:391–397, November 1987.
142. C. M. Lisse, D. J. Christian, K. Dennerl, K. J. Meech, R. Petre, H. A. Weaver, and S. J. Wolk. Charge Exchange-Induced X-Ray Emission from Comet C/1999 S4 (LINEAR). *Science*, 292:1343–1348, May 2001.
143. C. M. Lisse, T. E. Cravens, and K. Dennerl. X-ray and extreme ultraviolet emission from comets. *Comets II*, pages 631–643, 2005.
144. C. M. Lisse, K. Dennerl, J. Englhauser, M. Harden, F. E. Marshall, M. J. Mumma, R. Petre, J. P. Pye, M. J. Ricketts, J. Schmitt, J. Trumper, and R. G. West. Discovery of X-ray and extreme ultraviolet emission from comet C/Hyakutake 1996 B2. *Science*, 274:205–209, October 1996.
145. D. C. Lis, J. Keene, K. Young, T. G. Phillips, D. Bockelée-Morvan, J. Crovisier, P. Schilke, P. F. Goldsmith, and E. A. Bergin. Spectroscopic observations of Comet C/1996 B2 (Hyakutake) with the Caltech submillimeter observatory. *Icarus*, 130:355–372, December 1997.
146. D. C. Lis, D. M. Mehringer, D. Benford, M. Gardner, T. G. Phillips, D. Bockelée-Morvan, N. Biver, P. Colom, J. Crovisier, D. Despois, and H. Rauer. New molecular species in comet C/1995 O1 (Hale–Bopp) observed with the Caltech Submillimeter observatory. *Earth Moon and Planets*, 78: 13–20, 1999.
147. A. J. Lovell, F. P. Schloerb, E. A. Bergin, J. E. Dickens, C. H. De Vries, M. C. Senay, and W. M. Irvine. HCO^+ in the coma of comet Hale–Bopp. *Earth Moon and Planets*, 77:253–258, 1999.
148. A. J. Lovell, F. P. Schloerb, J. E. Dickens, C. H. De Vries, M. C. Senay, and W. M. Irvine. HCO + Imaging of Comet C/Hale–Bopp 1995 O1. *ApJL*, 497:L117–L121, April 1998.
149. J. Manfroid, E. Jehin, D. Hutsemékers, A. Cochran, J.-M. Zucconi, C. Arpigny, R. Schulz, and J. A. Stüwe. Isotopic abundance of nitrogen and carbon in distant comets. *A&A*, 432:L5–L8, March 2005.

150. C. G. Mason, R. D. Gehrz, T. J. Jones, C. E. Woodward, M. S. Hanner, and D. M. Williams. Observations of unusually small dust grains in the coma of comet Hale–Bopp C/1995 O1. *ApJ*, 549:635–646, March 2001.
151. J. W. (ed.) Mason. *Comet Halley*. Ellis Horwood, 1990.
152. D. J. McComas, J. T. Gosling, S. J. Bame, J. A. Slavin, E. J. Smith, and J. L. Steinberg. The Giacobini-Zinner magnetotail – Tail configuration and current sheet. *JGR*, 92:1139–1152, February 1987.
153. J. A. M. McDonnell, P.L. Lamy, and G.S: Pankiewicz. Physical properties of cometary dust. In R. L. Newburn, M. Neugebauer, and J. Rahe, editors, *ASSL Vol. 167: IAU Colloq. 116: Comets in the Post-Halley Era*, pages 1043–1074, 1991.
154. K. J. Meech and J. Dvoren. *Comets II*, Chapter Physical and chemical evolution of cometary nuclei. University of Arizona Press, 2005.
155. R. Meier, P. Eberhardt, D. Krankowsky, and R. R. Hodges. The extended formaldehyde source in comet P/Halley. *A&A*, 277:677, October 1993.
156. R. Meier, P. Eberhardt, D. Krankowsky, and R. R. Hodges. Ammonia in comet P/Halley. *A&A*, 287:268–278, July 1994.
157. R. Meier, T. C. Owen, D. C. Jewitt, H. E. Matthews, M. Senay, N. Biver, D. Bockelée-Morvan, J. Crovisier, and D. Gautier. Deuterium in Comet C/1995 O1 (Hale–Bopp): Detection of DCN. *Science*, 279:1707, March 1998.
158. G. Mie. *Ann. Physik*, 25:377, 1908.
159. M. Min, J. W. Hovenier, A. de Koter, L. B. F. M. Waters, and C. Dominik. The composition and size distribution of the dust in the coma of Comet Hale Bopp. *Icarus*, 179:158–173, December 2005.
160. A. Morbidelli. Comets and their Reservoirs: Current Dynamics and Primordial Evolution. In K. Altwegg, W. Benz and N. Thomas, editors, *Trans-Neptunian Objects and Comets*, Vol. 35, pages 79–164, 2008.
161. M. Müller, S. F. Green, and N. McBride. An easy-to-use Model for the Optical Thickness and Ambient Illumination within Cometary Dust Comae. *Earth Moon and Planets*, 90:99–108, March 2002.
162. M. J. Mumma, M. A. Disanti, N. dello Russo, M. Fomenkova, K. Magee-Sauer, C. D. Kaminski, and D. X. Xie. Detection of Abundant Ethane and Methane, Along with Carbon Monoxide and Water, in Comet C/1996 B2 Hyakutake: Evidence for Interstellar Origin. *Science*, 272:1310–1314, May 1996.
163. M. J. Mumma, M. A. Disanti, N. dello Russo, K. Magee-Sauer, E. Gibb, and R. Novak. Remote infrared observations of parent volatiles in comets: A window on the early solar system. *Advances in Space Research*, 31:2563–2575, June 2003.
164. M. J. Mumma, H. A. Weaver, H. P. Larson, M. Williams, and D. S. Davis. Detection of water vapor in Halley's comet. *Science*, 232:1523–1528, June 1986.
165. F. M. Neubauer. Giotto magnetic-field results on the boundaries of the pile-up region and the magnetic cavity. *A&A*, 187:73–79, November 1987.
166. H. F. Newall. The spectrum of the daylight comet 1910a. *MNRAS*, 70:459, March 1910.
167. R. L. Newburn and H. Spinrad. Spectrophotometry of 25 comets - Post-Halley updates for 17 comets plus new observations for eight additional comets. *AJ*, 97:552–569, February 1989.

168. M. B. Niedner and J. C. Brandt. Interplanetary gas. XXII – Plasma tail disconnection events in comets – Evidence for magnetic field line reconnection at interplanetary sector boundaries. *ApJ*, 223:655–670, July 1978.
169. G. Notesco, A. Bar-Nun, and T. Owen. Gas trapping in water ice at very low deposition rates and implications for comets. *Icarus*, 162:183–189, March 2003.
170. G. Notesco and A. Bar-Nun. Enrichment of CO over N_2 by Their Trapping in Amorphous Ice and Implications to Comet P/Halley. *Icarus*, 122:118–121, July 1996.
171. J. H. Oort. The structure of the cloud of comets surrounding the Solar System and a hypothesis concerning its origin. *Bull. Astron. Inst. Neth.*, 11:91–110, January 1950.
172. T. Owen. The isotope ratio 12C/13C in comet Tago-Sato (1969g). *ApJ*, 184:33–44, August 1973.
173. E.N. Parker. *Interplanetary Dynamical Processes.* New York: Interscience Publishers, 1963.
174. D. Prialnik, J. Benkhoff, and M. Podolak. *Comets II*, Chapter Modelling the structure and activity of comet nuclei. University of Arizona Press, 2005.
175. D. Prialnik. A Model for the Distant Activity of Comet Hale–Bopp. *ApJL*, 478:L107–L110, April 1997.
176. J. Rahe, C. W. McCracken, and B. D. Donn. Monochromatic and white-light observations of Comet Bennett 1969i /1970II/. *A&AS*, 23:13–35, January 1976.
177. H. Rauer, C. Arpigny, H. Boehnhardt, F. Colas, J. Crovisier, L. Jorda, M. Kueppers, J. Manfroid, K. Rembor, and N. Thomas. Optical observations of comet Hale–Bopp (C/1995 O1) at large heliocentric distances before perihelion. *Science*, 275:1909–1912, 1997.
178. H. Rauer, J. Helbert, C. Arpigny, J. Benkhoff, D. Bockelée-Morvan, H. Boehnhardt, F. Colas, J. Crovisier, O. Hainaut, L. Jorda, M. Kueppers, J. Manfroid, and N. Thomas. Long-term optical spectrophotometric monitoring of comet C/1995 O1 (Hale–Bopp). *A&A*, 397:1109–1122, January 2003.
179. H. Rauer, R. Wegmann, H. U. Schmidt, and K. Jockers. 3-D MHD simulations of the effect of comoving discontinuities in the solar wind on cometary plasma tails. *A&A*, 295:529–550, March 1995.
180. Weiler M. Hainaut O. Jehin E. Sterken C. Rauer, H. *A&A*, submitted, 2006.
181. C. Reylé and D. C. Boice. An S_2 Fluorescence Model for Interpreting High-Resolution Cometary Spectra. I. Model Description and Initial Results. *ApJ*, 587:464–471, April 2003.
182. K. Richter, M. R. Combi, H. U. Keller, and R. R. Meier. Multiple scattering of hydrogen Lyα radiation in the coma of comet Hyakutake (C/1996 B2). *ApJ*, 531:599–611, March 2000.
183. S. D. Rodgers, H. M. Butner, S. B. Charnley, and P. Ehrenfreund. The HNC/HCN ratio in comets: Observations of C/2002 C1 (Ikeya-Zhang). *Advances in Space Research*, 31:2577–2582, June 2003.
184. S. D. Rodgers, S. B. Charnley, W. F. Huebner, and D. C. Boice. *Comets II*, Chapter Physical processes and chemical reactions in cometary comae. University of Arizona Press, 2005.
185. S. D. Rodgers and S. B. Charnley. HNC and HCN in Comets. *ApJL*, 501: L227–L230, July 1998.
186. S. D. Rodgers and S. B. Charnley. On the origin of HNC in Comet Lee. *MNRAS*, 323:84–92, May 2001.

187. D. G. Schleicher, S. M. Lederer, R. L. Millis, and T. L. Farnham. Photometric behavior of Comet Hale–Bopp (C/1995 O1) before perihelion. *Science*, 275:1913–1915, 1997.
188. D. G. Schleicher, R. L. Millis, and P. V. Birch. Narrowband photometry of Comet P/Halley: Variation with heliocentric distance, season, and solar phase angle. *Icarus*, 132:397–417, April 1998.
189. F. P. Schloerb, M. J. Claussen, and L. Tacconi-Garman. OH Radio Observations of Comet p/ Halley. *A&A*, 187:469, November 1987.
190. F. P. Schloerb, C. H. De Vries, A. J. Lovell, W. M. Irvine, M. Senay, and H. A. Wootten. Collisional quenching of OH radio emission from comet Hale–Bopp. *Earth Moon and Planets*, 78:45–51, 1999.
191. F. P. Schloerb and W. Ge. Submillimeter molecular line observations of cometary Levy (1990c). In A. W. Harris and E. Bowell, editors, *Asteroids, Comets, Meteors 1991*, pages 533–536, December 1992.
192. F. P. Schloerb, W. M. Kinzel, D. A. Swade, and W. M. Irvine. Observations of HCN in comet P/Halley. *A&A*, 187:475–480, November 1987.
193. M. Schmidt-Voigt. Time-dependent MHD simulations for cometary plasmas. *A&A*, 210:433–454, February 1989.
194. H.-U. Schmidt, R. Wegmann, W. F. Huebner, and D. C. Boice. Cometary gas and plasma flow with detailed chemistry. *Comparative Physical Communication*, 49:17–59, 1988.
195. G. Schwehm and P. Murdin. Rosetta. *Encyclopedia of Astronomy and Astrophysics*, November 2000.
196. Z. Sekanina, D. E. Brownlee, T. E. Economou, A. J. Tuzzolino, and S. F. Green. Modeling the nucleus and jets of Comet 81P/Wild 2 based on the stardust encounter data. *Science*, 304:1769–1774, June 2004.
197. L. E. Snyder, P. Palmer, and I. de Pater. VLA Observations of Formaldehyde Emission from Comets Halley and Machholz 1988J. *PASP*, 101:882, October 1989.
198. L. A. Soderblom, D. C. Boice, D. T. Britt, R. H. Brown, B. J. Buratti, R. L. Kirk, M. Lee, R. M. Nelson, J. Oberst, B. R. Sandel, S. A. Stern, N. Thomas, and R. V. Yelle. Imaging Borrelly. *Icarus*, 167:4–15, January 2004.
199. A. Stawikowski and J. L. Greenstein. The Isotope Ratio $^{12}C/^{13}C$ in a Comet. *ApJ*, 140:1280, October 1964.
200. K. S. Swamy. *Physics of Comets*. Worlds Scientific, 1997.
201. P. Swings. Complex structure of cometary bands tentatively ascribed to the contour of the solar spectrum. *Lick Observatory Bulletin*, 19:131–136, 1941.
202. M. Tarenghi and T. L. Wilson. The ALMA Project. In L. I. Gurvits, S. Frey, and S. Rawlings, editors, *EAS Publications Series*, pages 423–430, 2005.
203. N. Thomas and H. U. Keller. Fine dust structures in the emission of comet P/Halley observed by the Halley Multicolour Camera on board Giotto. *A&A*, 187:843–846, November 1987.
204. A. J. Tuzzolino, T. E. Economou, B. C. Clark, P. Tsou, D. E. Brownlee, S. F. Green, J. A. M. McDonnell, N. McBride, and M. T. S. H. Colwell. Dust measurements in the coma of comet 81P/Wild 2 by the dust flux monitor instrument. *Science*, 304:1776–1780, June 2004.

205. V. Vanysek. Carbon isotope ratio in comets and interstellar matter. In A. H. Delsemme, editor, *IAU Colloq. 39: Comets, Asteroids, Meteorites: Interrelations, Evolution and Origins*, pages 499–503, 1977.
206. V. Vanysek. Isotopic ratios in comets. In R. L. Newburn, M. Neugebauer, and J. Rahe, editors, *ASSL Vol. 167: IAU Colloq. 116: Comets in the Post-Halley Era*, pages 879–895, 1991.
207. H. A. Weaver, T. Y. Brooke, G. Chin, S. J. Kim, D. Bockelée-Morvan, and J. K. Davies. Infrared Spectroscopy of Comet Hale–Bopp. *Earth Moon and Planets*, 78:71–80, 1999.
208. H. A. Weaver, P. D. Feldman, J. B. McPhate, M. F. A'Hearn, C. Arpigny, and T. E. Smith. Detection of CO Cameron band emission in comet P/Hartley 2 (1991 XV) with the Hubble Space Telescope. *ApJ*, 422:374–380, February 1994.
209. H. A. Weaver, M. J. Mumma, H. P. Larson, and D. S. Davis. Post-perihelion observations of water in comet Halley. *Nature*, 324:441–444, December 1986.
210. R. Wegmann, H. U. Schmidt, W. F. Huebner, and D. C. Boice. Cometary MHD and chemistry. *A&A*, 187:339–350, November 1987.
211. R. Wegmann. Large-scale disturbance of the solar wind by a comet. *A&A*, 389:1039–1046, July 2002.
212. M. Weiler, H. Rauer, J. Knollenberg, L. Jorda, and J. Helbert. The dust activity of comet C/1995 O1 (Hale–Bopp) between 3 AU and 13 AU from the Sun. *A&A*, 403:313–322, May 2003.
213. M. Womack, A. Homich, M. C. Festou, J. Mangum, W. T. Uhl, and S. A. Stern. Maps of HCO^+ Emission in c/1995 O1 (Hale–Bopp). *Earth Moon and Planets*, 77:259–264, 1999.
214. L. M. Woodney, J. McMullin, and M. F. A'Hearn. Detection of OCS in comet Hyakutake (C/1996 B2). *PSS*, 45:717–719, June 1997.
215. M. C. H. Wright, I. de Pater, J. R. Forster, P. Palmer, L. E. Snyder, J. M. Veal, M. F. A'Hearn, L. M. Woodney, W. M. Jackson, Y.-J. Kuan, and A. J. Lovell. Mosaicked images and spectra of J = 1 -> 0 HCN and HCO+ emission from Comet Hale–Bopp (1995 O1). *AJ*, 116:3018–3028, December 1998.
216. S. Wyckoff, S. C. Tegler, and L. Engel. Ammonia abundances in four comets. *ApJ*, 368:279–286, February 1991.
217. L. M. Ziurys, C. Savage, M. A. Brewster, A. J. Apponi, T. C. Pesch, and S. Wyckoff. Cyanide chemistry in comet Hale–Bopp (C/1995 O1). *ApJL*, 527:L67–L71, December 1999.

Acknowledgments

Kuiper Belt and Comets: An Observational Perspective

I thank Nick Thomas for inviting me to Saas Fee and NASA and NSF for supporting my research on the comets and Kuiper Belt Objects for many years. Sean Andrews, Audrey Delsanti, Nader Haghighipour, Pedro Lacerda, Rita Mann, and Bin Yang kindly read the manuscript.

Comets and Their Reservoirs: Current Dynamics and Primordial Evolution

I thank N. Thomas and W. Benz for their invitation to present a series of lectures at the 35th Saas-Fee advanced course, from which this chapter has been derived. I also thank R. Jedicke and D. Jewitt for their invitation to re-present the same lectures at the Institute for Astronomy of the University of Hawaii. I am grateful to all the colleagues with whom I had stimulating discussions on comet dynamics and the formation of the cometary reservoirs, in particular L. Dones and H. Levison. I also acknowledge that I re-cycled pieces of text originally written by M. Brown and Ph. Claeys in papers that we made together on the Kuiper belt and on the Late Heavy Bombardment. I am in debt to all those who read carefully the draft of this manuscript and gave valuable suggestions for improvements: D. O'Brien, W. Benz, P. Michel and, particularly, the reviewer J. Horner. Finally, I wish to devote this chapter to the memory of Michel Festou. This great French expert on comets was particularly aware of the importance of dynamics for understanding the role of these icy objects in the framework of Solar System formation. My discussions with him greatly motivated me – originally an asteroid person – to know more about the primordial evolution of the outer Solar System.

Comets

I thank Nicolas Thomas and the organizers of the Saas Fee lectures for inviting me and giving me the opportunity to show the field of comet science to young students. My warmest thanks goes to Philipp Eigmüller, Jörg Knollenberg, Michael Weiler and the team in Bern for careful reading of the manuscript.

Index